Conserving Active Matter

BARD GRADUATE CENTER

CULTURAL HISTORIES OF THE MATERIAL WORLD

Cultural Histories of the Material World is a series centered on the exploration of the material turn in the study of culture. Volumes in the series examine the ways human beings have shaped and interpreted the material world from a broad range of scholarly perspectives and show how attention to materiality can contribute to a more precise historical understanding of specific times, places, ways, and means.

Peter N. Miller, Series Editor

Other books in the series include:

Antiquarianism and Intellectual Life in Europe and China, 1500–1800
Peter N. Miller and François Louis, Editors

The Sea: Thalassography and Historiography
Peter N. Miller, Editor

Cultural Histories of the Material World
Peter N. Miller, Editor

Ways of Making and Knowing: The Material Culture of Empirical Knowledge
Pamela H. Smith, Amy R. W. Meyers, and Harold J. Cook, Editors

The Anthropology of Expeditions: Travel, Visualities, Afterlives
Joshua A. Bell and Erin L. Hasinoff, Editors

Ex Voto: Votive Giving Across Cultures
Ittai Weinryb, Editor

In Space We Read Time: On the History of Civilization and Geopolitics
Karl Schlögel

The Art of the Jewish Family: A History of Women in Early New York in Five Objects
Laura Arnold Leibman

The Museum in the Cultural Sciences: Collecting, Displaying, and Interpreting Material Culture in the Twentieth Century
Peter N. Miller, Editor

Object—Event—Performance: Art, Materiality, and Continuity since the 1960s
Hanna B. Hölling, Editor

Conserving Active Matter

Edited by Peter N. Miller and Soon Kai Poh

Bard Graduate Center
New York City

Peter N. Miller is dean and professor at Bard Graduate Center.
Soon Kai Poh is a conservator and Andrew W. Mellon Foundation Postgraduate Fellow at Bard Graduate Center.

Distributed by the University of Chicago Press. This book may be purchased in quantity for educational, business, or promotional use. For information, please e-mail marketing@press.uchicago.edu.

Designed by Jocelyn Lau
This book is set in ITC New Baskerville.
Printed in the United States of America

Library of Congress Control Number: 2021951108
ISBN 978-1-941792-32-2 (hardcover)
ISBN 978-1-941792-33-9 (e-book)

A catalogue record for this book is available from the British Library.

10 9 8 7 6 5 4 3 2 1

To Mariët, for asking the right question and then trusting us to answer it our way

Contents

4. Materials

Preface: Report on a Research Project

Abstract

Conserving Active Matter *draws together the main lines and interim conclusions of a five-year research project embedded in a ten-year effort to reimagine the relationship between conservation knowledge and the humanistic study of the material world—Cultures of Conservation. The effort to conserve things is part of the human struggle with the pervasive activity of matter. For as long as people have made things and kept things, they have cared for them and repaired them. The field of conservation developed in Europe and the United States and then spread around the world. Today's conservator uses a variety of tools and categories developed over the last 150 years to do this work. But in the next decades, new kinds of materials and a new scale of change will pose unprecedented challenges. Thinking through the lens of "active matter," as understood by philosophers, historians, materials scientists, conservators, and those who work on Indigenous artifacts, this project raises questions and establishes new lines of inquiry for the future rethinking of conservation and the human sciences of the object.*

The Beginning of Conserving Active Matter

The project called Conserving Active Matter was undertaken at Bard Graduate Center under the auspices of Cultures of Conservation, a ten-year, $2.1-million initiative largely funded by the Andrew W. Mellon Foundation, that had as its aim connecting conservators and conservation scientists on the one hand and the material culture–oriented human sciences (art history, history, anthropology, archaeology) on the other. It drove a teaching curriculum, research events, exhibitions, fellowships, and publications. The leading idea was that humanists were asking more and better questions than they ever had before about the way objects could, in principle, function as historical evidence, and conservators and conservation scientists were giving better and more precise answers to questions about the "what," "when," "how," and "where" of objects than ever before, but the usual siloing effect of disciplines and institutions worked to keep them apart. Putting the two groups together was no utopian venture but something concrete and attainable, especially in the context of the existing interdisciplinary environment at Bard Graduate Center.

Conserving Active Matter was designed from the outset as a research project. It shaped the second phase of Cultures of Conservation. But it was inspired by an accident, the kind of accident that was prepared for in advance. On Tuesday, October 18, 2016, at lunchtime, Wolfgang Schäffner, codirector of the excellence cluster "Image. Knowledge. *Gestaltung*" at the Humboldt University of Berlin spoke at Bard Graduate Center about the work of the cluster. Included in his talk was a discussion of "active matter"—a term he used to designate materials that were innately active or were designed to be active—and their impact on the future of architecture and design at every scale. In the question period, a visiting research fellow from Berkeley, Beth Piatote, asked about parallels in the traditions of Indigenous North Americans to the view of matter just presented. The very next lunchtime, Wednesday, October 19, Robert van Langh, head of the Department of Conservation and Science at the Rijksmuseum, spoke at Bard Graduate Center about the importance of research for art conservation. He sketched for his audience a scenario in which all treatment would be suspended for a generation while conservators researched their objects. For those in the audience for both talks, they seemed to speak to one other: What would a world of active matter and materials mean for art and object conservation?

Up to that moment, the concept of active matter had reached academic publics through two projects: one at MIT and one at the Humboldt University in Berlin.[1] An "Active Matter Summit" at MIT in 2015 featured three days of talks by materials scientists and was later published.[2] By contrast, the work of the excellence cluster "Image. Knowledge. *Gestaltung*" directed by Schäffner and Horst Bredekamp, spoke more directly to BGC thinking with its effort to present active matter from the perspectives of art, architecture, and design.[3] An exhibition organized by the cluster at the Martin-Gropius-Bau in Berlin in 2016, *+Ultra,* presented the nineteenth century's technologies as built upon a division between blunt matter and spirit (as animating force) that then shaped the twentieth's horizon of possibility. At the end of the eighteenth century, Immanuel Kant had stated what came to be the received opinion on this question: "The possibility of living matter [*lebenden Materie*]," he wrote, "(the concept of which contains a contradiction, because lifelessness, *inertia,* constitutes its essential characteristic) cannot even be conceived."[4] According to Schäffner, however, "We are currently standing on the threshold of a new era in which knowledge will become a state of matter itself, while matter, on the other hand, will present itself as an active code." What this opened up, he argued, was a "new design horizon."[5]

Wood provided the Berlin group with their best example. Wood is matter that is active, sometimes notoriously so. But wood is also the ur-matter. In Greek it is *hulê,* or construction wood, taken by Aristotle and generalized to refer to all substance that was not form. When translated into Latin as *materia,* whose original meaning was timber (it is used that way in Vitruvius), we gain our familiar "material." Architect Achim Menges is currently trying to harness this primordial activity for future building design: for instance, the hygroscopic workings of wood offer the possibility of self-powering ventilation systems.[6]

BGC asked how thinking in terms of matter's "natural" or "intended" activity would affect thinking about professional conservation, with its ambivalent, if no longer utterly agonistic, relationship to material change. The first phase of Cultures of Conservation had as its main goal the integration of conservation knowledge into the tool kit of the material culture scholar. Conserving Active Matter continued this path-widening conversation, but it also turned this widened perspective back on the field of conservation itself, asking how new research questions in the humanities and material sciences might expand conservation thinking and practice.

As an interdisciplinary, advanced humanities institute whose faculty come from history, art history, philosophy, anthropology, archaeology, and most recently, chemistry departments, BGC is well positioned to explore the potential meaning of active matter for conservation through the different lenses constituted by history, philosophy, materials science, and Indigenous ontologies of activity. The four working groups that were created to explore along these axes would probe the meaning of active matter from their different vantage points and also reflect on its usefulness for those looking out from those vantage points. This meant thinking about both conservation and active matter in broader cross-disciplinary perspectives but also in terms of the actual people (makers, users, conservators, custodians, researchers, curators, museum visitors, originating communities) who are stakeholders in shaping the field of professional conservation (as well as its more vernacular aspects). If "conservation" includes all this, "activity" can too.

Conceiving of the project in this way raised two challenges. The first was establishing a new field of questions at the junction of two distinct and up to that point unrelated spheres of inquiry, "active matter" (in the sense of materials science used by the teams at MIT and Humboldt University) and "conservation" (in the narrower field of professional art and object conservation). Moreover, we were interested in exploring the potential to expand the usage of the pair of terms beyond their immediate application to wider conceptions of activity in matter and of the care and repair of material things. This meant, in practice, getting the attention of people who might be interested in these themes (both narrow and expanded) and convincing them that there was a worthwhile question to be explored at their junction. The second challenge was created by the very means employed to deal with the first one, namely, that the fourfold approach of the working groups, each with its own focus, risked encouraging siloed inquiry. To see the methodological challenge of the project as ultimately balancing the centrifugal and the centripetal is, then, to place it in the dynamics of "comparison."

As I've discussed in an earlier volume in this series, the realm of comparative scholarship has had a liminal status in a humanities structure that remains anchored in discipline.[7] Discipline *is* crucial to teaching skills. And in all fields of inquiry disciplines do excel at solving the questions they acknowledge as significant. But that's just it: What about the questions they do not acknowledge as significant? What about questions forced upon disciplines by the unruly character of the world? Here disciplinary scholarship often produces lackluster research, what Thomas Kuhn thematized long ago in terms of "normal science," because of the decreasing ambitiousness of the questions being asked. Comparative scholarship, while not necessarily linked to interdisciplinarity—you could have comparative history within history, comparative literature within literature, comparative politics within politics—has tended in the direction of interdisciplinarity.[8] Standing outside a discipline's questions and horizons makes comparison exciting but also diminishes its probative force. And, sociologically speaking, it puts practitioners outside, if not at risk of marginalization from, sources of power and patronage, most of which remain tied to discipline, whether institutional or professional. For all these reasons, urging transformation from the perspective of the comparative is not an obviously winning proposition. And yet, as we are all aware, at least from time to time, it is only by comparison that we can make sense of something new, only by comparison that some hidden details emerge, only by comparison that obscured connections are revealed. In short, the whole clumsy epistemology of "splitting" and "lumping" from Charles Darwin through Max Weber and on to us has evolved out of the struggle of discipline-based scholarship with comparison. It was only much later that Roger Caillois, the lapsed surrealist, articulated the premise of a "diagonal" science that compared but in nonlinear ways, and thus stood somewhere between lazy lumping and dogmatic splitting.[9] In Conserving Active Matter, the four working groups operated relative to each other, and to the overall theme, in this kind of diagonal fashion. Although we eschewed the working-group distinctions in the layout of the concluding exhibit and its digital publication, we deemed it appropriate to maintain them in the structure of this volume as it is a report from a research project that took this shape.

How to Use This Book

Between 2017 and 2020, the four working groups scheduled workshops (see Appendix: Events of the Research Project Conserving Active Matter) with invited guests to explore the idea of active matter from their different perspectives. The working-group events, and the scholarship that flowed from them into this volume, spoke in very different disciplinary and professional languages. For example, the conservators and conservation scientists presented work in detail. The philosophers also spoke in detail—but very different details. The same was true for those addressing Indigenous approaches to activity and the historians exposing presences of activity in European and Asian epistemologies (including Kabbalah and Buddhism). Some speakers focused on the activity of *matter*—the core of the project inherited from the Berlin and MIT groups—and others addressed activity in terms of human agents who *activate* the matter at hand through their questions. The speakers did not always engage with the arguments or conclusions of the groups working on either side. Humanists did not always talk about either object "conservation" or material "activity," and the scientists and conservators did not always address the historical and philosophical implications of what they were talking about. Some of this talking past survives in the published book.

But this is nothing to be surprised about. In practice, with work on a very new subject that involves bringing two unrelated fields and undefined terms into conversation, one cannot expect people to be immediately comfortable working in those two areas and synthesizing them. What one is doing in a project (book and exhibition) of this sort is bringing a faceted presentation of the particular fields into a common space and then, at a "second order," stepping in to draw the possible contours of a new question that builds convincingly on some of that local detail. The goal is to persuade a wider audience of the worthiness of the larger question. So, in fact, having discrete pieces of work that bring out specific arguments in nugget-rich detail is desirable, even though—and sometimes *because*—the collected, commissioned scholarship does not try to do the work of comparison at the local level. This may not be everyone's vision of comparative scholarship—indeed, we might all like new fields of interdisciplinary learning to burst fully formed from their creators' institutional context—but the sociology of jump-starting new research agendas has its own logic. Getting people to do the comparative work at the beginning of a field's existence is very difficult because, aside from those doing the commissioning, it is to be expected that everyone else is commanded by their existing research paradigms.

One model that has evolved to pursue this kind of work is grant-based and involves paying postdoctoral scholars to pursue the work of the visionary project directors. But with more junior scholars one is even less likely to find scholarship based firmly in two disciplines simply because of the time it takes to master even one. Where our project has left us, by design, I need repeat, is with a series of discrete inquiries at the local level—philosophy, materials, Indigenous ontologies, history—that at the "section" and "volume" level provide us with the rudiments for a new way of looking at conservation through the lens of activity. We, the editors of this volume and each of its constituent sections, have done our best to tie things together. You, the readers, will have to participate in this work of "closing the circuits," "connecting the dots," and framing the implications. And if, twenty or even thirty years from now, it is not considered worth commenting on conservators curating exhibitions or historians describing themselves as conservators or scholars producing work that resists easy categorization as art historical, conservation, curatorial, or historical, then this necessarily imperfect beginning will have done its work.

To help us get to this future, the volume begins with two introductions, one written by a conservator and the other by a historian. The conservator, who also curated the eponymous exhibition, presents Conserving Active Matter as an experiment in showing how a conservator can think with humanistic disciplines in order to broaden the understanding of what it means to conserve and to be a conservator. The historian, who also formulated the project Cultures of Conservation, presents Conserving Active Matter as an experiment in showing how a historian can think with conservation knowledge in order to broaden the understanding of what history is and how to be a historian. The two approaches may or may not meet at some equidistant point; what they do is provide an example of what can be gained by the shift in perspective they each represent.

Where once upon a time Euro-American conservation practice viewed activity as a threat to be resisted, conservators now are more comfortable managing—even on occasion facilitating—change. We hope that our volume will encourage this tendency and will provide intellectual resources for those already inclined to view activity with less anxiety, if not more sympathy. Bringing in the perspective of Indigenous ontologies, referring to practices in Japan, and, in fact, showing that even within European traditions discomfort with activity was a choice among alternatives, will, we hope, provide conservators of the future with the ability to formulate a more complex set of responses to what surely will be a more complex set of challenges.[10] We believe that looking through the lens of active matter will help contribute to the future of conservation thinking among both conservators and humanists.

On July 13, 2017, on page A11, the *New York Times* reported the results of a startling experiment. Scientists had inserted digital video with its pixels recoded in combinations of adenine, guanine, thymine, and cytosine into a bacterium's genome, taking advantage of the fact that all DNA contains a large quantity of "junk." This had the effect of turning one species into a storage device for another (bacteria as external "hard drive"). Even more startling, when the bacterium reproduced, it passed along that digital code to its descendants with enough fidelity to be retranslated into a recognizable video image. Life—active matter—is doing the conserving, and active matter—the now-living code—is conserved.

Self-driving cars, smart textiles, self-regulating buildings, and artworks that change themselves are already with us. Our twenty-first-century future will turn all these from science fiction to humdrum. What will be the impact of this sea change on the field of conservation?

Or on our notion of preservation more broadly? Of course, less dramatic versions of this process have been with us for a long time. The human body itself could be said to pose the most acute example of "active matter"—and its conservation has been a challenge to philosophers and medical doctors from diverse cultures for millennia. But, in terms of scale and audaciousness, the twenty-first century will likely pose an especial challenge to conservation because of active matter. If we add to this the impact of planetary climate change on cultural heritage, it is clear that the field of conservation will be addressing new questions on an unprecedented scale. This volume, its companion, and the accompanying exhibition project are contributions toward this future thinking.

—Peter N. Miller

NOTES

This volume captures the state of the research project at the point it took on a curatorial form. A separate publication will reflect the knowledge created by the exhibition process. For a parallel to this introduction, see Giard, "Histoire d'un recherche," esp. xvi–xxiii. I am grateful to Aaron Glass for a careful reading of an earlier draft of this essay and some very good suggestions, many of which I have incorporated.

1. Like-minded initiatives, such as Lee, *Material Alchemy*, aim at this same target through a vernacular design focus on synthetic design and bio-inspiration.
2. Perkins, *Active Matter.*
3. BGC had been collaborating with Bredekamp in one of his predecessor projects, The Technical Image, which resulted in a postdoctoral fellowship, exhibition, and publication at BGC in 2012 (Samuel, *Islands of Benoît Mandelbrot*) as well as a translation project (Bredekamp, Dunkel, and Schneider, *Technical Image*).
4. Kant, *Critique of the Power of Judgment,* II.2, § 73, 265.
5. Schäffner, "Immateriality of Materials," 23, 26. See also Schäffner, "Interdisziplinäre Gestaltung."
6. In Doll, Bredekamp, and Schäffner, *+Ultra*, see Schäffner, "Immateriality of Materials," 25; Michael Friedman and Karin Krauthausen, "Inspired Mechanics: Active Matter as Machine and Structure," 169–70; Peter Fratzl, "The Bioinspired Design of Materials," 177. For Menges's work,

see Reichert, Menges, and Krieg, "*HygroSkin—Meteorosensitive Pavilion.*" For a deeper look at wood from this perspective, see Eder et al., "Wood and the Activity."

7. Miller and Louis, introduction.
8. For instance, *Comparative Studies in History and Anthropology.*
9. For this notion, see Caillois, "New Plea." An inspiring model for this kind of cross-cutting study, albeit more sideways than diagonal, is Eggert, *Securing the Past.*
10. Conservation thinking in Japan and India will be part of *Conserving Active Matter: Exhibition and Afterthoughts,* also to be published by Bard Graduate Center in this series.

BIBLIOGRAPHY

Bredekamp, Horst, Vera Dunkel, and Birgit Schneider, eds. *The Technical Image: A History of Styles in Scientific Imagery.* Chicago: University of Chicago Press; New York: Bard Graduate Center, 2015.

Caillois, Roger. "A New Plea for a Diagonal Science." In *The Edge of Surrealism: A Roger Caillois Reader,* edited by Claudine Frank, 343–47. Durham, NC: Duke University Press, 2003.

Doll, N., H. Bredekamp, and W. Schäffner, eds. *+Ultra: Knowledge & Gestaltung.* Leipzig: E. A. Seemann, 2017.

Eder, M., W. Schäffner, I. Burgert, and P. Fratzl. "Wood and the Activity of Dead Tissue." *Advanced Materials* (2020): https://doi.org/10.1002/adma.202001412.

Eggert, Paul. *Securing the Past: Conservation in Art, Architecture and Literature.* Cambridge: Cambridge University Press, 2009.

Giard, Luce. "Histoire d'un recherche." In *L'invention du quotidien. 1. Arts de faire,* by Michel de Certeau, i–xxx. 2nd ed. Paris: Gallimard, 1990.

Kant, Immanuel. *Critique of the Power of Judgment.* Edited by Paul Guyer. Translated by Paul Guyer and Eric Matthews. Cambridge: Cambridge University Press, 2007.

Lee, Jenny, ed. *Material Alchemy.* Amsterdam: BIS, 2014.

Miller, Peter N., and François Louis. Introduction to *Antiquarianism and Intellectual Life in Europe and China,* 1500–1800, edited by Peter N. Miller and François Louis, 1–24. Ann Arbor: University of Michigan Press, 2012.

Perkins, Schuyler, ed. *Active Matter.* Cambridge, MA: MIT Press, 2017.

Reichert, Steffen, Achim Menges, and Oliver David Krieg. "*HygroSkin—Meteorosensitive Pavilion.*" In *Fabricate,* edited by Fabio Gramazio, Matthias Kohler, and Silke Langenberg, 272–79. London: UCL Press, 2014.

Samuel, Nina, ed. *The Islands of Benoît Mandelbrot: Fractals, Chaos, and the Materiality of Thinking*. New York: Bard Graduate Center, 2012.

Schäffner, Wolfgang. “Immateriality of Materials.” In Doll, Bredekamp, and Schäffner, *+Ultra*, 23–31.

———. “Interdisziplinäre Gestaltung: Einladung in das neue Feld einer Geistes- und Materialwissenschaft.” In *Haare hören—Strukturen wissen—Räume agieren. Berichte aus dem Interdisziplinären Labor Bild Wissen Gestaltung*, edited by Horst Bredekamp and Wolfgang Schäffner, 199–213. Bielefeld: Transcript, 2015.

Acknowledgments

Conserving Active Matter has been a collaborative endeavor through and through. The editors thank our colleagues who have given this project their time and thought. From 2017 to 2020, they led the four working groups: Ivan Gaskell (Philosophy), Aaron Glass (Indigenous Ontologies), Jennifer L. Mass (Materials Science), and Ittai Weinryb (History). In the final two years of the project, Gaskell, Glass, Mass, and Meredith Linn joined the volume's editors as the curatorial committee that planned the exhibition accompanying this book. During these years, Soon Kai Poh led the weekly curatorial meetings.

The editors thank the members of the local steering group of Cultures of Conservation, which at one time or another included Lisa Bruno, Jim Coddington, Joseph Godla, Marco Leona, Judith Levinson, Michele Marincola, Ian McClure, and Glenn Wharton. Further thanks are due to the visiting fellows who were with us for *Conserving Active Matter*, Jessica Walthew and Alicia Boswell, and the members of the wandering seminar of 2018–19, who participated in many fruitful conversations: Joy Bloser, Lara Kaplan, Vicki Parry, and Nicholas Pedemonti. We also thank our hosts for these visits: Sarah Scaturro, Carol Mancusi-Ungaro, Sarah Barack, Roberta Olson, Valerie Paley, and Margie Hofer. Many of our collaborators already named here remained engaged with *Conserving Active Matter* as it came together as an exhibition; Bloser, Coddington, Leona, Levinson, Marincola, Scaturro, Walthew, and Wharton, joined by Robert van Langh, advised as a scientific committee to the exhibition.

Laura Minsky, associate director of the research institute, has our gratitude for the typical precision and care with which she organized the six project meetings held between 2017 and 2019. Heather Topcik, director of research collections, as always, is responsible for integrating research interests like conservation into the institution's holdings. Dan Lee, director of publishing, is again responsible for the manner in which these ideas reach the public. The editors also thank Kat Atkins, managing editor, Jocelyn Lau, designer, and Florence Grant, copyeditor, for their indispensable and meticulous work on this volume.

Soon Kai Poh thanks Peter N. Miller and Bard Graduate Center for the incredible opportunity to contribute to the ambitious vision that is *Conserving Active Matter* in the early stages of his career. He is grateful to mentors and colleagues, many of whom are named above, and numerous others who are too many to name here, for their generosity of spirit and mind in sharing their valuable experiences and perspectives with him.

Peter N. Miller thanks especially Francesca Casadio, Jim Coddington, Robert van Langh, Marco Leona, Ellen Pearlstein, and Glenn Wharton for their ongoing contributions as conversation partners. *Conserving Active Matter* was the last act of a ten-year project prompted and mostly funded by the Andrew W. Mellon Foundation. For those ten years he learned a great deal by watching the Arts and Cultural Heritage group in action. He thanks Alison Gilchrest and, latterly, Abby Choudhury for their guidance and collaborative spirit.

Finally, the book is dedicated to Mariët Westermann, who, while vice president of the Andrew W. Mellon Foundation, gave this project its wings. Without her initial push, none of this would have happened. Like so many others in this country and around the world who have benefited from her vision and leadership, it is left to us to say: thank you.

Introduction: Conserving Active Matter and the Conservator

Soon Kai Poh

Abstract

The notion of active matter is inscribed into conservation on multiple levels: from its processes and procedures to its conventions and normative structures, including its organization as a profession. Indeed, the practice of conservation is, at its core, about providing context to the activity of matter. Yet the underlying assumptions surrounding active matter that are manifest through the relationships that conservators share with activity are less often examined. How might an interdisciplinary group of materials scientists, historians, philosophers, Indigenous scholars, scholars of Indigeneity, and, not least, conservators provide insight into the fundamental questions of conservation? From the perspective of a conservator at the beginning of his career, Conserving Active Matter *is a pertinent reminder that things are active, materially, temporally, and in relation to people. Might we envision a future in which conservation plays a central role at the nexus of intellectual inquiry? Conservation presents opportunities to reconsider how all of us—not just conservators—participate in and contribute to the relationships between people and things that sustain and keep us.*

I write as a conservator, and one at the beginning of his career. On first encounter, the notion of "active matter" seemed unproblematic and requiring no further elaboration. After all, the role of a conservator is conventionally associated with addressing the implications of activity in matter, especially those related to material changes, whether realized in the form of restorative or preventive measures.

In fact, it is hardly controversial to suggest that a view of matter as "active" is central to the discipline and moreover that it is operative on multiple levels. The processes of conservation as they have been established in the present, writ large, are exactly about responding to the expression of such activity as it unfolds in time, doing so through examination and documentation procedures, condition assessments, and preventive or interventive actions. By examining and documenting items, conservators observe and describe their activity in textual or photographic terms, or through other forms of imaging, isolating them at a discrete moment in time. Condition assessments are moreover performed against certain preconceived expectations of activity, largely associated with a scientific understanding of materials and their potential for change. Put differently, the work of conservation provides context to identified activity. Deciding to intervene in various ways may be viewed as a manipulation of said activity, through actions taken to alter its course. In real terms, conservation represents a direct and ongoing interaction with the activity of matter.

As a profession, conservation (at least in the Euro-American context) remains largely organized in materials-based specializations. As mentioned earlier, the scientific knowledge of the activity of materials remains a core paradigm of conservation and is foundational to the way it is structured. Today, conservators are educated on the basis of such classifications and continue to largely identify as specialists on the basis of their understanding of and facility with specific materials (such as wood, metals, stone, paper, textiles, etc.) or their commonly associated composites, such as painted surfaces. Indeed, conservators are generally acknowledged for their acute sensibility toward the material nature of things—one of the factors that make their contributions to material culture studies particularly valuable. In what follows, I use my recent training as a conservator, and my subsequent immersion in an interdisciplinary humanities institute, as a lens through which to look at the essays in this volume—and through them, the practice of conservation more broadly.

Material

Activity, for conservators, if it isn't already clear, has a distinct material bias. Among the most prominent narratives about activity that conservators can relate to are those that conceive of it in material terms, particularly as described through scientific analysis and empirical inquiry and as addressed by the conservator's intervention. Over the past decades, conservators and scientists who work with cultural heritage have gained an ever better grasp of materials and their mechanisms of change over time, and new technological developments continue to enable questions to be asked and answered with increasing accuracy and precision or, at the very least, resolved to more satisfactory ends for the purposes of conservation. These center on the identification and greater understanding of materials found in things at various stages of their transformation through time and changing environmental conditions. This type of research will likely continue to account for the work of the conservation community into the future, but its intellectual foundation presupposes a static and unchanged, core material interpretation of activity. This research is cumulative. A survey of current conservation literature will readily show that notwithstanding the research into the (accelerating) profusion of newly developed materials used for the making of things in the twentieth and twenty-first centuries, conservators continue to grapple with the questions raised by the activity of "old" materials from centuries past, as the contributions by cultural heritage scientists in this volume attest.

Marc Walton and his coauthors address the complexity underlying the preservation of holograms, which were first developed in the early 1960s. While it may seem that one could address the underlying material interface in the form of silver halide emulsions, in order to preserve these otherwise immaterial "sculptures" of light, Walton et al. demonstrate that the key challenge here is documentation, to capture the specific spatial geometry of their light sources that may now be lost, as, for example, is the case for Stephen Benton's *Engine no. 9* series. The caveat with such an approach is, of course, that the generation of such metadata presents its own problems of storage and preservation. Moreover, there is perhaps no better argument against a material-centric view of activity than the act of documenting light with light-dependent techniques: the tools of preservation themselves may no longer be strictly material in nature.

In her discussion of metal soaps found in centuries-old oil paint, Francesca Casadio offers an alternative intellectual framework for activity that challenges the centrality of "materials." Notwithstanding the attention expended on this research and the significant unresolved questions that remain, Casadio suggests that what is really at stake here extends beyond these established lines of inquiry: that the study of metal soaps offers an opportunity not just to examine activity at various temporal and physical scales but also to contextualize meaning across spectra of values.

What about other ephemeral, performative, and virtual "materials" that conservators are increasingly called upon to preserve? Furthermore, will preexisting frameworks for thinking about what constitutes deterioration, loss, function, and existence continue to be useful tools with which conservators evaluate their activity? As Walton et al. and Casadio have demonstrated, perhaps material-centric approaches to activity deserve reexamination. Might expanding how we conceive of material activity to include its contextualization, for example, through processes of documentation, augment the conservator's toolbox of the future? As a result, in addition to being concerned with materials, might we more deliberately frame our actions as participatory in time?

Time

A prevailing narrative of conservation has been that of activity as phenomena to be "remedied." Present in the popular imagination as providers of remedial care to largely, if not exclusively, artifactual things, conservators and their work are often tied to such physical actions as polishing silver, washing garments, mending documents, removing dust and stains, consolidating paint, rebinding book manuscripts, joining ceramic fragments—the list goes on and on. Conservators take these actions on the "front lines" of activity, in pursuit of a reversion to a perceived former or original state insofar as it has been desirable to do so, even while the impossibility of such a goal is clear: at least for the moment, humans are not capable of manipulating time itself. The shift in language toward the acceptance of "retreatability" in lieu of "reversibility" as a desirable outcome in conservation reflects the significance of communicating how conservation may be less about "turning back time" than it is about the process and implications of attempting and purporting to do so.[1]

Yuriko Saito writes about activities of repair, or remedying activity, in precisely this mode. Saito describes what it means to live in a world where repairing is constant, where repair actions mediate the way in which we aesthetically, personally, and emotionally relate to the material world. It may be tempting to interpret the examples of textile mending and ceramic repair that Saito describes as vernacular or colloquial and set apart from what has become codified as professional conservation practice, but I suggest that doing so is a lost opportunity for recognizing conservation as a form of reparative activity: labor that is similarly concerned with issues of (im)permanence, (im)perfection, and (in)visibility. Taking inspiration from Saito, how might conservators redescribe their work in new ways, in a world characterized by repair?

Arguably, the paradigm of reversibility itself arose in part out of the observation of residues and vestiges of past efforts to conserve items, prohibiting later interventions or causing further change or damage. Moreover, all conservation in practice operates with the acknowledgment that there is no substitute for acting upon the actual item; in spite of rigorous testing on suitable material substitutes, there is nothing quite like conserving the item itself.

As a result, best practices in conservation have evolved to privilege materials and techniques that grant a certain degree of control over their application and (future) removal. While attributes such as "stable" or "archival" are more often invoked to describe materials used in conservation including (but not limited to) cleaning agents, adhesives, fillers, and paint, I suggest that it is in fact "control" that guides their use and selection, insofar as they interact with the item being conserved through the conservator's hand in an expected manner. An aspiration for stability or permanence in reality lurks within the sense of control. After all, conservators are intimately familiar with false promises of stability, being expert witnesses to the contrary. Preventive conservation practices are often thought of by conservators to be indirect, even though managing environmental parameters such as temperature, relative humidity, and light levels can have a significant impact on their physical condition.[2] The conservator's hand may be invisible but is still acting in an item's history. While often posited as strategies that minimize future deterioration and loss—that is, instances that may require remedial action in the future—arguably these same measures can be also be characterized as forms of systemic control. In other words, conservation balances our expectations of activity and our ability to manage different materials as they interact over time.

Jennifer L. Mass elaborates on this further, noting that activity is inseparable from time. The expected lifetimes that we ascribe to items of cultural heritage and that we seek to define for modern high-performance materials stem from the same underlying motivation to find this balance. As Mass points out, this balance has conventionally been tipped in favor of permanence, particularly for items that have been preserved beyond their presumed intentions—a notion that has increasingly become untenable today. Even the ways in which it has been pathologized for certain inorganic materials through their description belies a bias toward permanence. It would seem that our expectations of activity outlast these items and materials themselves, to which Mass adds that they may now extend beyond material boundaries to include their narrative potential.

Sherri Irvin's description of Zoe Leonard's preference that *Strange Fruit* exemplify decay through actively deteriorating fruit peels, rather than having the work's conservator, Christian Scheidemann, masterfully preserve the mere appearance of decay in these items, speaks to a similar predilection. Irvin explicates our expectations of decay by considering two other case studies, Kara Walker's *A Subtlety or the Marvelous Sugar Baby, an Homage to the unpaid and overworked Artisans who have refined our Sweet tastes from the cane fields to the Kitchens of the New World on the Occasion of the demolition of the Domino Sugar Refining Plant* and Marc Quinn's *Self*, proposing that the expressive import of decay is that of the potential for psychosomatic response. Notably, "activity" has been selected for its neutrality in the framework of this discussion. But much as decay may be construed as a negative form of activity, might completeness, perfection, and the manner in which they are produced be considered a positive kind? After all, conservation with regard to material activity operates between these poles. Here, we are reminded of Saito's caution: an unchanging world without repair (and decay) is one where we lose the possibility of reckoning with our own mortality. Perhaps we might consider the expressive import of conservation to be, in fact, an active reminder of this possibility, and of the human accomplishment born of trying to overcome it. By conserving, we reassure ourselves of our own activity in this world.

Chris McGlinchey narrates the collaborative efforts of scientists, conservators, and exhibition staff at the Museum of Modern Art to negotiate control over activity in the reperformance of Tania Bruguera's *Untitled (Havana 2000)*, as its matter, sugarcane bagasse, rots and decays. As McGlinchey relates, the significant lengths to which the team endeavored to upend conventional expectations of a rotting material by its desiccation ultimately required a manufactured scent to be added, among other accommodations, for the work to be presented according to Bruguera's vision. Moreover, instituting such control necessarily grew to involve the performers and the gallery space within its orbit. Together, these conservation actions to simulate decay further highlight the significance of Irvin's contribution above and receive greater context in Carolyn Korsmeyer's essay discussed in the following pages.

How might conservators conceive of time? The determination of an item's history remains a productive area of interdisciplinary inquiry in conservation, and conveniently, the oft-described tripartite relationship between scientific analysis, (art) historical research, and the conservator's intuition and connoisseurship come together in writing an item's biography. Conservators understand each conservation action to be fundamentally additive and the conserved item to be an accumulation of all actions performed by it or on it—not just conservation actions. Obviously, this further complicates their task. Can an item perpetually be in a combination of "states" at any given time? What are the metaphysical boundaries of any historical state of an item or its constituent parts? The possibilities are endless, and the trajectory of each item is largely unique. It is this reality that conservators are responding to. While there have been some recent attempts to apply philosophical ideas to sharpen these distinctions via discrete case studies, conservators recognize that much as philosophers have long thought about these issues, they may be less concerned with the specific circumstances that conservators are confronted with: a gap that the contributions from philosophers in this volume have sought to bridge.

Ivan Gaskell and A. W. Eaton investigate the concept of active matter itself, laying the philosophical groundwork for conceiving of activity, change, and the nature of (conservation) interventions that influence them. If conservation is a way of providing a context to activity, then Gaskell and Eaton's considerations are more than theoretical: they are a road map for doing so and a reminder that the occasion for conserving any item fundamentally entails inquiry into the nature of its activity.

Carolyn Korsmeyer explicates the differences in the aesthetic processes belying the construction of age—in real things, in replicas, and in referential shams. As Korsmeyer points out, one sees how these categories are blurred when one considers items that have been conserved. This complex territory is familiar to conservators who balance aesthetic options so that the conserved item remains, perceptibly, a real thing. Yet the ways in which conservators have described and thought about these decisions have been less precise, which, I suggest, hampers a deeper public appreciation for the performance of conservation. A clarification of these processes not only lets conservators more astutely engage with this ongoing debate within their discipline but also allows them to more effectively convey their work to the public. Might doing so lead to more active public engagement with what Aloïs Riegl termed "age-value" and, in turn, allow conservators to work with greater latitude?[3] All of us—not just conservators and philosophers—participate in these processes.

If conservators are fundamentally working *in* time, how might we learn from historians—as a group trained to think *in* time—whose work is concerned with the interplay of events and material evidence? As Ittai Weinryb writes, the motivation to conserve may be understood from the ur-moment of creation; for as long as people have created things, they have preserved them. Defining and contextualizing the role of the maker's intent articulated at the point of creation remains an active area of debate in conservation. Weinryb aptly captures the spaces within which historians and conservators alike work, described as dissonances between temporal, epistemological, perceptual, and perspectival binaries. What emerges are convincing parallels between the work of the historian and conservator, both contextualizing activity within these spaces.

This is exemplified by the hermeneutic challenge of historical inquiry that André Laks identifies in his essay exploring how we might find historical precursors of active matter in the work of Presocratic thinkers. One might consider conservation as the transposition of Laks's project into spatial terms insofar as the accumulations of the past inevitably characterize and shape a conservator's understanding of a specific item.

The recent turn by conservators to preserve time-based media that are ephemeral, performative, and/or virtual, such as those described earlier, has led them to consider new frontiers of activity. If the goal of historical research is to preserve the accumulations of the past, and the goal of conservation to preserve things, does the challenge of conserving these "things" that are in fact events not lead conservators back into the province of the historians? Moreover, as Hanna Hölling and Peter N. Miller suggest, can we not think of all things as events, unfolding over relatively longer or shorter timescales?[4]

People

Conspicuously absent in this discussion thus far has been significant mention of the role of people in contextualizing activity—not just conservators, but all the others present within the orbit of the items being conserved. Even as conservation continues to be preoccupied with issues of materials and time, over the past decades there has been increasing interest in the recentering of the discipline on people and communities. This turn raises important questions: Whose expectations of activity do conservators respond to? Who performs conservation, and for whom do conservators conserve? Here we ask not just *what* is active, but *who.* Notably, these questions are complex and rarely the preserve of individuals or even a single group of people; more often they concern pluralities who may or may not agree in their perspectives. As a result, we have seen the development of values-based decision-making models and significance-based assessments that seek to identify appropriate conservation action by taking into account the voices of relevant stakeholders. These conversations have most prominently been sparked by the challenges to preexisting assumptions about material activity that are routinely faced by conservators working with Indigenous materials and with modern and contemporary works. As the authors below point out, these efforts largely remain siloed in their normative specializations within conservation. A broader shift in emphasis within the discipline has not happened yet. The question is: Why not? or perhaps, Why should it not?

Kelly McHugh describes some of these challenges in the context of her role as a conservator at the Smithsonian Institution's National Museum of the American Indian. She focuses on *Always Becoming*, made in 2007 by Nora Naranjo-Morse, an ephemeral, site-specific, and ever-changing contemporary sculptural installation. McHugh's reflections on the care and stewardship of Indigenous items at the museum leads her to suggest that these conservation treatments were "inspired by and conducted for people, not materials": a significant shift in the conservator's worldview.

Lee Palmer Wandel guides us through the changing ontological relationships between bread and wine, and the body and blood of Christ, over time, as practiced and accepted by various churches. Wandel discusses the process of transubstantiation and the significant lengths to which early modern theologians went in order to account for errors, defects, and accidents that could thwart the transformation of inanimate, not just to animate, but to holy and sacred. Here, the active negotiation of the significance of transubstantiation is a keen reminder that such ontological determinations are not parenthetical, but fundamental to the conservation of religious relics, inspired objects, deities, and other living beings who may persist in material items. Furthermore, these efforts remind us that their conservation may similarly hinge upon sustaining processes that enable animation and life, whether doing so occurs materially through the circumstances of their housing and presence at institutions or through requisite mental, emotional, or physical preparation prior to any interaction.

For Aaron Glass, what is operative in these conversations is the acknowledgment of Indigenous ontologies about activity and how they may necessitate new approaches to the care of their belongings. From my own perspective, and as Glass suggests, to marginalize these concerns as solely relevant to Indigenous items is to relinquish the potential that lies in learning from the broader relational significance of framing the conservation of things alongside that of people and communities. Certainly, there is need for caution so as not to extrapolate, functionalize, or worse, fetishize these perspectives. It is important, and urgently so, for us to seek broader representation of underrepresented communities and peoples within the field of conservation (see Rose Evans's contribution). In fact, that the means by which one acquires the skills, knowledge, and sensitivity to conserve are not equitably accessible to all has created problems of diversity, equity, and inclusion in the field that continue to bear reckoning. But the present reality is that conservators will continue to find themselves at the forefront of bridging gaps of identity between Self and Other in their work and in the experiences of other people. Conservation in the future may well benefit, as is the aspiration of the project underpinning this volume, not just from a broader array of tools with which to address activity, but from a richer array of concepts and more flexibility in wielding them.

Rose Evans describes these opportunities, reflecting on her career over the past three decades as a conservator of Māori descent trained within a Western or Euro-American scientific framework. As Evans points out, the embedding of Māori values within the dominant *Pākehā* (Western) system is crucial to the care of Māori *taonga*, underscoring the importance of institutional commitments to cultivating conservators of Māori descent that are generally absent in Aotearoa (New Zealand) today.

As Evans demonstrates, however much one may consider the "activity" of matter, the personhood of the conservator is equally relevant. Within a paradigm of people-centric activity, neither conservators nor their labor should remain invisible or stripped of subjectivity, recalling themes that Saito introduces in her essay. That is, one might consider conservators as participants, if not "activators," in the conservation of things. Their putatively necessary critical distance from the item being conserved, until now veiled in objectivity or described in normalized codes of ethics, has recently come under scrutiny. The apparatus of conservation, its conventions and procedures, has itself become the site of inquiry. Put together, these impulses implicate the conservation of things within a broader network of relationships that is concerned with the sustenance of meaning(s).[5]

Spike Bucklow contextualizes the activity of rood screens in medieval churches in this manner, seamlessly weaving the material and social histories of oak, oil, chalk, and flint into an ecological framework that considers the interactions of the inanimate with animate beings and of the numerous human participants to whom modern conservators confronted with the preservation of these items necessarily trace their interactions back. Further, Bucklow reminds us that such material and social ecologies may well take on different forms over time and that the differences between premodern and modern frameworks of navigating these relations may not be overlooked.

Marco Leona and Henry D. Smith II offer still another vision of such an approach. Their extensive study of Meiji-period Japanese woodblock prints focuses on the material identification of the changing colorants used at this time. Through their analysis, Leona and Smith contextualize what they term the semiotic activity implicated by shifts and preferences in the use of dyes and pigments for the creation of these prints. They suggest that the binary notions of cost, natural/artificial, or indigenous/imported, were relevant for observers at different times in different ways, complicating a coherent narrative. In short, the activity of these prints extends beyond their color palettes and concerns historical networks of relations revealed by their investigation.

As all the contributions to this volume attest, in various forms, while conservation has conventionally operated under prevailing notions of activity in largely fixed material and temporal terms, it is clear that such a path grows increasingly unsustainable as we look to the future. The refamiliarization of conservation as a human impulse—one centered on people and communities, and not least on conservators themselves, has the potential to refocus conservation on relationships as much as on things.

Where Is the Future of Conservation?

It is not by coincidence that the preceding sections address fundamental questions of conservation that form the organizational basis for the accompanying exhibition sharing the title of this volume. The exhibition was curated by a disciplinarily diverse team including the two volume editors and the four authors of the section introductions, joined by Meredith Linn, a historical archaeologist also on the faculty at Bard Graduate Center. The exhibition traces the conservator's relationship to activity, proposing that to ask, What is conservation? is to consider how things are active and who acts on objects—when and why. But what of the future of conservation?

As the other contributors of this volume have also remarked to varying degrees, the norms of conservation today are, after all, products of conventions that trace their roots to the late nineteenth and early twentieth centuries, when the field was formed. If we are to consider the future of conserving active matter, we might begin with reconsidering the binary oppositions of inert/active and dead/living matter that Guido Giglioni suggests have become unwieldy in the present. They will, surely, only continue to do so in a world in which such distinctions turn out to be either coincidental or fluid. Giglioni traces a history of Western perspectives on living matter from ancient Greece to the early modern period. It is a reminder that such ontological "relational interdependencies" from the long-forgotten past may yet help us in reorienting conservation—contextualizations of activity—in the future.

Jamie Jacobs, a ritual custodian of the Tonawanda Ceremonial Longhouse, shares the history and the significance of wampum to the Haudenosaunee people, where conventional Euro-American conservation definitions of "material" and "activity" are similarly incompatible. Indeed, the scientist or conservator may consider the calcium carbonate that is found in chalk, as relayed by Bucklow, to be chemically indistinct from that found in the shells from which wampum are made, but from its formation to its use, these two forms of calcium carbonate are quite different. Jacobs notes that wampum belts are essential to spiritual life and the transmission of cultural memory as mnemonic devices that aid in the preservation of stories and events. As Jacobs relates, the responsibility of caring for sacred wampum belts is a significant one, not least because they are considered living beings and ancestors. The "scientific care" that conservators may provide remains relevant to the physical preservation of things, but here it is clear that there are instances where it may be insufficient for conservators to do so. Conservators' role of providing context to activity may nonetheless lead them to find that their very act of doing so is inappropriate. Far from implying that conservators' material expertise will become obsolete in the future, I suggest that the manner in which it is applied must be carefully considered. It must be remembered that conservators have traditionally been among the strongest advocates for the repatriation of items, which is itself a form of conservation.

What about instances where Indigenous belongings have yet to be repatriated? Sven Haakanson, a Sugpiaq scholar, narrates his work to preserve the intangible knowledge of making Alutiiq *angyaat* by reviving it through the study of historical models of these open boats presently held in museum and research collections. Haakanson offers that repatriation may take the form of returning knowledge to its living community in the interim and that doing so entails a form of active conservation that engages communities. This model of conservation contextualizes activity as a hybrid of material and people-centered concerns. More importantly, it entails the joint effort of "conservators" as presently understood, and Haakanson and his collaborators, whom we may consider conservators too. Collaboration in conservation typically brings together stakeholders representing different perspectives. I wonder whether broadening the definitions of what constitutes "conservation" and, moreover, who we might consider "conservators" could provide greater legitimacy to the overarching objective of preservation. It might also facilitate the productive intermixing of expertise and ontological perspectives and broaden out the narrow delineation of conservation that has prevailed until now. Might this be one way of finding the intersectionality that Casadio proposes in her essay?

Alva Noë leads us plainly to a remarkably profound expression of conservation: that to conserve is to "make it possible for us to care about the work of art." In other words, that to conserve is to provide context to activity—of material things, of choreographed dance, and of ourselves. Conserving active matter collapses these preceding notions of material, time, and people into the "dynamic singularity" that Noë borrows from Susan Hurley. Indeed, Noë posits that conservation isn't just about maintaining organizational logic—one might think of this as the sustenance of internal coherence, the balance of (visual) elements in a painting or sculpture, perhaps—but also about maintaining the ability to organize ("meanings" or "values" as they have been established in conservation), and of being organized (the possibility of plural futures embedded in the motivations for "reversibility" and "retreatability").

And so, what of the future of conservation? The reality is that conservators hardly work in a vacuum. On the contrary, they work alongside a variety of collaborators, from cultural heritage scientists, preparators, artifact handlers, facilities and security personnel, educators, exhibition designers, curators, art historians, artists, and not least, communities of relevant stakeholders, just to name a few. Some of these interactions take place under the auspices of institutions such as the museum, a collecting, exhibiting, and researching structure, with its accompanying, and purely conventional, guidelines and requirements. The conventional discourse that describes conservation as a reactive discipline (reacting to activity) has relegated conservators to a narrowly defined relationship with things—focused more on their physical properties and less on their embedded intellectual challenges—in the spaces of the conservation laboratory or studio. If we are to take seriously the implications of conserving active matter, might this organizing principle be transformed? Where will conservators find their place, physically and metaphorically, in the future?

The refamiliarization of conservation with active matter presented here impacts not just the discipline of conservation but equally the people with whom conservators collaborate. In many instances, including those described in this volume, it has been precisely such collaborations that have paved the way for this project. Recursively, then, it is perhaps crucial that shifts in the conservation of active matter receive broad support and recognition within these communities too. In some ways, this is the grand project of conservation hermeneutically writ large.

Let us end by considering how the people who work with conservators will change how they do so, into the future. Might we envision a future in which conservation plays a central role at the nexus of inquiry? This volume presents one attempt at doing just this, and I have intentionally structured my essay to reflect how the interests of all the contributors may be woven into a cross-disciplinary narrative about how one might consider conserving active matter. If the aforementioned institutions exist to ensure the preservation of items such that people may continue to experience them, could their underlying guidelines be instead led by conservation? Could we imagine conservation as Noë's California condor, "[i]n its nexus with the wider world" and where its "activity is world involving"? Collectively, the efforts of this volume, its accompanying exhibition, and the Cultures of Conservation initiative at Bard Graduate Center show the promise of interdisciplinary discourse centered on conservation. We may well envision other models for doing so in the future that further elide disciplinary distinctions and that embrace a greater diversity of histories, cultures, and philosophies. If these projects represent the first steps toward such possibilities, I eagerly anticipate the next.

Conserving active matter is a reminder that things are active, materially, temporally, and in relation to people. The work of conservators has been one that provides context to this activity, necessarily involving themselves in this process. In its evolution thus far, the field of conservation has been shaped by the challenges posed by the expanded categories of things under conservators' care that grows with them. But in the future, responding to this activity may not exclusively be the province of conservators, who rely on an increasing group of constituents to do this work. What might proactive conservation look like? I suggest that there remains untapped potential for leadership in the conservator's expertise in thinking about activity. The questions of conservation are not means to an end but present opportunities to reconsider how all of us—not just conservators—participate in and contribute to the relationships between people and things that sustain and keep us. The conservators of the future are those who will orchestrate these discussions around just how we might do so.

NOTES

1. This is compounded also by the reality that certain conservation interventions colloquially understood to be "reversible" are deemed unsuitable and disqualified because they do not provide adequate or substantive support, in which case "reversibility" becomes an ever more aspirational and idealistic goal.
2. "Preventive Conservation—ICOM-CC."
3. Riegl, "Modern Cult of Monuments."
4. Hölling, "Aesthetics of Change"; Miller, "Introduction: Conserving Active Matter and the Historian," in this volume.
5. Muñoz Viñas, *Contemporary Theory of Conservation.*

BIBLIOGRAPHY

Hölling, Hanna. "The Aesthetics of Change: On the Relative Durations of the Impermanent and Critical Thinking in Conservation." In *Authenticity in Transition: Changing Practices in Art Making and Conservation*, edited by Erma Hermens and Frances Robertson, 13–24. London: Archetype, 2016.

Muñoz Viñas, Salvador. *Contemporary Theory of Conservation.* London: Routledge, 2011. First published 2005.

"Preventive Conservation—ICOM-CC." International Council of Museums–Committee for Conservation (ICOM-CC). Accessed June 2021. http://www.icom-cc.org/36/working-groups/preventive-conservation/.

Riegl, Aloïs. "The Modern Cult of Monuments: Its Character and Its Origin." Translated by Kurt W. Forster and Diane Ghirardo. *Oppositions*, no. 25 (Fall 1982): 21–51.

Introduction: Conserving Active Matter and the Historian

Peter N. Miller

Abstract

What contribution can a historian make to conservators' thinking about active matter? The historian might be able to bring a long-lens perspective to the problem of activity by locating current discussions in the millennium-rich global literature on ruins and ruination. Everywhere we find ruins we are seeing a conservation discourse. The historian can also help ground the current discussion of active matter by identifying forgotten or overlooked presentations of the activity of matter, demonstrating their connection to one another, and suggesting reasons for the eclipse of this approach. Recovering the history of active matter offers a bridge to non-European traditions of thought in which matter's activity was never doubted. This turns the table on the presumed normativity of European—in fact, one strand only of European—views of matter. With this comes the possibility of a new kind of conservation thinking that is both global and true to Europe's own diverse heritage. Finally, reapproached through the lens of quantum physics, active matter offers a way to engage profitably with a range of questions that have been generated by the quantum revolution. A quantum theory of the object (QTO) could anchor conservation within a human sciences of "becoming" and "process."

> All things are full of gods
> —Thales of Miletus

Before we begin, we need to object: Is there not something fundamentally paradoxical about conserving active matter? Don't the terms pull in different directions? We will leave the paradox for later and the definitions for a moment. Instead a story, and a very familiar one, that introduces conserving active matter not as an abstraction but as something important to us all: near at hand, and close to heart.

At the end of book 5 of *The Odyssey*, the hero has washed ashore on the island of the Phaeacians. Exhausted from his swim, he hauls himself up the beach and collapses in a close-grown thicket, but not before burying himself under a pile of dead leaves to protect against cold.

> As when a man buries a burning log in a black ash heap
> in a remote place in the country where none live near as neighbors,
> and saves the seed of fire, having no other place to get a light
> from, so Odysseus buried himself in the leaves.[1]

"Saving" the fire from going out, saving his life from expiring—this, surely, is an act of conservation. But the fire, like the sleeping body, remains active. It is "banked," covered with ash, slowed, stabilized, managed. But it is alive.

In what follows I will speak as a historian who lives in an institute with philosophers, art historians, archaeologists, anthropologists, and materials scientists and has been immersed for ten years in the questions of conservators and cultural heritage scientists. "Conserving," "active matter," and "conserving active matter" are not terms that historians think much about. Talking with conservators and with my colleagues from other disciplines, I've come to see their importance—not just for conservators but also for historians. Nor are these but the inevitable benefits of a cross-disciplinary conversation that, like a fascinating dinner party conversation, may leave a happy feeling the next day, but not much else of substance. No, as I hope to show, these terms take us to very important issues, ones that historians could—perhaps even *should*—address.

Activity, Ruins, Conservation

Wolfgang Schäffner would have us pay attention not only to the history of materials but to the way that this particular history intertwines with other histories. Walter Benjamin made the association for all time between cast-iron architecture and the *passages* of Paris (and his imagined alter ego, Carl Roseman, took this one logical step further and annexed the architecture of New York's SoHo as well). Schäffner points out how those nineteenth-century materials shaped the vision of the new field of conservation as it was then developing. With iron shaping its horizon, it is no surprise that conservation would view the stability of matter as a good. (There is also a long history behind this, as we will see below.) In the same way, if conservation were beginning right now as a field, it might more intuitively respond to the pervasiveness of activity.[2]

But, as it stands, "activity" is not how conservation as a profession has mapped its territory. The International Council of Museums–Committee for Conservation in 1984 defined the work of "preservation" as "action taken to retard or prevent deterioration of or damage to cultural properties by control of their environment and/or treatment of their structure in order to maintain them as nearly as possible in an unchanging state."[3] The American Institute for Conservation defines the profession of conservation as one dedicated to the "preservation" of cultural heritage and "preservation" as work "that minimize[s] chemical and physical deterioration and damage and that prevent[s] loss of informational content."[4] Granted that these kinds of professional definitions are the supertankers of the world of ideas: slow moving and hard to maneuver. The fact that "slowing," "stabilizing," "minimizing," and even "managing" change—once upon a time the goal was "arresting change"—are, of course, ways of dealing with activity without daring to speak its name makes our point.[5] Key terms related to activity are heavily value laden. Change is "degradation." The change set in motion by the material constitution of the object is "inherent vice."

But maybe things are beginning to change. *Conservation of Cultural Property—Main General Terms and Definitions*, published by the Republic of Italy in 2012, defines "alteration" as "change in condition, beneficial or not, intentional or not"; "aging" as "natural alteration over time"; and "deterioration" as "gradual change in condition that reduces significance or stability."[6] Though even here, we have to note, "activity" still roams undefined.

Ivan Gaskell and Anne Eaton put "activity" under the analytical philosopher's loupe. They see it as something that can be imposed on matter from without, as something that comes from within, and as something that comes from within but can be triggered by something coming from without. Activity could interact with matter to create new matter, could interact with matter to destroy matter, and could trigger random behavior by that matter. What they do not find is biologically inactive material that is yet capable of generating activity from within itself. Gaskell and Eaton point to the frontier between "biologically living" and nonliving active matter as marking both the zone of greatest uncertainty and where the future of active matter will play out. Achim Menges and the Berlin group presented pine cones as a readily grasped example of the activity of nonliving matter—dead cells opening and closing. Even in this case, however, Gaskell and Eaton note that there remains the triggering work of moisture in the air, however minimal. That they would argue that only a god can create from nothing might, in practice, only mean that for everyone and everything else, activity has a context.

Philosophers look at the natural activity of matter—always declining from a higher energy state to a lower one—and see it as bound up with our understanding of life itself. Humans, for example, have tended to experience activity as a negative: closely related to decay, dissolution, and death, though Yuriko Saito's exploration of the Japanese *wabi* aesthetic reminds us that this needn't be considered a universal disposition. For anyone encountering change, but especially for those who do not like its effects, conservation can be an assertion of the values of life. So, medicine, whether practiced by a doctor or a shaman, is a form of conservation. History, whether practiced by a poet or a priest, is a form of conservation. Architecture is a form of conservation. And so on. If entropy is the background condition of all things, then the effort of "conservation" to slow its course could be another way of describing the work of *cultura*, the Latin term for the human intervention in the natural world. This intervention suggests that "culture" and "conservation" have long been connected. Maybe, even, we could formulate a strong argument that culture is actually *made* by acts of conservation, both material and mental. That's how important conservation is.

"Active matter" encourages a new way of thinking about conservation by putting the emphasis squarely on activity and not its evaluation. Another thing we learned from Conserving Active Matter was that in the world of contemporary art conservation these questions were already being asked—but without the language of "active matter" to connect them to other fields of inquiry. Chris McGlinchey casts his essay not only as a first-person report from the field but as a report from the field set like a jewel into a fragment of a *Bildungsroman*. He is the young scientist blithely oblivious to change, even when pointed out to him by conservators, who then grows through experience into the wise head who can reflect on a complexity he never imagined, but that he has come no longer to perceive as a threat: "The tolerance for authenticity can be variable and arguments for what exactly is authentic, complex." With works such as Tania Bruguera's *Untitled (Havana 2000)*, the activity of matter, which had stood as a threat to the human and the work of humans, and had been either repressed or denied, was now embraced. What would this mean for a professional practice that was dedicated to defending the prospect of permanence against the sharp, eating teeth of time?

Once we stop to think about it, we see that the impermanence of things and the corresponding need to conserve are a fundamental human experience. We could even describe hominins as the tribe that conserves. A million years ago in what is now South Africa, a hand ax was made by our ancestors. And not long after, it was reknapped, or repaired, to maintain its usefulness.[7] But the significance of the encounter with activity, and of the preservationist response, can perhaps be most clearly discerned in the endeavor to preserve not tools only, but memory itself. The erection of menhirs and dolmens across preliterate Europe and of pyramids just on the other side of literacy in ancient Egypt are the biggest, and oldest, representations of this desire to defeat the force of change, decay, entropy, or as we might say, matter's activity. Mesopotamia's kings, building with mud brick that they knew eventually to decay, sought futurity by stamping their names on the bricks and leaving in the foundations of the temples they erected but knew would collapse texts urging their successors to rebuild them and perpetuate their names. Around the same time, in Middle Kingdom Egypt, a poet set out to argue that words made for a more enduring monument than stone because nothing lasted longer than a human memory that could be passed along from generation to generation.[8] Alain Schnapp reminds us that Pindar's Sixth Pythian Ode resumes this theme, for which Horace provided Europe, at least, with its canonical formulation: "I erect a monument more permanent than bronze" (Exegi monumentum aere perennius). Our records from one of the other ancient civilizations, China, show the same feelings. Stephen Owen sees the *Analects* of Confucius as the moment when the past became distant and needed to be sought after. "He said, 'I am not someone who was born knowing it; I love the past and am someone who seeks for it earnestly.'"[9] The "poems of memory" sprung from this. They are efforts at conservation—yet efforts that reinscribe loss within every attempt at remembrance.

> Hold with the past, don't lose the past:
> If you lose the past, your will easily breaks;
> If you lose the past, even the sword snaps;
> If you lose the past, the zither too laments.
> And the Master's tears for the loss of the past
> In those days fell streaming in torrents.[10]

I knew about ruins, and I knew about their study. But I never realized that the discourse of ruins, which continued unabated into the nineteenth century, at which point industrial modernity, rather than antiquity, became its focus, was a way to discuss the dialectic of matter's apparent stability but inherent mobility. Carolyn Korsmeyer's essay takes up their challenge and explores the implications of different kinds of ruins. Alva Noë's essay goes a step further, asking whether death—or "ruination"—is not just a different way of talking about living. In wrenching around our perspective, Noë brings us back to Buddhism. Lurking in the background of Yuriko Saito's discussion of Japanese aesthetics of degradation (*wabi*) is the very different ontology, or understanding of Being, that came from Buddhism, even though it was a fourteenth-century Japanese Buddhist priest who acknowledged that "people *commonly* regret that the cherry blossoms scatter or that the moon sinks in the sky, and this is *natural*."[11]

Which means that wherever we find ruins depicted or invoked, whether in image or in word, we find ourselves in a conservation discourse. And since there is such an expansive literature of ruin, from all places of human habitation and, seemingly, from all times, we need to acknowledge that preservation is a fundamental human feeling, experience, need.

If we look closely into this discourse of ruins, and even if we never stray from its high road—documents like Hermannus Posthumus's 1536 *Landscape of Ruins* or Sir Thomas Browne's 1657 *Hydriotaphia, Urn Burial* or Percy Bysshe Shelley's 1818 "Ozymandias"—what we find is that the "ruins lesson" is, in the end, always about time's passage.[12] Leonardo da Vinci, he of flying machines and armored vehicles, was a contemporary of the Frisian Hermannus Posthumus and like him a reader of Ovid's *Metamorphoses*, the same lines from which they both quote: "O time, devourer of things! O envious age! Thou dost destroy all."[13] William Shakespeare, he of Hamlets and Falstaffs and Pucks, was also a reader of *Metamorphoses.*

> When I have seen by Time's fell hand defac'd
> The rich-proud coat of outworn buried age; . . .
> Ruin hath taught me thus to ruminate—
> That Time will come and take my love away.[14]

Or:

> To ruinate proud buildings with thy hours,
> And smear with dust their glittering golden towers;
> To fill with worm-holes stately monuments,

> To feed Oblivion with the decay of things
> . . . .
> to spoil antiquities of hammered steel
> And turn the giddy round of fortune's wheel.[15]

Sherri Irvin reflects on a class of contemporary artworks whose subject is ruination and that *exemplify* their subject by ruining themselves in the course of their display-life. If Fluxus works from the 1960s expose some *aporie* of active matter, they do so by exemplifying transience, fragility, and lack of self-regard. But Irvin's objects of inquiry are not "about" transience. They are about decay. And they *are* decay. Irvin's analysis helps us understand the astonishing staying power of the discourse of ruins. Like Zoe Leonard's *Strange Fruit*, but at a very, very much slower pace, the ruins of some man-made things allow for an intergenerational philosophical experience. Their power lies in the fact that, like the neutron star that remains after a supernova, when all else—people, history, even myth—is gone, time remains. And unlike those early photographs whose necessarily long exposure had the effect of making activity vanish, ruins magically do the reverse: they accentuate it. Hence, Marguerite Yourcenar's essay on conservation bears the title *That Mighty Sculptor, Time* (1984). How important to us is this struggle with *Tempus Edax Rerum*? We could even write an equation to state its worth in monetary terms. It is the cost of climate control (structures, equipment & utilities) + insurance + security + conservation services (people & equipment) ÷ number of objects in a given collection.

Just beyond, and a constant threat to the strongest of minds, lurks nostalgia, or the interiorization of this experience of time's passage registered on external things. As the famous archaeologist Sir William Matthew Flinders Petrie wrote, in a blurring of archaeology and subjectivity, "We dwell upon those portions of the past, those days at Athens, or Florence, or in the Forum, as on a treasure; they are a portion of our life crystallized into the structure of our thoughts—a haven of the imagination."[16] This is Robert Pogue Harrison's very point when he talks about the "humic" layer as holding "in its conserving element the unfinished story of what has come to pass." Burial, he writes, does not mean only inhumation; "it means to store, preserve, and put the past on hold." And, once so preserved, "dead etymons, latent meanings, and lateral connotations lie buried in the roots and phonemes of our living words, where they carry on an active afterlife. Our psyches are the graveyards of impressions, traumas, desires, and archetypes that confound the law of obsolescence."[17]

Historians work on time all the time. But sometimes they forget to think about what that means. Listening to conservators can bring back the philosophical significance of what we are doing. So, for instance, while Cesare Brandi's *Teoria del restauro* (1963) is the classic text of modern art restoration, he will not be read by historians. And yet his chapter titled "The Role of Time in a Work of Art" could easily be read by history students as a lesson in what makes historical evidence. For example, he describes the phenomenon of patina as "the sedimentation of time on the work." So bound up are they that Brandi forbade the removal of patina or any other trace of the object's passage through time.[18] Conservation treatment occurs, he went on, in "the second historicity" of the object. This is what Yourcenar means when she writes in the very first sentence of her essay, "On the day when a statue is finished, its life, in a certain sense, begins."[19] And with this we are teetering on the very brink of seeing conservation as a way of exploring "the social life of things."[20]

Active Matter in the History of Thought

Introduced by conservators to the problem of the activity of matter, the historian might naturally be tempted to wonder whether there was a history of this way of thinking. What said historian might find is some scattered comment on "hylozoism" designed to wrap up in a nice, neat package a whole series of thinkers who in a post-Newtonian age have come to seem heterodox and marginal and just wrongheaded. And yet.

Thales of Miletus's famous comment about the pervasiveness of "gods" is a very old and very poetic way of talking about the pervasiveness of activity in a surrounding material world that need not be considered distinct from our subjectivity. In his essay for this volume, Guido Giglioni outlines a trajectory of more recent thinkers who struggled with the reality of an interpenetration of matter and spirit, from Aristotle's *hulê* (fourth century BCE), Galen's *dunamis* (second century CE), Paracelsus's matrices (d. 1541), and Petrus Severinus's *spiritualis materia* (1571) through to Francis Glisson's *vita materiae* (1677). Alongside the philosophical and medical tradition of animated matter there is the equally robust alchemical tradition.[21] Nor should we underestimate its appeal at the heart of phenomena like the revival of antiquity. When Marsilio Ficino brought the *Corpus Hermeticum* from the ancients to his fellow Florentines in the age of Lorenzo the Magnificent, he also passed along Hermes's catechism: "Is God then in matter, O Father? Where could matter be placed if it existed apart from God?"[22] A version of this same question lurked in the many pages and almost endless debates of the councils of the early church before returning with a kind of vengeance in the Reformation-era conflict over the Eucharist. Lee Palmer Wandel reminds us how central to Protestantism was the rejection of "transubstantiation," with its inherent doctrine of living matter. The sharper split between matter and life effected by Martin Luther plays in the language of theology the role René Descartes would play, philosophically, in the seventeenth century, in fostering a hard dualism. The "spirit of capitalism," to look to a context not explored in our volume, could turn out to have as much to do with dualism and its discontents as with double predestination. Rabbi Judah Loew of Prague's Golem was not a commentary on these contemporary debates, but the mass of clay that became animate when a sacred text was placed in its mouth also belongs to the rich history of active matter.

The "new science" is often identified as the moment when dualism hardened into a victorious orthodoxy that squeezed the activity out of matter, but we should be careful about seeing this as an "either/or." While it was Galileo Galilei's enemies who feared for the "imperfections" of God's creation made visible by his telescopic observation of the moon in 1610, it was the scientist who embraced activity. In the words of Francesco Sagredo from the first day of the *Dialogue Concerning the Two Chief World Systems* (1632): "I can only hear with greatest reluctance that the properties of the unchangeable and unchanging are considered something noble and perfect and, in contrast, mutability is considered something imperfect. I consider the earth to be extremely noble precisely because of this, of the changes that take place on it, and the same is true of the moon, of Jupiter and other planets."[23]

Baruch Spinoza's blurring of ontology and activity could have been a comment on Thales. *Deus sive natura*, "God or Nature," he wrote, and with this flourish dualism was undone, or would have been, if Christianity, which still guarded over ontology like an only child, had been able to let it pass. But it couldn't, and the marginalization of Spinoza and his followers only solidified the orthodoxy of dualism.[24] We would do no violence to the past to redescribe the various controversies sparked by the pulsar-like appeal of Spinoza at different times and in different places, such as the "pantheism" controversy in late eighteenth-century Germany, György Lukács's recruiting of Spinoza (and Friedrich Wilhelm Joseph Schelling) to the anti-dualist goal aimed at by Karl Marx, or the Louis Althusser–Gilles Deleuze reading of Spinoza in late twentieth-century France, as chapters in the history of thinking active matter.[25] In fact, the passage in which Kant defined active matter as an impossibility, cited above from his *Critique of Judgment,* came in a section devoted to attacking Spinoza.

Yet, however we have gotten there, if we look to the nineteenth century in Europe on the eve of conservation's professionalization, it is clear that a dualist ontological orthodoxy had arisen—and was already taken for granted. We can see this clearly when we delve with those critics of this regime who, looking back, point to what they take to be their nemeses. For Goethe, it is Newton who is the great upholder of dualism. For Schelling, it is Kant. For Marx, it is Hegel. For Heidegger, it is Descartes. "Active matter" as a term of art may have been brought to our attention by materials science, but paying attention to it—another argument for the value of interdisciplinary conversation—makes visible a major fault line in the history of knowledge.

Johann Wolfgang von Goethe's struggle with the philosophy of Isaac Newton is famously played out in terms of his color theory. But it is in his notion of morphology that he strikes out most decisively against a rigid dualism of forces acting and acted upon. Goethe sees a complex interplay between internal and external agents of causation, much as Galen and the Galenists had struggled to explain the individuation of matter as life.[26] The most profound taking up of Goethe's term was Ernst Haeckel's *Generelle Morphologie der Organismen* (1866), which began with a long citation of Goethe on its flyleaf, and others introducing the parts of the book and at the head of *every* chapter. Haeckel deploys morphology in support of Darwinism—but also to attack dualism and defend a conception of active matter.[27] He finds in Anaximander, Heraclitus, and Empedocles the teaching that "nowhere in the world exists absolute rest, and that all standing-still is but apparent, we are compelled everywhere to assume a perpetual change of matter, a constant variation of form."[28] For Schelling, the whole burden of Johann Gottlieb Fichte's interpretation of Kant's transcendental defense of the possibility of scientific knowledge reified a split between the thinking subject and everything else. Much of Schelling's early work, through his 1809 treatise on freedom, is about peeling back idealism so as to recover a unity of thinking and feeling that was lost with the dualism of Descartes and Kant. Marx accused Georg Wilhelm Friedrich Hegel of "failing to overcome the duality of thought and being, of theory and practice, of subject and object," and his dialectic, conceived of as an instrument for "the ending of a rigid confrontation of rigid forms is enacted essentially *between the subject and the object.*"[29]

Finally, Martin Heidegger's magnum opus of 1927, *Being and Time*, points the finger of blame at Descartes. It is his attempt to defend the certainty of the thinking being ("Cogito ergo sum") that for Heidegger crystallizes the path wrongly taken. But for Heidegger the problem goes even deeper and even further back. Indeed, one could argue that Heidegger's self-proclaimed *Destruktion* of Western philosophy is about unravelling the historical foundations of dualistic thinking. He calls it the history of Being, and in this narrative Being loses its way with the departure of philosophy from Ionia and "Presocratic" thinkers such as Anaximander and Heraclitus. In this volume, André Laks explores the notion of "Presocratic" and then grapples with the way "Becoming" has been attached to the legacy of these Ionian thinkers, sometimes at the expense of their own dualism. The story of active matter before the twenty-first century's "Active Matter" can, then, be run from Thales's gods-drenched landscape of the early sixth century BCE to Hermann Usener's discovery in late nineteenth-century Bonn that there were gods for every moment of need (*Augenblicksgötter*).[30]

Friedrich Nietzsche, a decade removed from his lectures on "Preplatonic Philosophy," went so far as to identify the struggle *against* active matter as coterminous with the broader history of (European) culture. His point is that modern (European) culture was the product of an enormous effort to neutralize and, thus, overcome a feeling of powerlessness vis à vis nature. Francis Bacon, Descartes, Newton, Kant—the whole edifice of "science" that put human beings in control was an effort at rewriting a still-fearful reality.

> Because it was thought for many thousands of years that *things* (nature, tools, belongings of all kinds) were also alive and animate, with the power to injure and to elude human purposes, the feeling of powerlessness among humans was much greater and much more common than it need have been. . . . But because the feeling of powerlessness and fear was in a state of almost perpetual excitation for so long a time . . . the feeling of power has become their strongest inclination; the methods they discovered to create this feeling just about constitute the history of culture.[31]

If Heidegger's was the high philosophical attack on Western dualist accounts of being, Georges Bataille's was the lower. From his perch atop the masthead of *Documents*, Bataille assailed Plato and the theory of ideas that exercised influence through his works and the teachings of his Academy. In his essay, "The Academic Horse," Bataille juxtaposed ancient Gallic and contemporary African depictions of horses to undermine the superiority of Greek art. Debunking idealism was also the subject of another essay, on gnosticism. Bataille identified in the "base materialism" of the gnostics a means of assaulting the abiding dualist ontology. "Base" invoked the Western hierarchy that saw all things material as base. But he then worked to subvert the hierarchy by showing that matter, even the basest matter, was not so passive after all. "In practice, it is possible to see as a *leitmotiv* of gnosticism the conception of matter as an *active* principle having its own eternal autonomous existence as darkness."[32]

If we return now from Bataille to Fluxus, we can see the latter's relationship to surrealism extending deep into its philosophical commitments. Its attack on the dualism underpinning "high" art is an ontological challenge that emerges clearly through the matrix of issues that confront conservators of Fluxus art.[33] Perhaps it is no coincidence that when giving an example for how "diagonal science" could connect the human and natural sciences, Roger Caillois chose active matter.[34] Jennifer Mass takes the biomedical language of conservation—"bronze disease," "ultramarine disease," "glass disease"—and gives its history from Pliny to now.[35] Conservation as a field was born in the same nineteenth century as modern medical positivism, and both were dedicated to purging an inherited and, they believed, mistakenly applied set of terms to describe the lives of inorganic things—until scientists discovered a few years ago that it actually *was* a fungus that created the copper oxalate alteration sometimes found on ancient Egyptian artifacts. It may not actually be the bronze that is animate, but this is a tale worthy of *Documents.*

Around the same time, but in the postrevolutionary Soviet Union, Lev Vygotsky developed cultural historical activity theory—essentially a representation of the philosophy of history of the young Marx ("Theses on Feuerbach," "German Ideology") in terms of activity. It had its roots, according to a recent student, in Goethe's rebellion against the abstraction and empiricism he associated with Kant and Newton. The "morphology" that he developed in his plant studies enabled him to skirt the dualisms that these predecessors had relied upon to establish certain knowledge. In this genealogy of activity, Hegel learned from Goethe, and Marx reformed Hegel. And if Stalin's purges extinguished cultural historical activity theory as historical interpretation, it did reemerge in the 1960s as a form of psychotherapy, and there is even the suggestion of a further, and wider, rehabilitation of activity.[36]

In France, in the 1930s and 1940s, André Leroi-Gourhan developed his own approach to material culture that was heavily influenced by the nineteenth-century naturalist and antiquary Georges Cuvier. He borrowed from Cuvier the notion of "mécanique vivante" to describe what he saw as a very blurred boundary between matter and activity. Georges Canguilhem, around the same time, also talked about how "'the first tools were no more than prolongations of human organs in motion. Flints, clubs, and levers prolong and extend the arms' organic movement of percussion.'"[37]

To end this very brief flying overview of alternative histories of matter, let us note Carlo Ginzburg's turn back to that same Cuvier, whom he describes "as a new species of antiquarian," in the new introduction to the second edition of his *Night Walkers* (*Benandanti*). Cuvier's use of morphology to identify species and their changing relationships leads Ginzburg back to Goethe, whose development of this analytical framework he describes as Goethe's real gift to the history of knowledge. In other words, once we start taking active matter seriously, we begin to see that others have too.[38]

Activity and Deprovincialization

From the perspective of Indigenous peoples, however, all this might seem like a long story about getting back on the highway after having taken a wrong turn. Historians listening to Indigenous voices might have come to see the robustness of active matter thinking before their very eyes.[39] As Aaron Glass observes in his introduction to the essays in this section of the volume, for Indigenous North Americans—and, with qualifications, this could stand for the broader category of Indigenous peoples worldwide—the narrative is flipped. Their conceptual vocabulary was, and remains, more like what we have been finding in the Presocratics or Spinoza or Schelling or even Heidegger. For Indigenous peoples, matter has long been viewed as possessing active properties. Sometimes forms, such as ritual objects, required periodic change over the course of their lives. Other objects were active because they were believed to situate spiritual power or agency. And then there were objects that were active, not because of *what* they were made of, but *how* they activated networks of humans.

Jamie Jacobs and Sven Haakanson give us two slightly different examples of humans activating matter. Jacobs explains that for the Haudenosaunee people wampum was valued for the hardness of its elemental calcium but that this hardness only mattered because of the messages invested in it by living people who, like those ancient Mesopotamian kings, sent them into the future confident that their descendants would eventually be on the receiving end. Haakanson's tale of the *angyaaq*, or open boat, of the Sugpiat people of Kodiak Island in the Aleutians is a tale of research, with archival and museum collections providing the past's knowledge, stored against malign fate, that enabled Haakanson to resurrect an extinct cultural form and, through that form, to reconstitute a human nexus and a social practice.

Rose Evans and Kelly McHugh focus on the kind of activation that can happen in a museum setting. Evans's memoir tells the story of how a conservator in New Zealand from within the Indigenous community could bring community values into the theory and practice of conserving that community's heritage. Back in North America, McHugh shows us how community values could be incorporated into the stewardship of Native objects, especially by museum conservation departments. For our purposes, the case study she chooses to illustrate the way principles are put into practice could not be more appropriate: a sculpture with the title *Always Becoming*. For conservators, she writes, it is a lesson in learning to accept "change as inherent and beautiful" and "implementing this Indigenous practice of 'active' or relational conservation" (267). We could think of this as defining "care" as a form of preservation.

Historical and anthropological approaches braid together. Ittai Weinryb, in his introduction to the essays from the History working group, presents a different active matter spectrum. It is composed of sourdough starter, a twelfth-century European depiction of God creating the world, and a thirteenth-century missionary account of fetish gods of the Black Sea steppes. From the perspective of the history of objects, we still find ourselves asking about the ways in which activity is both in the thing and in the human relation to the thing. Bruno Latour's actor-network theory might come to mind here for some; Lewis Hyde's *The Gift*, for others.[40] Henry D. Smith II and Marco Leona argue in their study of artificial colors in later nineteenth-century Japan that the "'activity' of materials must always go beyond materiality itself, and take into account the changing meanings assigned to those materials, and the ways in which they are chronicled historically" (378). We can read this as an argument for "collaborative conservation" of a different kind. When dealing with a deeper past for which there are no living interlocutors, the conservator can partner with those able to make the past speak in languages that we can still understand if we listen carefully, whether these be historians, anthropologists—or poets. To return a last time to Robert Pogue Harrison, he suggests that conservation needs to work along the full spectrum of meaning making. "The dead speak from beyond the grave as long as we lend them the means of locution; they take up their abode in books, dreams, houses, portraits, legends, monuments, and graves as long as we keep open the places of their indwelling."[41]

For Spike Bucklow, the dead speak from within the very walls of churches. He finds active matter where we might now expect it—in a medieval East Anglian partnership of physical labor and communal ideal—and where we might not—in the limestone (chalk and flint) walls that brought an ocean's activity from deep time to stand side by side with living medieval Englishfolk. In all these forces upholding the church's walls Bucklow sees a "poetic boundary crossing" and "the natural interactions between relatively fluid materials." He argues that if conservation attached itself to chemistry at the start of the twentieth century, in the twenty-first its partner and guide will be ecology. And as we look to this moment from the vantage point provided by active matter, we can see how historians can learn the same lesson from listening to conservators of kayaks and conservators of rood screens.

If the differences that separate "fine art" and "ethnographic" conservation disappear when viewed through the prism of activity and active matter, then maybe there are also lessons to be learned from how other, distant cultures think about conservation. This means more than simply adding Indigenous American or Japanese or Indian ontologies of matter to a Greco-Euro-American "default" mode. Moving from different ontologies to different conservation practices means allowing objects to live different kinds of social lives. Outside Europe there were cultures of conservation whose grounding ontologies were, and remain, nondualist. Looking out from the workshops and studios of Taos, Kyoto, and Delhi, it is Euro-American conservation that could seem exceptional. One of the immediate outcomes of our project in colliding ontologies and crossing diagonals is the "provincialization" of Western conservation history.[42] Looking at this, the historian might wonder whether there are similar lessons to be learned in the way the past can be grasped from those not operating within the modern, university-driven practice of history.[43]

Recognizing what is provincial in what is so easily taken as the default mode, the assumption that conservation simply is what happens in Euro-American contexts, enables us to identify the *de-provincializing* turn in conservation—"collaborative conservation" coming from the "ethnographic fringe" to change the way Old Master paintings, or ancient sculpture, get conserved—with the same move Bruno Latour described as anthropology "coming home from the tropics" in *We Have Never Been Modern* (1991). The publications of the past two decades devoted to alternative, nondualist approaches to ontology only make clearer the stakes of his argument.[44] For Latour, the claim of "modern" is built on a dualism insistently dividing modern from premodern even at the expense of ingesting a host of paradoxes, but also a modern "us" from a distant, premodern "them" that fails by *not* being dualist.[45] Without being aware of, or intending to situate, conservation at the core of either modernity or his proposed nonmodernity, Latour's argument does just that. He describes "intentionality," a key term in conservation theory and practice, as a "notion" that transforms a distinction, a separation, a contradiction, into "an insurmountable tension between object and subject." The moderns looked out and "were always struck by the diffuse aspect of active or spiritual forces in other so-called premodern cultures. . . . Spirits and agents, gods and ancestors, were blended in at every point." The future, he writes, involves taking this sphere into account. A conservator of Indigenous materials, such as Kelly McHugh in this volume, could have written the following sentence of Latour's: "[T]he paradigm of the knowledge of objects has to be replaced by the paradigm of mutual understanding between subjects capable of speech and action."[46]

And time itself is at the heart of one of Latour's key examples of the paradoxical nature of the claim to be modern. He notes that while "the moderns indeed sense time as an irreversible arrow," they also "want to keep everything, date everything, because they think they have definitely broken with their past." The more dramatic the ruptures—Latour nods to the political revolutions of the eighteenth and nineteenth centuries—"the more they save; the more they capitalize, the more they put on display in museums." Conservation and collection become, like ruins, ciphers for transience; archives, for all their vastness, indices of absence. "Maniacal destruction," he writes, "is counterbalanced by an equally maniacal conservation." In the very next sentence, Latour recruits Leopold von Ranke and the passions of archival scholarship to this desperate psychological crabwalk. The past reconstituted "as it actually was" is both nevermore and ever insurgent. "Historical reconstitution and archaism are two symptoms of the moderns' incapacity to eliminate what they nevertheless have to eliminate in order to retain the impression that time passes."[47]

Objects As the Poetry of Time

Latour's recruiting Ranke for a lesson about conserving makes me ask: Can historians learn about time from conservation? Historians have only recently begun to take objects seriously as historical evidence. Conservators know a lot about objects. But it isn't from how conservators currently think about objects that historians stand to learn the most. It's from how conservators *could* think about objects.

Karen Barad focuses our attention on the question of ontologies in conflict. Writing about space, Andrew Pickering notes that "a taken-for-granted ontology of neutral space (the 'featureless container of events') fosters a certain class of 'modern' sciences and of engineering projects, and these reflect the ontology back to us as self-evidently true." Change the ontology and you change "how we think," not only "about space, but about how we go on in the world."[48] Similarly, Latour names one part of his critique "Variable Ontologies."[49] It is in the disagreement between Werner Heisenberg and Niels Bohr that Barad sees that conflict of ontologies. Where Heisenberg argued that measuring disturbs the existing reality, Bohr's point was that measurement fails in the face of a deep indeterminacy. Barad concludes that "[w]hat he is doing is calling into question an entire tradition in the history of Western metaphysics: the belief that the world is populated with individual things with their own independent sets of determinate properties."[50] With Bohr, individuals do not exist, and thus intentionality does not exist; "the very binary between 'interior' and 'exterior' states needs to be rethought."[51] Barad tells us that she wants to understand the implications of quantum physics for the epistemology and ontology of the old Newtonian world. As we have noted, the field of conservation has relied, since the nineteenth century, on that same Newtonian epistemology and ontology. What are implications of a different ontology—whether from "Indigenous peoples" or materials scientists—for conservation knowledge and practice?

Let me begin my answer with a provocation. If we follow the argument of this essay to its natural end point, we arrive at what we might call the *quantum theory of the object* (QTO). In *The Order of Time*, the physicist Carlo Rovelli describes matter from the perspective of quantum physics.

> The hardest stone . . . is in reality a complex vibration of quantum fields, a momentary interaction of forces, a process that for a brief moment manages to keep its shape, to hold itself

> in equilibrium before disintegrating again into dust, a brief chapter in the history of interactions between the elements of the planet, a trace of Neolithic humanity, a weapon used by a gang of kids, an example in a book about time, a metaphor for an ontology, a part of a segmentation of the world that depends more on how our bodies are structured to perceive than on the object of perception—The world is not so much made of stones as of fleeting sounds, or of waves moving through the sea.[52]

Rovelli's point is that from a quantum perspective, nature is not about objects (or subjects) marked off from one another; rather, it describes a world of relations, where the reality lies in the relations themselves. "Individual objects *are* the way in which they interact. And since they interact with us, then we as observers are part of this same reality, not just when we are doing work in the laboratory, but all the time."[53]

In this world, because objects are relations, they are also events.

> The quantum world is more tenuous than the one imagined by the old physics; it is made up of happenings, discontinuous events, without permanence. It is a world with a fine texture, intricate and fragile as Venetian lace. Every interaction is an event, and it is these light and ephemeral events that weave reality, not the heavy objects charged with absolute properties that our philosophy posited in support of these events.[54]

"Heavy objects"—elsewhere Rovelli refers to them as "entities with definite properties or unique facts." Properties are "bridges between objects," and "objects are nodes where bridges meet." Rovelli is aware of, and calls attention to, the way in which quantum theory challenges Western philosophical presuppositions about subjects and objects.[55] But he might have been addressing conservators directly when he wrote, "The unambiguous description of an object includes the objects to which it manifests itself"—and we are among those other objects.[56]

Rovelli began from quantum physics and we from conserving active matter. But we have arrived at the same place. Conservators, like historians, may think that the things they hold in their hands are made of matter. But what Rovelli teaches us, so succinctly, is that matter is itself a chronograph. When all else is gone, as with distant ruins, what remains is time.

What might this QTO look like? Rovelli actually starts us down this path. "Take an object," he begins. "This chair that I see in front of me. It is real and stands before me, objectively, no doubt about it. But what does it mean, exactly, that this whole is an object, an entity, a chair, real?" A chair is defined by function and has characteristics, but always relative to us observers. Thus, color, he explains, is not *of* the chair, but "comes from the encounter between the frequencies of light reflected from the surfaces of the chair and the particular receptors in human retinas. It is not about the chair; it is a story between light, retina and reflection." We could, Rovelli continues, think similarly about what holds a chair together of its parts. "If we look for the chair in itself, independently of external relations, and especially of its relations to us, we struggle to find it."[57]

One relation of the chair that Rovelli does not consider is time. If an object is just duration, then instead of talking about media (wood, metal, ceramic) we would talk about time signatures—how they are active, their way of activity, and their rate of activity until they are active no more. Like muscles in the human body, one kind of object might be classified as "fast"—these could include holograms, as in Marc Walton, Pengxiao Hao, Marc Vermeulen, Florian Willomitzer, and Oliver Cossairt's essay—or "slow"—such as the metal soaps described by Francesca Casadio, which do not "cease to be in flux" even centuries later—and others very, very slow, like the boulder overhanging a hillside-hugging corniche. This problem seems to have fascinated the children's book author William Steig, as he examines it from two different vantage points, the tragic and the comic, in *Sylvester and the Magic Pebble* and *Solomon, the Rusty Nail.* It was also significant for Nelson Goodman, who drew no metaphysically significant distinction between objects and events, treating both as entities of the same kind: an object would simply be a "monotonous" event; an event would be an "unstable" object. George Kubler groped toward finding time in objects; he ended up, instead, finding many different ways of placing objects in time.[58] Another way of saying this is that it comes down to the way we conventionally measure and frame process. Arguably, we humans have tended to make our own lives and self-understanding the measure of the processes we label events.[59]

With time replacing matter, and longer time framing become acceptable, things can become events, and as they become events, they also become historical, that is, entangled with all the distinct components we associate with events. As Casadio points out in her story about the "lives" of lead soaps in painted canvases, with this step the convenient but oversharp dualisms by which we navigate our world in motion fall away. So, in Rovelli's account, the object is also the "gang of kids" who might once have used it as a weapon, it is "an example in a book about time," it is "a metaphor for ontology," and so on. Without a hard line between subject and object, active and passive, intentional and accidental, there is also a flow between foreground and background, text and context, cause and effect.[60]

Some historians, reading these paragraphs, might think about Friedrich Engels's re-presentation of dialectical materialism: "'that the world is not to be comprehended as a complex of ready-made *things*, but as a complex of *processes*.'"[61] Others, who know something about Buddhism, might think here, with Rovelli, about Nāgārjuna.[62] For still others, what Fernand Braudel argued seventy years ago might come to mind. His *The Mediterranean and the Mediterranean World in the Age of Philip II* (1949; 2nd edition 1966; translated into English 1972) is organized into three very different scales of retelling: that of the very, very slow, the slow, and the relatively fast: geological time, cyclical economic time, and human time, the last only looking like a conventional "event." This is much closer to the quantum conservator's approach—or Cesare Brandi's way of doing history—than, say, that of a political or military or diplomatic historian.

At this point the boundary between conservation and history begins to blur. But so, too, that between conservation and "historic preservation." Recent developments in urbanistics have recast the city as an event rather than an object. What this has meant is imagining different uses for streets, spaces, and institutions at different times of the day for different types of residents. In Mexico City, for example, Laboratorio para la Ciudad tried to find ways to make space more malleable and adaptable to the changing needs of citizens. As this happens more and more, we will be drawn further into thinking about objects, even mega-objects such as cities, as events.[63]

If we might have earlier casually described conservators as historians of the object, seen in terms of active matter—the quantum realm—this becomes a formal definition: like historians, conservators will be telling stories about events. Like conservators, historians will recognize that events come in different shapes, sizes, and materials.

The QTO emphasizes activity and change, relations and process, and eliminates any metaphysical separation between mental and physical worlds. It reinforces the sovereignty of time in the realm of conservation and thus makes clear the priority of time to treatment. Change out the word "becoming" in the following sentence for "conservation"—a perfectly legitimate move in the spirit of Rovelli—and a vast field for human understanding opens before us.

> Becoming is also the mediation between past and future. But it is the mediation between the concrete, i.e. historical past, and the equally concrete, i.e. historical future. When the concrete here and now dissolves into a process it is no longer a continuous, intangible moment, immediately slipping away; it is the focus of the deepest and most widely ramified mediation, the focus of decision and of the birth of the new.[64]

In this vast field, Robert van Langh's trial balloon about postponing treatment for a generation in order to amass more knowledge about the object takes us to new meaning. Conservation as a whole, not just the project called Conserving Active Matter, is about research. And research is driven by questions asked in the present for the sake of the future. Van Langh's suggestion may sound utterly fantastical, yet it closely follows the serious suggestion of the British philosopher and archaeologist R. G. Collingwood. In a speculative essay dated February 1939 devoted to sketching the principles of a future archaeological practice, Collingwood proclaims that no physical labor should be performed by an archaeologist before

> he can answer the question "what historical problems lead you to that site, why do you think you can solve them there, and how exactly do you mean to go about it?"; unless, every time he orders a new trench to be opened or even a single shovelful of earth to be moved, he is prepared to explain, in terms of historical questions and their possible answers, exactly why he is doing it; unless his record shows that, instead of nibbling away at his site like a small boy with a cake until nothing is left, he is capable of saying "now I have answered the questions I came here to answer, and we are going home."[65]

Finally, the QTO also has implications for the person of the inquirer, whether historian or conservator or archaeologist. Yannis Hamilakis argues that archaeology was born in the same nineteenth-century moment as photography (and, we might add, the natural sciences and the museum) and that it worked to produce a clearly defined scientific discipline by excluding everything that complicated this picture.[66] The view of matter as dumb and passive, patiently waiting to be acted upon—like the cast-iron architecture studied by Benjamin—also evacuates the agency of the human actor (photographer, archaeologist, conservator, historian).

Wanting now to extend archaeology's purview into those very excluded complexities, Hamilakis reaches for his Bergson. In the work of Henri Bergson, as readers of Proust well know, "every perception of a present moment is replete with memories. . . . Past and present thus are not successive moments in a line but rather coexist side by side." Crack open the meaning of time, he writes, and the human inquirer finds herself grappling with all those hard-to-hold memories. Bergson for archaeologists leads Hamilakis to envision an archaeology reborn "as a multi-temporal, corporeal, and sensorial practice." Can we see the implications of Bergson for conservators in what Hamilakis imagines for archaeologists? "Such a new discipline," he continues, "will be attentive to the sensorial lives of things as multi-temporal entities. It will be an undisciplined discipline, not only because of the risky and unpredictable nature of the sensorial, but also because of the activation of the political lives of matter."[67]

Turning back to physics and the QTO, Anthony Aguirre has explained that, from this perspective, objects are collections of information applied to atoms, atoms which themselves are "'made' of a quantum state, which is itself a sort of informational entity." And information is itself something that lives as much in the mind of the observer as outside it. "Familiar properties like color, utility, and beauty," he writes, "are human constructs that no single atom has, and even more primitive notions like motion, solidity, locality, or change are ways that observers or agents choose to break up and describe the cohesive whole that is reality." But at the same time, Aguirre cautions, what we might think of as *our* ideas and *our* thoughts are shared across people, space, and time. His real target is dualism: "*objective* and *subjective* should perhaps not be considered two sides of a coin, but rather two ends of a continuum."[68]

The antidualism of the QTO sends us to Schelling, again, and his 1802 *Lectures on University Studies.* In the eighth, "On Philosophy and the Positive Sciences," he argued against the blinkering impact of professional disciplinization on the project of knowing and the project of being human. He argued that "[h]istory is neither a purely rational process subject to the concept, nor is it purely irrational; rather it combines necessity in the whole with the appearance of freedom in the individual."[69] Making sense of the causal relationship between external conditions and internal impulsion is a riddle that has to be seen as a whole, whether we are talking about the lives of rocks or parliaments or proletarians or paintings.

This is hard stuff. To return one last time to Rovelli. The concluding sentence in the passage we cited reads, "The world is not so much made of stones as of fleeting sounds, or of waves moving through the sea." This isn't physics or chemistry or history. It's poetry. Giving a lecture, "Conservation of the Future," at BGC in March 2015 as part of Cultures of Conservation, Paul Eggert began with a reading and interpretation of Thomas Hardy's "The Self-Unseeing."[70] Hardy was an architect who did historic preservation before he returned to his native Dorset and wrote novels that were a form of historic preservation. The poem moves with economy between these vocations in the way that past and present can live together only in the realm of art. It's why Goethe the poet was attracted to morphology, and Haeckel the morphologist to Goethe.[71]

It is at the very end of Goethe's *Faust Part II* that we are introduced to Lynkeus, whose namesake was the far-seeing lookout on the mythical *Argo.* We meet him on his high tower, looking into the distance, admiring nature's beauty and proud of his vision. But then he sees material destruction, too—the burning of a hut and its inhabitants—and laments that his ability to see means also that he cannot avoid seeing loss. Nor has he the power to preserve all that he sees. The tension cannot be resolved, only recast in a whisper of melancholy.

> Something lovely to behold
> Has vanished like an old tale told.[72]

Is this the conservator of the future? With talent and training to look closely, the conservator can see ever more deeply into the wonder of the world and also comprehend its ceaseless activity. But to grasp this whole and give it voice, the new Lynkeus will need the soul of a poet.

NOTES

I am grateful to Francesca Casadio, Jim Coddington, Ivan Gaskell, Aaron Glass, Soon Kai Poh, Sarah Scaturro, Jessica Walthew, Marc Sebastian Walton, and Glenn Wharton for reading and commenting on earlier versions of this essay.

1. Homer, *Odyssey*, lines 488–91.
2. Kishkik, *Manhattan Project*; Schäffner, "Active Matter," 4–6.
3. "Restoration" involved "action taken to make a deteriorated or damaged

artifact understandable, with minimal sacrifice of aesthetic and historic integrity." "The Conservator-Restorer."

4. "Conservation Terminology."
5. As is almost always the case, change is coming from what is often perceived as the periphery of the field. From the perspectives of First Nations and contemporary art, activity is a more common category to think with. See, for instance, Clavir, *Preserving What Is Valued*; Muñoz Viñas, *Contemporary Theory of Conservation.*
6. *Conservazione dei beni culturali*, 9.
7. Chazan, "Technological Trends," 719. I thank Nathan Schlanger for pointing me in Chazan's direction.
8. For this see Schnapp, *Histoire universelle des ruines*, chap. 1, and the chapters in Schnapp et al., *World Antiquarianism*, devoted to ancient Egypt and Mesopotamia, 121–59.
9. Confucius, *Analects*, 7.19, quoted in Owen, *Remembrances*, 13.
10. This is taken from "Autumn Meditations" by Meng Chiao (751–814), quoted in Owen, *Remembrances*, 18.
11. Kenkō, *Essays in Idleness*, 115, quoted in the chapter by Yuriko Saito in this volume.
12. Stewart, *Ruins Lesson.*
13. Leonardo, *Thoughts on Art and Life*, no. 101; Ovid, *Metamorphoses*, 15:234–36: "Tempus edax rerum, tuque, invidiosa vetustas, omnia destruitis vitiataque dentibus aevi paulatim lenta consumitis omnia morte!"
14. Shakespeare, "Sonnet 64."
15. Shakespeare, *Rape of Lucrece*, lines 944–47, 951–52.
16. Flinders Petrie, *Methods and Aims in Archaeology*, 189.
17. Harrison, *Dominion of the Dead*, x, xii.
18. Brandi, *Teoria del restauro*, 27, 35, 44.
19. Yourcenar, *That Mighty Sculptor, Time*, 57–62.
20. Kopytoff, "Biography of Objects."
21. See, for example, Chang, "Alchemy as Studies."
22. Quoted in Yates, *Giordano Bruno*, 34.
23. Galilei, *Dialogue*, 58–59, translation slightly emended.
24. Israel, *Radical Enlightenment*, part 2, "The Rise of Philosophical Radicalism"; Mulsow, *Enlightenment Underground.*
25. For example, Lukács followed an account of a world transformed by dialectic with the following statement: "It might look as if this would take philosophy back to the great system-builders of the beginning of the modern age. The identity, proclaimed by Spinoza, of the order to be found in the realm of ideas with the order obtaining in the realm of

things seems to come very close to this point of view. The parallel is all the more plausible (and made a strong impression on the system of the young Schelling) as Spinoza, too, found the basis of this identity in the object, in the substance." Lukács, "Reification and the Consciousness," 142–43.

26. For Goethe's notion of morphology and, in particular, its utility for historical thinking, see Miller, "Tentative Morphology," 74–77; Miller, "Goethe and the End."
27. Very clearly, in the foreword: "Die vorliegenden Grundzüge der 'generellen Morphologie der Organismen' unternehmen zum ersten Male den Versuch, diesen heillosen und grundverkehrten Dualismus aus allen Gebietstheilen der Anatomie und Entwickelungsgeschichte völlig zu verdrängen, und die gesammte Wissenschaft von den entwickelten und von den entstehenden Formen der Organismen durch mechanisch-causale Begründung auf dieselbe feste Höhe des Monismus zu erheben, in welcher alle übrigen Naturwissenschaften seit längerer order kürzerer Zeit ihr unerschütterliches Fundament gefunden haben." (The existing basic features of the "general morphology of organisms" undertake for the first time the attempt to completely drive out this hopeless and fundamentally wrong dualism from all areas of anatomy and evolutionary history, and raise up the entire science of the developed and emerging forms of organisms through mechanical-causal justification to the same fixed height of the monism in which all other natural sciences have found their unshakable foundation for a longer or shorter period of time.) Haeckel, *Generelle Morphologie*, 1:xiv–xv.
28. "Professor Haeckel on Darwin," 537.
29. The words are those of Lukács in "What Is Orthodox Marxism?" 16, and "Reification and the Consciousness," 142.
30. For Usener's notion of *Augenblicksgötter*, see his *Götternamen*, chap. 16.
31. Nietzsche, *Dawn*, 21–22.
32. Bataille, "Bas matérialisme," translated in Bataille, *Visions of Excess*, 47. This paragraph draws on my "Visions of Juxtaposition."
33. On Fluxus and conservation, see Hölling, *Object—Event—Performance.*
34. Caillois, "New Plea," 346: "Productive exchanges between the human sciences and the natural sciences can be established and developed. The dialogue should be even broader and include the physical sciences. Crystal, for example, has properties akin to those of living matter. For one thing, it can scar over a break through heightened regenerative activity, in much the same way that a lobster regenerates its claw or a Saurian its tail; for another, it can gradually eliminate foreign bodies accidentally trapped in its well-ordered, homogeneous substance."

35. For more, see Mass's contribution to the website accompanying the *Conserving Active Matter* exhibition, 2022, https://exhibitions.bgc.bard.edu/cam/.
36. Blunden, *Cultural Historical Activity Theory*, 2–3, 31.
37. Cited in Schlanger, *André Leroi-Gourhan on Technology*.
38. Ginzburg, "Medals and Shells."
39. Robert Macfarlane, in the English Faculty at Cambridge, has done just this already. *Underland*, 111–12.
40. I am grateful to Jim Coddington for pointing this out to me.
41. Harrison, *Dominion of the Dead*, 153.
42. Burawoy, "Provincializing the Social Sciences," 508–25; Chakrabarty, *Provincializing Europe*.
43. Miller, *History and Its Objects*, 14–15.
44. A list of such publications could include Graham Harman's *Tool-Being: Heidegger and the Metaphysics of Objects* (2002), Andy Clark's *Natural Born Cyborgs* (2003), Jane Bennett's *Vibrant Matter* (2010), and Terrence Deacon's *Incomplete Nature* (2012).
45. Latour, *We Have Never Been Modern*, 11–12, 53, 99.
46. Ibid., 58, 128, 60.
47. Ibid., 69. Latour animates what Pierre Nora had already noted: "'As traditional memory fades, we feel obliged religiously to accumulate the testimonies, documents, images, and visible signs of what was, as if this ever-proliferating dossier should be called on as evidence in some tribunal of history. Hence the inhibition against destroying, the retention of everything.'" Preface to *Rethinking France*, xxv–xxvi.
48. Pickering, "Space," 2, 7.
49. Latour, *We Have Never Been Modern*, 85.
50. Barad, *Meeting the Universe Halfway*, 19.
51. Ibid., 23.
52. Rovelli, *Order of Time*, 99. A similar account of "the event" is in Dewey, *Logic of Inquiry*, 222–23: "On the other hand, a mountain, which to the layman is a standing symbol of permanence, is to the geologist the scene of a drama of birth, growth, decay and ultimate death. Unless the difference between existential change as barely existential and as subject-matter of judgment is borne in mind, the nature of event becomes an inexplicable mystery. *Event* is a term of judgment, not of existence apart from judgment. The origin and development the Appalachian Mountain Range is an event, and so is the loosening and rolling of a particular pebble on a particular ledge on a particular foot-hill. There may be a situation in which the latter sort of episode is much more important in judgment than is the history of long duration: as, for example, when a

rolling pebble is the 'cause' of a sprained ankle. In the story of the cyclical weathering of the mountain, the roll of the pebble would hardly be *an* event at all; it would be but a specimen, unnoted in itself, of a *kind* of thing that is significant only *en masse.* An event is, strictly that which comes out; that which issues forth; the net outstanding consequence, the eventuation. It involves a teleological concept; it is capable of description-narration only in terms of a delimiting beginning, an interval and a termination."

53. Rovelli, *Helgoland*, 74.
54. Ibid., 85.
55. Ibid., 142–45, and the notes there.
56. Ibid., 88, 117.
57. Ibid., 145–46. Compare this with Heidegger's very similar description of a chair in *Prolegomena*, § 5.a, 37–40.
58. George Kubler, *The Shape of Time: Remarks on the History of Things* (New Haven, CT: Yale University Press, 1962), 84–85, 96–97.
59. Steig, *Sylvester and the Magic Pebble*; Steig, *Solomon, the Rusty Nail*; Nelson Goodman, *Structure of Appearance.* For this language as a way of framing performance art, see Hillier, "Objects as Events." I thank Ivan Gaskell for this reference.
60. Everyone has their favorite theorist of entanglement. No bibliography would be complete without Thomas, *Entangled Objects*; Hodder, *Entangled*; Deleuze and Guattari, *Thousand Plateaus.*
61. Friedrich Engels, "Ludwig Feuerbach and the End of Classical German Philosophy," quoted in Lukács, "Reification and the Consciousness," 199–200.
62. Rovelli, *Helgoland*, 149–58. See also Nāgārjuna, *Mulamadhyamakakarika.*
63. For more, see the website of Laboratorio para la Ciudad, accessed August 11, 2021, https://labcd.mx/.
64. Lukács, "Reification and the Consciousness," 204.
65. Collingwood, *Principles of History*, 65.
66. Oliver Wendell Holmes talked about photography in just this way in 1859: "Form is henceforth divorced from matter. In fact, matter as a visible object is of no great use any longer, except as the mould on which form is shaped." Quoted in Solnit, "Annihilation of Time and Space," 17.
67. Hamilakis, *Archaeology & the Senses*, 122.
68. Aguirre, *Cosmological Koans*, 350, 352.
69. Schelling, *Lectures on University Studies*, 75–76.
70. Eggert, "Conservation and the Future."
71. It may also explain why the most poetic of our era's great historians cares about morphology. Ginzburg, "Medals and Shells."
72. Goethe, *Faust: Part Two*, 11336–37.

BIBLIOGRAPHY

Aguirre, Anthony. *Cosmological Koans: A Journey to the Heart of Physical Reality.* New York: W. W. Norton, 2020.

Barad, Karen. *Meeting the Universe Halfway: Quantum Physics and the Entanglement of Matter and Meaning.* Durham, NC: Duke University Press, 2007.

Bataille, Georges. "Le bas matérialisme." In *Documents,* edited by Denis Hollier, 2 vols., 2:1–8. Paris: Jean-Michel Place, 1991.

———. *Visions of Excess: Selected Writings, 1927–1939.* Edited by Alan Stoekl. Minneapolis: University of Minnesota Press, 1985.

Blunden, Andy. *Cultural Historical Activity Theory.* Leiden: E. J. Brill, 2010.

Brandi, Cesare. *Teoria del restauro.* Turin: Einaudi, 1977. Originally published 1963.

Burawoy, Michael. "Provincializing the Social Sciences." In *The Politics of Method in the Human Sciences: Positivism and Its Epistemological Others,* edited by George Steinmetz, 508–36. Durham, NC: Duke University Press, 2005.

Caillois, Roger. "A New Plea for a Diagonal Science." In *The Edge of Surrealism: A Roger Caillois Reader,* edited by Claudine Frank, 343–47. Durham, NC: Duke University Press, 2003.

Chakrabarty, Dipesh. *Provincializing Europe: Postcolonial Thought and Historical Difference.* Princeton, NJ: Princeton University Press, 2000.

Chang, Ku-ming (Kevin). "Alchemy as Studies of Life and Matter: Reconsidering the Place of Vitalism in Early Modern Chymistry." *Isis* 102 (2011): 322–29.

Chazan, Michael. "Technological Trends in the Acheulean of Wonderwerk Cave, South Africa." *African Archaeological Review* 32 (2015): 701–28.

Clavir, Miriam. *Preserving What Is Valued: Museums, Conservation and First Nations.* Vancouver: UBC Press, 2002.

Collingwood, R. G. *The Principles of History and Other Writings in the Philosophy of History.* Edited by W. H. Dray and W. J. van der Dussen. Oxford: Oxford University Press, 1999.

"Conservation Terminology." American Institute for Conservation. Accessed January 26, 2021. https://www.culturalheritage.org/about-conservation/what-is-conservation/definitions.

"The Conservator-Restorer: A Definition of the Profession." International Council of Museums–Committee for Conservation. 1984. http://www.icom-cc.org/47/about/definition-of-profession-1984/#.YOdnjRNKi_J.

Conservazione dei beni culturali: Principali termini generali e definizioni. UNI EN 15898, Luglio 2012.

Deleuze, Gilles, and Félix Guattari. *A Thousand Plateaus: Capitalism and Schizophrenia.* Translated by Brian Massumi. Minneapolis: University of Minnesota Press, 1987. Originally published 1980.

Dewey, John. *Logic of Inquiry.* New York: Henry Holt, 1938.

Eggert, Paul. "Conservation and the Future." Talk delivered at the seminar series Conservation Conversations, Bard Graduate Center, New York, March 31, 2015. https://www.bgc.bard.edu/research-forum/articles/30/conservation-and-the-future.

Flinders Petrie, William Matthew. *Methods and Aims in Archaeology.* London: Macmillan, 1904.

Galilei, Galileo. *Dialogue on the Two Chief World Systems.* 2nd ed. Translated by Stillman Drake. Berkeley: University of California Press, 1959.

Ginzburg, Carlo. "Medals and Shells: On Morphology and History, Once Again." *Critical Inquiry* 45 (2019): 380–95.

Goethe, J. W. von. *Faust: Part Two.* Translated by David Luke. Oxford: Oxford University Press, 2008

Goodman, Nelson. *The Structure of Appearance.* Cambridge, MA: Harvard University Press, 1951.

Haeckel, Ernst. *Generelle Morphologie.* Berlin: Georg Reimer, 1866.

Hamilakis, Yannis. *Archaeology & the Senses: Human Experience, Memory, and Affect.* Cambridge: Cambridge University Press, 2013.

Harrison, Robert Pogue. *The Dominion of the Dead.* Chicago: University of Chicago Press, 2003.

Heidegger, Martin. *Prolegomena to a History of the Concept of Time.* Translated by Theodore Kisiel. Bloomington: Indiana University Press, 1992.

Hillier, Susan. "Objects as Events Extended over Time: The Nature of Representation." In *Thinking about Art: Conversations with Susan Hillier,* edited by Barbara Einzig, 168–82. Manchester: Manchester University Press, 1996.

Hodder, Ian. *Entangled: An Archaeology of the Relationships between Humans and Things.* Hoboken, NJ: Wiley-Blackwell, 2012.

Hölling, Hanna B., ed. *Object—Event—Performance: Art, Materiality, and Continuity Since the 1960s.* New York: Bard Graduate Center, 2022.

Homer. *The Odyssey of Homer.* Translated by Richard Lattimore. New York: Harper and Row, 1967.

Israel, Jonathan. *Radical Enlightenment: Philosophy and the Making of Modernity 1650–1750.* Oxford: Oxford University Press, 2001.

Kishkik, David. *The Manhattan Project: A Theory of a City.* Stanford, CA: Stanford University Press, 2015.

Kopytoff, Igor. "The Biography of Objects." In *The Social Life of Things,* edited by Arjun Appadurai, 64–92. Cambridge: Cambridge University Press, 1986.

Latour, Bruno. *We Have Never Been Modern.* Cambridge, MA: Harvard University Press, 1991.

Leonardo da Vinci. *Thoughts on Art and Life.* Translated by Maurice Baring. Boston: Merrymount Press, 1906.

Lukács, György. "Reification and the Consciousness of the Proletariat." In *History and Class Consciousness,* 83–222. Cambridge, MA: MIT Press, 2000.

———. "What Is Orthodox Marxism?" In *History and Class Consciousness,* 1–26. Cambridge, MA: MIT Press, 2000. First published 1971.

Macfarlane, Robert. *Underland: A Deep Time Journey.* New York: W. W. Norton, 2019.

Miller, Peter N. "Goethe and the End of Antiquarianism." In *For the Sake of Learning: Essays in Honor of Anthony Grafton,* edited by Anne Blair and Anja-Silvia Goeing, 897–916. Leiden: E. J. Brill, 2016.

———. *History and Its Objects: Antiquarianism and Material Culture Since 1500.* Ithaca, NY: Cornell University Press, 2017.

———. "A Tentative Morphology of European Antiquarianism, 1500–2000." In *World Antiquarianism,* edited by Alain Schnapp, 67–87. Los Angeles: Getty Research Institute, 2012.

———. "Visions of Juxtaposition: Peiresc/Bataille, Monuments/Documents." In *The Museum in the Cultural Sciences: Collecting, Displaying, and Interpreting Material Culture in the Twentieth Century,* edited by Peter N. Miller, 223–32. New York: Bard Graduate Center, 2021.

Mulsow, Martin. *Enlightenment Underground: Radical Germany 1680–1720.* Charlottesville: University of Virginia Press, 2015.

Muñoz Viñas, Salvador. *Contemporary Theory of Conservation.* London: Routledge, 2004.

Nāgārjuna. *Mulamadhyamakakarika.* Translated in J. L. Garfield, *The Fundamental Wisdom of the Middle Way: Nāgārjuna's "Mulamadhyamakakarika."* Oxford: Oxford University Press, 1995.

Nietzsche, Friedrich. *Dawn: Thoughts on the Presumptions of Morality.* Translated by Brittain Smith. Afterword by Keith Ansell-Pearson. Stanford, CA: Stanford University Press, 2011.

Nora, Pierre. *Rethinking France: Les lieux de memoire.* Vol. 1, *The State.* Chicago: University of Chicago Press, 1999. First published 1984.

Owen, Stephen. *Remembrances: The Experience of the Past in Classical Chinese Literature.* Cambridge, MA: Harvard University Press, 1986.

Pickering, Andrew. "Space: The Final Frontier." In *Collection-Laboratory-Theater: Scenes of Knowledge in the 17th Century,* edited by Helmar Schramm, Ludger Schwarte, and Jan Lazardzig, 1–8. Berlin: De Gruyter, 2005.

"Professor Haeckel on Darwin, Goethe, and Lamarck." *Nature* 26 (September 28, 1882): 533–41.

Rovelli, Carlo. *Helgoland: Making Sense of the Quantum Revolution.* Translated by Erica Segre and Simon Carnell. New York: Riverhead Books, 2021.

———. *The Order of Time.* New York: Riverhead Press, 2018. First published 2017.

Schäffner, Wolfgang. "Active Matter." In *23 Manifeste zu Bildakt und Verkörperung,* edited by Marion Lauschke and Pablo Schneider, 1–10. Berlin: De Gruyter, 2017.

Schelling, Friedrich Wilhelm Joseph von. *Lectures on University Studies.* Translated by E. S. Morgan. Edited and with an introduction by Norbert Guterman. Athens: Ohio University Press, 1968.

Schlanger, Nathan. *André Leroi-Gourhan on Technology: Material Civilization, Operational Behavior, and the Evolution of Technicity.* New York: Bard Graduate Center, 2022.

Schnapp, Alain. *Une histoire universelle des ruines.* Paris: Le Seuil, 2020.

Schnapp, Alain, with Lothar von Falkenhausen, Peter N. Miller, and Tim Murray, eds. *World Antiquarianism: Comparative Perspectives.* Los Angeles: Getty Research Institute, 2013.

Shakespeare, William. *Rape of Lucrece.* Folger Shakespeare Library. Accessed August 10, 2020. https://shakespeare.folger.edu/shakespeares-works/lucrece/the-poem/.

———. "Sonnet 64." Poetry Foundation. Accessed August 10, 2020. https://www.poetryfoundation.org/poems/45096/sonnet-64-when-i-have-seen-by-times-fell-hand-defacd.

Solnit, Rebecca. "The Annihilation of Time and Space." *New England Review* 24 (2003): 5–19.

Steig, William. *Solomon, the Rusty Nail.* New York: Farrar, Strauss, Giroux, 1985.

———. *Sylvester and the Magic Pebble.* New York: Windmill Books, 1969.

Stewart, Susan. *The Ruins Lesson: Meaning and Material in Western Culture.* Chicago: University of Chicago Press, 2020.

Thomas, Nicholas. *Entangled Objects: Exchange, Material Culture, and Colonialism in the Pacific.* Cambridge, MA: Harvard University Press, 1991.

Usener, Hermann. *Götternamen. Versuch einer Lehre von der religiösen Begriffsbildung.* Bonn: Friedrich Cohen, 1896.

Yates, Frances A. *Giordano Bruno and the Hermetic Tradition.* London: Routledge and Kegan Paul, 1964.

Yoshida, Kenkō. *Essays in Idleness: The Tsurezuregusa of Kenkō.* Translated by Donald Keene. New York: Columbia University Press, 1967.

Yourcenar, Marguerite. *That Mighty Sculptor, Time.* New York: FSG, 1993.

PART 1

Philosophy

ONE

Introduction: Active Matter—Some Initial Philosophical Considerations

Ivan Gaskell and A. W. Eaton

Abstract

In the near absence of discussion of conservation by philosophers, this chapter addresses two fundamental philosophical questions underlying Conserving Active Matter. The first is conceptual, the second normative. First, in considering what active matter might be in terms of a viable concept, it identifies two possible views: a taxonomic view, proposing distinctions among types of matter in relation to activity, and a spectral view that conceives of active matter as being at one end of a set of continua pertaining to reactive matter. Second, this chapter poses a series of questions about normative considerations of active matter. These include the following: When is material change something to resist or repair, and when is it something to embrace and facilitate? On what bases and according to what norms can those responsible for material things make these determinations? The purpose of this chapter is not to advance any determinative theory concerning active matter and conservation, but simply to serve as a prolegomenon offering an initial map of the terrain in philosophically consistent terms.

> Ah! how hard a thing it is to tell what a wild,
> and rough, and stubborn wood this was, which
> in my thought renews the fear!
> —Dante, *Inferno*, 1:4–6

Two philosophical questions lie at the heart of the Conserving Active Matter research project. First: What is *active matter*? Second: Under what conditions is active matter something to value or disvalue? Like all philosophical questions, these do not have definite, or even straightforward, answers. Accordingly, we could not hope to answer them in this brief introductory chapter. Instead, we set out to map the conceptual terrain in which these questions are located with the hope of offering routes through the dark wood.

We begin with an attempt to sharpen the very concept of active matter. First, we acknowledge some of the various stipulative definitions used by scientists, and, second, we compare and contrast active matter with neighboring concepts involving material change. We then explore the various kinds of change of which matter is capable. This investigation will be part conceptual, part empirical. We then turn to some central normative questions: When is matter's activity something to resist or repair, and when is this activity something to embrace and perhaps even facilitate? Again, our aim will not be to provide answers but, rather, to sharpen the questions and set the stage for contributions by four fellow philosophers: Sherri Irvin, Carolyn Korsmeyer, Alva Noë, and Yuriko Saito.

Active Matter As Concept

The term "active matter" is used by some scientists, technologists, and designers in a variety of sometimes overlapping senses. For instance,

> Active matter describes systems whose constituent elements consume energy and are thus out-of-equilibrium. Examples include flocks or herds of animals, collections of cells, and components of the cellular cytoskeleton. When these objects interact with each other, collective behavior can emerge that is unlike anything possible with an equilibrium system.[1]

In other usage, the term can apply to "physical materials that can assemble themselves, transform autonomously, and sense, react, or compute based on internal and external information."[2] Introducing a journal issue devoted to active matter, two physicists and a biomedical engineer state, "Active matter includes living and non-living systems that have in common that they contain energy-consuming and force-generating microscopic constituents that drive emergent dynamic properties on larger scales."[3] Our discussion of active matter is responsive to, but is not constrained by, these and other doubtless locally useful stipulative definitions.

Our own discussion begins with the less controversial component of the conceptual dyad: matter. We take matter to be constituted of atomic particles, either independently (in the case of elements) or in molecular combinations. Matter is that which occupies space. Since the seventeenth century, philosophers have commonly captured this notion with the concept *extension*. Matter is that which is extended.[4]

But what does it mean to say that matter is, or could be, active? We identify two conceptually distinct yet promising approaches to this question: a taxonomic view of active matter and a spectral view of active matter. As we explain below, on a taxonomic view, active matter is but one species of a broader category of *changing matter*, whereas on a spectral view, active matter lies at one end of a complex continuum of *reactive matter*.[5]

First, we propose a taxonomic view of active matter, where active matter is distinct, or mostly distinct, from other subsets of changing matter, where "changing" means a progressive difference in a thing's material features.[6] These other subsets include, but are not limited to, *passive* and *reactive matter*. Passive matter would be matter subject to change without responding or initiating action in return, for example, a lump of damp clay subject to changes in form by manipulation owing to its malleability or ductility. In the case of reactive matter, an external stimulus is the motor of change, but in this case the reactive matter participates in the change, as in the case of reagents added to a system to bring about, or test for, a chemical reaction, one example being hydrochloric acid when used in the industrial production of polyvinyl chloride (PVC), a common synthetic plastic polymer.

Following this line of thinking, *activity* connotes a thing's direct participation in its own processes of change. We might, then, think of active matter as matter capable of spontaneous change in the sense of *self-generating* and not dependent on any external stimulus. Spontaneous change can be of three kinds—which, by the way, also apply to the spectral view of active matter, to be discussed below. One is change that leads to replication, proliferation, or the emergence of new constituents of the matter concerned. The second is change that is self-consuming, whereby matter feeds upon itself degeneratively, leading eventually to dissolution. The third is nondirectional, or random. Living matter is an example of matter subject to the first kind of change. Stars, including our own sun, are an example of the second kind. An example of random change is the process of change that underlies evolution, however directional it may appear in hindsight. Let us look more closely at the first kind.

Although living matter may be active matter in a broad sense, one assumption motivating this project is that there exists a kind of directional active matter that is nonliving in the biological sense. This biologically nonliving active matter is what we shall discuss henceforth as "generative active matter." When considering this together with degenerative active matter and random active matter, we shall use the term "active matter" without qualification.

Defining living matter is itself a topic of considerable contention within both philosophy and biology. Biologist Radu Popa, for instance, offers three hundred or so definitions of life.[7] Whatever living matter might be, generative active matter, in contradistinction to it, comprises things other than entities that are capable of "self-reproduction with variations," as biologist Edward Trifonov has defined life.[8] The key thought here is that generative active matter is biologically nonliving but active in the sense that the changes it exhibits are not generated externally by other factors, whether physical or chemical, acting upon it.

Active matter on the taxonomic view—whether generative, degenerative, or random—excludes biologically living matter but may include matter that is living in other senses. These include being animate or endowed with personhood. We refer here to living things recognized as such in the knowledge systems of peoples other than those of the European diaspora. Our colleagues discuss examples in this volume and elsewhere in this project under the rubric "Indigenous ontologies."

We are hard-pressed to identify matter that is generatively or randomly active, but nonliving in the biological sense. What could this self-generating changing stuff be? Candidates might be the fabric MycoTEX, produced from fungal mycelium by Dutch clothing designer Aniela Hoitink and her company, NEFFA, or the "programmable water-based biocomposites for digital design and fabrication across scales" named Aguahoja, produced by the Mediated Matter research group at the Massachusetts Institute of Technology led by founder and associate professor of media arts and sciences, Neri Oxman.[9] Yet neither MycoTEX nor Aguahoja appears to be self-generating beyond the bounds of life. Generative active matter, in the sense we have defined it, would seem, at present, to be conceptual but not yet actual matter.

In contrast, *changing matter* that spontaneously degenerates appears to exist in actuality. It is of two kinds. The first, an example of which is the sun, is self-contained but complex, comprising numerous constituents that react with one another whether chemically or in nuclear and physical senses. The second kind of degenerative active matter is simple, as in the case of the unstable isotopes of single elements that undergo spontaneous nuclear decay. We shall say more about isotopes below.

Thus far we have been talking about active matter on the taxonomic view, but there is another way to think of it, namely, a spectral view that conceives of active matter as being at one end of a set of continua pertaining to reactive matter. These continua include, but are not invariably confined to, the extent of change, the rate of change, and the frequency of change. Our idea is that reactive matter admits of degrees along these and other lines and that a spectral view reserves active matter to describe not a distinct species of matter that differs in kind from others, but for matter that, for example, reacts often and quickly and changes dramatically.

We do not mean to suggest that these continua apply only to active matter on the spectral view. Active matter conceived as a taxon of changing matter also admits of degrees: active matter, like its neighbors passive and reactive matter, can change more or less quickly, more or less dramatically, and so on. But on the taxonomic view, these variables are not characteristic of the taxon, whereas on the spectral view, it is precisely the degree of change along these various continua that characterizes active matter.

We have just discussed extent of change, rate of change, and frequency of change as variables applying to changing matter. Earlier, when considering the taxonomic view, we examined another variable, namely, *direction of change*. We shall now return to direction of change, and also add remarks on regularity of change. Direction of change encompasses movement toward a given state, such as generative changes including growth and development, and movement away from a given state, such as degenerative changes, for instance, decay and deterioration. Direction of change also encompasses change that is not directed toward any state in particular, such as change that is random. *Regularity of change* encompasses change that is consistent and predictable, or irregular, stochastic, and unpredictable. Some processes of material change can exhibit aspects both regular and irregular, as in the formation of microscopic crystals when chlorides in the patina of copper-containing alloys interact with oxygen and water to produce the condition known as bronze disease. All these variables—extent of change, rate of change, frequency of change, direction of change, and regularity of change—admit of degrees and cut across both the taxonomic and spectral views of material change.

This map of the conceptual terrain accounts for what we take to be the principal ways of thinking about how matter changes. But how do these conceptual possibilities meet, or fail to meet, the empirical world? How exactly does changing matter in our known world actually change? Of course, a full discussion of actual material change is not best addressed by philosophers, but no discussion of this issue would be complete without a brief exposition of how changing matter actually undergoes change. We leave it to scientists and others who deal with actual changing matter to determine whether the empirical world inclines us toward a taxonomic understanding of material change or toward a spectral one.

Scientists usually speak of three basic types of change to matter: chemical, physical, and nuclear. Chemical change occurs when atoms and molecules interact—bond or rebond—to form new materials. A change is physical when the matter subject to change retains its underlying atomic or molecular identity. Nuclear change is a change in nuclear structure, which occurs when an atomic nucleus splits (fission), when two or more nuclei combine (fusion), or when an unstable nucleus loses energy by radiation. Many physical changes can be easily reversed, whereas chemical and nuclear changes either cannot be reversed or cannot be reversed without extraordinary effort. A shift in the energy environment of matter can bring about physical or nuclear change, as in the case of exposure of matter to radiation in the form of waves or subatomic particles, or the collision of subatomic particles. A shift in the energy environment can result in changes in the *state* of matter.

Matter exists in a variety of states. Matter has likely existed, can actually exist, and might theoretically exist in over twenty states, but we draw attention to those four that most people experience in everyday life: solids, liquids, gases, and plasmas. External conditions affect the state of any given matter. H_2O (water) can freeze to form solid ice yet remain the same in chemical terms as when it was liquid. Ice and water are two states of H_2O.[10]

A specific type of nuclear change is isotopic. Elements—matter not reduceable to further massive atomic constituents—exist in a variety of isotopes. An isotope is a form of an element that contains the same number of protons as other forms of the same element but that has a different number of neutrons in its nucleus. Therefore, each isotope of any given element differs from other isotopes of the same element in relative atomic mass but not in chemical properties. There are, for instance, seventy known possible isotopes of gold, only one of which, ^{197}Au, can be counted stable in any practical sense. The radioisotopes of gold decay at various rates, with half-lives ranging from 300 nanoseconds to 186 days, though they invariably retain seventy-nine protons in each of their atomic nuclei.[11]

Whatever the changes any matter undergoes—whether chemical, physical, or nuclear—that matter is perpetually, actually, or potentially changing. We would be hard pressed to identify any matter in the actual known universe that is perfectly static, which means that material stasis is a mere conceptual possibility, which is worth entertaining insofar as it may help us to find a route through the dark wood.

Normative Considerations

We have just been outlining various conceptual possibilities for understanding what active matter is. We now turn to explicitly normative questions. We consider there to be three sets of questions of particular relevance. We shall do no more than mention two of them, and then give greater attention to the third.

First, What is the relationship between the values or disvalues placed on material change, on the one hand, and the directionality often attributed to change (as we discussed in the first section above), on the other? Is to attribute a direction or telos to material change already to value or disvalue it, at least implicitly?

Second, Can aging and degradation improve a material thing, and if so, under what conditions? What sort of improvement might this be, for instance, aesthetic, practical, or ethical? What might the implications of such questions be for conservation and other institutionalized forms of practice?[12]

Our third set of questions will be the focus of the rest of this chapter. As in the previous section, our aim is not to provide answers but instead to sharpen the questions, thereby setting the stage for the contributions by Sherri Irvin, Carolyn Korsmeyer, Alva Noë, and Yuriko Saito. These questions include the following: When is material change, however conceived, something to resist or repair, and when is it something to embrace and perhaps even facilitate? On what bases and according to what norms can those responsible for material things make these determinations? What value do some communities find in an ideal of material stasis? Why do these communities take elaborate pains to try to ensure material stasis through preservation, restoration, and conservation? What value do some communities find in material change, even change understood as degradation and decay? Why do they lend privilege to things that have changed or that continue to change? And when some communities ascribe to both of these values in respect of different things or, in some instances, even of the same things, what are the bases for these differentiations? Are there reasons aesthetic, practical, ethical, or some combination of these, or other reasons entirely?

We refer to *communities* in order to foreground the fact that various peoples at various times and in various places have valued material change or material stasis for a variety of reasons and often in markedly different ways. We mean community as an elastic concept that can encompass groups as diverse as practitioners of Japanese tea ceremony, European romantic poets with a love of ruins, Māori guardians of living *taonga* (treasured ancestral items), and the international community of professional conservators.

Let us proceed by briefly exploring aspects of this last example. The practice of object conservation has long had material stasis as its overarching ideal. This means that the fundamental condition of matter—namely, its inevitable susceptibility to some degree of change—provides the central challenge that all professional conservators face. Especially, though not exclusively, the ideal for items in museum collections is that they should be preserved in perpetuity. In museums, perpetuity is a value with broader applications than to collections alone. Donors' lawyers often insist that the terms of a gift to a museum, whether of funds or of objects, should ensure the public acknowledgment of the donor's act in perpetuity. No parties involved believe that "in perpetuity" applies in any literal sense. This is so whether concerning curators charged with the overall care of museum objects, conservators responsible for their physical well-being, or directors intent on keeping existing donors placated and encouraging new potential benefactors. When asked how long "in perpetuity" might actually be, a prominent art museum director responded, "Thirty years."[13]

Similarly, though less cynically, conservators know that material stasis in an item subject to treatment is unattainable and that one of their roles is to minimize instability. Furthermore, conservators have largely abandoned talk of *reversibility of treatment*, that is, the notion that one can treat an item in such a way as to erase that treatment in the future, should it fail or be superseded or become otherwise undesirable. Instead, conservators have now adopted the norm of *retreatability* (or *removability*), according to which one treats a material thing so as to ensure its amenability to further treatment.[14] This is evidence of an increasing acceptance in contemporary conservation practice of material instability as chronically inevitable and an understanding that conservation can be no more than an attempt to manage change, not arrest it, let alone reverse it. Conservation scientists can do a great deal to analyze and account for the specific changes that materials undergo and can help conservators to formulate treatments that can decelerate or compensate for change. Some of their considerations are to be found in the "Materials" section of this volume.

Final Remarks

This concludes our mapping of the terrain of active matter, setting out what we take to be fundamental philosophical questions under two headings, conceptual and normative. Even as we ask what active matter might be, and under what conditions it is something to value or disvalue, we realize that we can do no more than indicate possible routes through the dark wood.

We should note that among the possible routes we have deliberately chosen not to take is the one usually discussed under the heading "agency."[15] Although humans, nonhuman animals, and other entities can affect one another in complex networks of action and reaction, the agency of humans differs from the agency sometimes ascribed to those others. It is only too easy to elide the distinctions among the various senses of agency, which is not a single predicate but a number identically termed. To do so can lead to the erroneous assumption that their respective properties are interchangeable. The changeability of changing matter and the activity of active matter may be affordances in psychologist James Gibson's sense, which ascribes complementarity of effect to constituents within an environment, but do not constitute agency of a kind analogous to the purposive agency of which humans are capable.[16]

The routes we have indicated, no less than the dark wood itself, may induce bewilderment in readers who expect this topic to be straightforwardly comprehensible. To dispel that bewilderment, the other philosophers who contribute to this volume—Sherri Irvin, Carolyn Korsmeyer, Alva Noë, and Yuriko Saito—in their own ways, and on their own terms, explore some of these routes further, as well as others as yet unmentioned. We are confident that clarification will follow. As Henry David Thoreau remarks in *Walden*: "All change is a miracle to contemplate, but it is a miracle which is taking place every instant."[17]

Postscript: Philosophers at Work on Conserving Active Matter

The philosophy working group collaborated in a loose manner. This is largely owing to the nature of the discipline and the working habits of its practitioners. While we philosophers do gather to exchange ideas in society meetings, symposia, and workshops, we tend to do our writing alone.

Philosophers in our working group assembled at Bard Graduate Center on two occasions for symposia dedicated to the Conserving Active Matter project. The first was the initial discussion in November 2017, at which all four sections were represented. Philosophers Justin Broackes (Brown University), Sherri Irvin, and Carolyn Korsmeyer participated. The fourth of the sectional symposia, on philosophy, was held in November 2019. Sherri Irvin and Carolyn Korsmeyer participated once again, joined by philosophers Erich Hatala Matthes (Wellesley College), Alva Noë, Yuriko Saito, and Elisabeth Schellekens (University of Uppsala). Also taking part were conservator Francesca Esmay (Solomon R. Guggenheim Museum), architectural theorist Rumiko Handa (University of Nebraska), and independent curator and art historian Jeffrey Weiss. Thanks to the generosity of the Objects Conservation Laboratory of the Museum of Modern Art, conservator Roger Griffith, and curator Paola Antonelli, we were able to hold a workshop for the speakers in the laboratory the day before the 2019 symposium. This was the first time that most of the philosophers among the participants had ever been in a conservation laboratory. The discussion was very lively, indeed, and led to new appreciations of challenges faced by conservators, scientists, curators, and—for the first time—philosophers.

We have said that philosophers tend to write in isolation, but we (the authors) have collaborated on this project from the beginning. We produced what eventually became this introduction thanks to the support of the Lichtenberg-Kolleg (Advanced Study Institute in the Humanities and Social Sciences) of the Georg-August University, Göttingen, which gave us a shared office for an extended period as fellows in the spring of 2017. This allowed us to spend many hours together puzzling over conceptual definitions, covering whiteboards with endless arcane scribbles. We subsequently built on this foundation by sharing a document online that we could write and revise together simultaneously while discussing it by telephone once we were both back in the United States.

What is the contribution of philosophy to the topic of this volume? While philosophers have addressed related topics, such as the definition of art, for generations, we know of almost no philosophical discussion of conservation prior to this introduction and the chapters that immediately follow it. David Carrier's 1985 article, "Art and Its Preservation," is a rare exception.[18] A very few conservators have appealed to philosophy in their own writings. Salvador Muñoz Viñas is a recent exception who in his influential *Teoría contemporánea de la restauración*, published in 2003, mentions the work of several philosophers.[19] The only conservator we can cite who has attempted to engage with a conservation issue in a philosophical register, and did so creditably, is Steven W. Dykstra in his 1995 article, "The Artist's Intentions and the Intentional Fallacy in Fine Arts Conservation."[20] The consequence of this near absence of philosophical discussion of conservation is that we and our colleagues have had to address it from scratch in this project.

NOTES

Ivan Gaskell and A. W. Eaton gratefully acknowledge the generous support of the Lichtenberg-Kolleg (Advanced Study Institute in the Humanities and Social Sciences) of the Georg-August University, Göttingen. The simultaneous provision of fellowships allowed them to work together on this project. They thank the director, Martin van Gelderen; the former executive director, Dominik Hünniger (now of the University of Hamburg); the present executive director, Kora Baumbach; and the former president of the university, Ulrike Beisiegel, for their support and friendship. They also thank Jane Whitehead and two anonymous reviewers for their editorial suggestions.

1. "Active Matter."
2. Tibbits, *Active Matter*, 11.
3. Das, Schmidt, and Murrell, "Introduction to Active Matter," 7185.
4. We are concerned with what scientists term *baryonic matter*: the product of baryons (composed primarily of protons and neutrons) that have not been annihilated by an encounter with antibaryons (antimatter particles). This is the ordinary matter of human experience. We acknowledge that not all philosophers subscribe to the definition of matter as that which is extended without qualification, for example, Bergson, *L'évolution créatrice*.
5. The taxonomic view proposes differences in *kind*; the spectral view proposes differences in *degree*.
6. We say "mostly distinct" so as to allow for the overlap of the different

subsets on the model of Ludwig Wittgenstein's *family resemblances.* Wittgenstein, *Philosophical Investigations,* part I, §§ 66–76, pp. 27–31.

7. Popa, *Between Necessity and Probability.* For debates in the philosophy of biology, see Weber, "Life."
8. Trifonov, "Definition of Life."
9. "MycoTEX Proof-of-Concept"; "Aguahoja."
10. Pure still water can be supercooled at standard atmospheric pressure to as low as about -40°C and stay liquid, prevented from solidifying by a lack of nucleation centers around which crystals form, afforded by tiny impurities or even vibrations. For recent research, see Kringle et al., "Reversible Structural Transformations." Our thanks to Daniel Lord Smail for drawing our attention to the supercooled state of water.
11. "Gold."
12. We address these questions in our contribution to the *Conserving Active Matter* website, 2022, https://exhibitions.bgc.bard.edu/cam/.
13. This was a jocular remark made to Ivan Gaskell in conversation; yet it reflects a complex English common-law doctrine, modified in various jurisdictions by statute, to inhibit *mortmain*: the prevention of the inclusion of provisions in a deed or a will that would continue to affect the ownership of property long after the person making those provisions is dead.
14. Muñoz Viñas, *Contemporary Theory of Conservation,* 183–89.
15. Discussion of agency is legion, relying directly or indirectly on the idea of *agencement* (usually poorly translated from French into English as "assemblage") in Deleuze and Guattari, *Milles plateaux.* Other influential exponents include Gell, *Art and Agency,* and Latour, *Reassembling the Social.* Elaborations of agency have come increasingly to resemble late medieval European attempts to retain the Ptolemaic account of the planets; see, for example, Bennett, *Vibrant Matter.* Notably, at the symposium "Conserving Active Matter" at Bard Graduate Center, New York (November 27–28, 2017), at which historians, Indigenous scholars, materials scientists, and philosophers spoke, the term "agency" was not used once in the nineteen papers delivered nor in discussion.
16. See, in particular, Gibson, *Ecological Approach.*
17. Thoreau, *Walden,* 11.
18. Carrier, "Art and Its Preservation."
19. Muñoz Viñas, *Contemporary Theory of Conservation,* passim.
20. Dykstra, "Artist's Intentions," 197–218.

BIBLIOGRAPHY

"Active Matter." Hagan Group, Martin A. Fisher School of Physics, Brandeis University (website). Accessed August 4, 2021. https://www.brandeis.edu/physics/hagan/research/active-matter/index.html#:~:text=Active%20matter%20describes%20systems%20whose,components%20of%20the%20cellular%20cytoskeleton.

"Aguahoja." Mediated Matter, MIT Media Lab. 2020. https://mediatedmattergroup.com/aguahoja.

Bennett, Jane. *Vibrant Matter: A Political Ecology of Things*. Durham, NC: Duke University Press, 2010.

Bergson, Henri-Louis. *L'évolution créatrice*. Paris: Alcan, 1907.

Carrier, David. "Art and Its Preservation." *Journal of Aesthetics and Art Criticism* 43, no. 3 (1985): 291–300.

Dante Alighieri. *The Inferno of Dante Alighieri*. Rev. ed. Translated by John Aitken Carlyle. London: J. M. Dent, 1909.

Das, Moumita, Christoph F. Schmidt, and Michael Murrell. "Introduction to Active Matter." *Soft Matter* 16 (2020): 7185–90.

Deleuze, Gilles, and Félix Guattari. *Milles plateaux*. Vol. 2 of *Capitalisme et schizophrénie*. Paris: Éditions de Minuit, 1980.

Dykstra, Steven W. "The Artist's Intentions and the Intentional Fallacy in Fine Arts Conservation." *Journal of the American Institute for Conservation* 35 (1996): 197–218.

Gell, Alfred. *Art and Agency: An Anthropological Theory*. Oxford: Clarendon Press, 1998.

Gibson, James J. *The Ecological Approach to Visual Perception*. Boston: Houghton Mifflin Harcourt, 1979.

"Gold." Periodic table of the elements. Radio Chemistry Society (website). Accessed August 4, 2021. https://www.radiochemistry.org/periodictable/elements/79.html.

Kringle, Loni, Wyatt A. Thornley, Bruce D. Kay, and Greg A. Kimmel. "Reversible Structural Transformations in Supercooled Liquid Water from 135 to 245 K." *Science* 369 (2020): 1490–92.

Latour, Bruno. *Reassembling the Social: An Introduction to Actor-Network-Theory*. Oxford: Oxford University Press, 2005.

Muñoz Viñas, Salvador. *Contemporary Theory of Conservation*. London: Routledge, 2011. First published as *Teoría contemporánea de la restauración*. Madrid: Editorial Síntesis, 2003. First English edition, Oxford: Elsevier Butterworth-Heinemann, 2005.

"MycoTEX Proof-of-Concept." NEFFA: Growing the Future of Fashion (website). Accessed August 4, 2021. https://neffa.nl/portfolio/mycotex/.

Popa, Radu. *Between Necessity and Probability: Searching for the Origin and Definition of Life.* Berlin: Springer, 2004.

Thoreau, Henry David. *Walden; or Life in the Woods* (1854). In *Walden and Other Writings of Henry David Thoreau,* edited by Brooke Atkinson, 3–312. New York: Modern Library, 1992.

Tibbits, Skylar, ed. *Active Matter.* Cambridge, MA: MIT Press, 2017.

Trifonov, Edward N. "Definition of Life: Navigation through Uncertainties." *Journal of Biomolecular Structure & Dynamics* 29, no. 4 (2012): 647–50.

Weber, Bruce. "Life." In *The Stanford Encyclopedia of Philosophy.* Summer 2018 ed. Edited by Edward N. Zalta. https://plato.stanford.edu/archives/sum2018/entries/life/.

Wittgenstein, Ludwig. *Philosophical Investigations.* 3rd ed. Edited by G. E. M. Anscombe and Rush Rhees. Translated by G. E. M. Anscombe. Malden, MA: Blackwell, 2001. First published in 1953.

TWO

The Expressive Import of Degradation and Decay in Contemporary Art

Sherri Irvin

Abstract

Many contemporary artworks include active matter along with rules for conservation that are designed to either facilitate or prevent that matter's degradation or decay. I discuss the mechanisms through which actual or potential states of material decay contribute to the work's expressive import. Nelson Goodman and Catherine Elgin introduce the concepts of literal *and* metaphorical exemplification, *which are critical to expression: a work literally exemplifies a property when it both possesses and highlights that property, and it metaphorically exemplifies a property when the properties it literally exemplifies bring that property powerfully to mind. I argue that the literal exemplification of actual or potential states of decay enhances a work's expressive power by stimulating our bodily and emotional responses to the physical potential of the work's active matter. Conservation practices, by affecting the properties the work literally and metaphorically exemplifies, are key to the expressive power of works that employ states of degradation and decay. The argument is illustrated through discussion of works by Zoe Leonard, Marc Quinn, and Kara Walker that literally exemplify actually or potentially decaying materials, as contrasted with works by Ai Weiwei and Sam Taylor-Johnson that represent, rather than exemplify, decaying materials.*

While new media artists play in the fourth dimension by creating technology-based works whose content unfolds over time, many artists of object-based works play in the fourth dimension by creating works designated for degradation or decay. Decay is, after all, one of the most fruitful processes of continual evolution that one can readily set in motion for an object-based work. There is, of course, a long tradition of artists representing or alluding to decay in visual artworks. But what is the distinctive expressive import of creating a work whose very substance is destined for dilapidation or collapse?

I'll begin by introducing the philosophical concept of *exemplification,* as discussed by Nelson Goodman and Catherine Elgin, and its implication that actual decay—not just depicted or simulated decay—might have particular expressive power. While Goodman and Elgin hold that exemplification is expressive, they say little about the psychological or phenomenological mechanisms that account for its distinctive expressive power. Attending to the bodily and emotional effects that decaying materials and structures have on us helps to explain the distinctive expressive power of these works.

Exemplification

Nelson Goodman and Catherine Elgin discuss exemplification, which is the situation in which an object both *possesses* a feature and *refers back to* that very feature. Goodman gives the example of a tailor's swatch: the swatch shows us the nature of the fabric it exemplifies by referring back to its own color and texture. While the swatch also possesses other features, such as size and shape, it does not refer to those features. The size of the swatch is not exemplified; it is merely incidental, giving us no insight into what the fabric is like.[1] Exemplification is a key expressive resource: as Goodman says, "Not all exemplification is expression, but all expression is exemplification."[2]

Elgin explains what is involved in an object's referring to its own properties. "An exemplar refers to certain of its properties: it exhibits them, highlights them, shows them forth, makes them manifest. . . . In highlighting some properties, an exemplar overshadows, marginalizes, or downplays others."[3] And, elsewhere: "To highlight, underscore, display, or convey involves reference as well as instantiation. An item that at once refers to and instantiates a feature may be said to *exemplify* that feature."[4] How does the item refer to its own features? By "present[ing] features in a context contrived to render them salient."[5] This can happen in more than one way. Some properties are salient because the context of use mandates our attention to them. Color and texture are salient for the tailor's swatch, while shape and size are not, because the swatch has a known function that involves revealing the former qualities and not the latter. The painted surface of an oil painting is salient, while the oil-stained reverse (however aesthetically interesting) is not, because of established expectations relative to the art form of painting. Properties can also become salient, naturally attracting our attention, because they are surprising or unexpected in a context. Georges Seurat's pointillism and Robert Rauschenberg's incorporation of three-dimensional nonart objects into his painted combines are salient because they distinguish these artists' works from their precursors in the art form of painting. There are, then, at least two quite different paths to exemplification: features may be salient because the expectations relevant to the context direct us to attend to them or because they stand out as unexpected and thereby naturally attract our attention.

Exemplification is relative to use. While a tailor's swatch normally exemplifies color and texture but not size or shape, it can be deployed to exemplify something it does not exemplify in typical uses, including the very property of being a tailor's swatch.[6] And in contexts where it exemplifies that property, it will also exemplify its own size and shape, since these properties are relevant to our understanding of what a tailor's swatch is like. An exemplar does not exemplify all the properties it possesses and may not exemplify those that appear most conspicuous, if the context conveys to us that these are not the ones most relevant given the current use.[7]

Zoe Leonard's *Strange Fruit* (1992–97; collection Philadelphia Museum of Art) exemplifies the property of being made from real fruit peels. This property is not merely possessed by the work but is highly salient in an art context, where the inclusion of unstable organic materials not only is historically uncommon but poses a notorious challenge from a conservation standpoint.[8] The decay of the materials is also perceptually salient: the objects are visibly subject to ongoing forces that have resulted in darkening, drying, shrinking, shriveling, and breakage. We can thus say that the work not only possesses but exemplifies the property of being in a state of decay.

Leonard could have represented or alluded to decaying fruit through a conventional art-making practice that would have been far less unwieldy for the context of a collecting institution—a still-life painting, say, or a realistic sculpture made from traditional art-making materials. Instead, she chose to make a work that directly incorporates decaying fruit, thereby exemplifying it. What is the expressive import of this choice?

One part of the answer lies in the concept of *metaphorical* exemplification. When a work literally exemplifies a certain property, this may make it especially apt to metaphorically exemplify further properties in a way that we experience as expressive. Goodman notes that a work may metaphorically exemplify properties it does not literally possess: "A picture literally possesses a gray color, really belongs to the class of gray things; but only metaphorically does it possess sadness or belong to the class of things that feel sad."[9] Elgin offers further examples: "[A]n experiment can metaphorically exemplify properties like power, elegance, panache, and promise; a painting, properties like electricity, balance, movement, and depth."[10] Narrative or representational works can metaphorically exemplify by way of what they literally represent or depict: "*A Doll's House* metaphorically exemplifies discontent; the figure in *Guernica*, grief."[11] *Guernica* represents people who are visibly in the throes of violent terror and grief; the marked salience of the states we are led to attribute to the characters leads Elgin to conclude that these states themselves are metaphorically exemplified by the work. The exemplification is metaphorical because the work does not *literally* possess the property of grief: a painting cannot grieve.

As Elgin's examples of painting imply, metaphorical exemplification may be secured through means other than representation. It appears, from the examples, that if a painting tends to give us a strong impression of or call to mind properties like movement and depth, these properties should be understood as metaphorically exemplified. I suggest that for contemporary object-based works, metaphorical exemplification is often fed by literal exemplification: the properties a work literally exemplifies—such as being made of certain materials—call to our minds properties that are metaphorically exemplified by the work.

Artists can make additional choices, beyond simply including materials in their works, that boost the salience of the materials or their states of decay, thereby shaping what the work exemplifies both literally and metaphorically. To understand this, let us look more closely at Leonard's *Strange Fruit.* Leonard, grieving the death from AIDS of her friend and fellow artist David Wojnarowicz, began gathering the peels of fruits that she and other friends had eaten. She used needle and thread to sew the pieces of peel back together, leaving the stitching quite visible. On some of the peels she added embellishments like zippers and buttons. To display the work, the reconstituted and decorated fruit peels are spread out in a seemingly random array on the gallery floor.

Strange Fruit literally exemplifies the property of being made of fruit in a way that is central to its content. Leonard's work alludes, through both its materials and its title, to the song "Strange Fruit," written by Abel Meeropol and first recorded in 1939 by Billie Holiday, which poignantly deplores lynching and other forms of racism; and it alludes to the use of the word "fruit" as a pejorative term for gay men. It also evokes theistic notions of the Garden of Eden and the Fall, whereby humans lost their privileged status through a choice to consume the forbidden fruit—often used as a metaphor for socially or religiously stigmatized sexual activity—and were condemned to earthly suffering. This complex of allusions invites us to contemplate the role of societal oppression in the AIDS epidemic.

Leonard could have secured such allusions by, say, making a painting that depicts decaying fruit. But the choice to make the work out of actual fruit peels allows Leonard to secure the allusive content in a direct and powerful way. It also allows her work to literally exemplify the fruit's degradation, which she understands as crucial to its meaning. When the work was acquired by the Philadelphia Museum of Art, she initially explored aggressive conservation measures with the well-known conservator Christian Scheidemann. But as Ann Temkin, the curator responsible for the work's acquisition, writes,

> Leonard surprised herself and found that she recoiled at Scheidemann's hard-won results. She realized that the appearance of decay was not enough for her, the metaphor of disappearance was insufficient. . . . [T]he mere pretense of deterioration was no longer persuasive. Leonard set herself a criterion of honesty and rejected the twenty-five preserved pieces.[12]

Leonard decided that it was essential to the work's expressive power that it directly embody deterioration rather than only alluding to it: these acts have different communicative resonance.

In prohibiting aggressive conservation measures for *Strange Fruit,* Leonard increased the salience of the fruit's decay.[13] When a literally exemplified property is made highly salient in this way, its potential for metaphorical exemplification is similarly heightened.

Bodily, Imaginative, and Emotional Responses

We now have the concept of exemplification in our repertoire. But is it really explanatory? Why does the actual presence of a material or property in the work have power that a depiction or simulation does not? If Leonard's decaying fruit peels are deployed in service of a metaphor, why couldn't that metaphor be activated just as effectively by, say, a vanitas painting?

The special powers of the fabric swatch are epistemic: the swatch shows us in detail what the fabric is like, including its weave, drape, transparency, and feel on the skin. But Zoe Leonard does not give us a sample of fruit peels in order to show us what fruit peels are like. The power is expressive, not epistemic, and this seems to require some explanation.

To answer, I'd like to bring together some reflections by Susanne Langer from the middle of the twentieth century. Langer suggested that we tend to understand space as organized by our kinetic possibilities within it. When a sculpture is present in the space, we see the space as organized by the kinetic possibilities we imagine for the sculptural object (even if we know it is not going to move), and this changes how we experience the space. She describes sculpture as "*virtual kinetic volume*" and suggests that the sculpture's implied kinetic nature tends to activate a connection between vision and touch.[14] She says,

> The intimate relationship between touch and sight which is thus effected by the semblance of kinetic volume explains some of the complex sensory reactions which sculptors as well as laymen often have toward it. Many people feel a strong desire to handle every figure. [Some] imagine the touch of stone or wood, metal or earth; they wish to feel the substance that is really there, and let their hands pass over its pure form.[15]

Langer's suggestion recalls remarks by Johann Gottfried Herder in the eighteenth century. Herder holds that we use vision as a substitute for touch in our apprehension of sculpture: the eye is guided to seek out the information that the hand desires. He suggests that the beauty of three-dimensional forms, and the pleasure they occasion, belong to touch: the form is felt (perhaps imaginatively), not seen, as beautiful.[16]

Langer invokes several interrelated processes. First, the presence of sculpture affects the way I experience space to be organized. Second, I have imaginative bodily experiences of what it would be like to touch the sculpture (and as we know from recent scientific findings, imaginative bodily experiences do indeed involve bodily, not merely mental, activation).[17] Third, I have longings associated with the sculpture. So the sculpture is entering my consciousness cognitively, imaginatively, emotionally, and somatically.

Langer argues further that imagination has a profound role both in shaping our perceptions of the world and in constituting our emotional lives. She says,

> It is sensation remembered and anticipated, feared or sought or even imagined and eschewed that is important in human life. It is perception molded by imagination that gives us the outward world we know. And it is the continuity of thought that systematizes our emotional reactions into attitudes with distinct feeling tones, and sets a certain scope for an individual's passions. In other words: by virtue of our thought and imagination we have not only feelings, but a *life of feeling.*[18]

I suggest that these various insights, brought together, help to explain how the actual presence of certain materials, including their actual or potential states of decay, has distinctive expressive power. The sculptural object has real physical potential, and thus offers real possibilities for us. Materials configured sculpturally reorganize our sense of possibility, spark imaginative bodily engagement, disturb us, and ignite our longings.

To see how these phenomena elevate the expressive potential associated with decaying materials, let us consider some examples. For his self-portrait busts, an ongoing series that began in 1991, Marc Quinn takes a direct cast of his head. Then he creates a mold that he fills with his own blood. The work is completed through a process of freezing.[19]

Now, blood is a liquid, and it has the natural tendencies of a liquid, to dissipate if uncontained. These tendencies lend themselves adequately to drawing or inscription—coagulation notwithstanding—but less well to sculpture.[20] Quinn's process wryly mimics what we do with bronze: render it temporarily liquid for casting, knowing that it will return to a highly durable solidity as it cools. Quinn, instead, uses the natural liquidity of blood and then suspends it in a solid state.

When we encounter the work, we know that this is wrong. The wrongness of it, the potential liquidity that is also the possibility of a bloodbath, of the artist's head dissolving into a pool of blood—this is present to us. Our knowledge that the object is constituted of frozen blood makes it hum with the potential of disaster. You can feel this even if you aren't aware that one of the works in this series may have been destroyed when power to the refrigeration unit was mistakenly shut off, even if you aren't thinking about the grisly failures of early attempts to cryogenically preserve human bodies for future revival.[21] We humans aren't very good at keeping things frozen; even our icebergs are melting.

Also present to us is the object's current actual state: the coldness, solidity, and ghastliness of this strangely shaped blood popsicle. And then there is our further contemplation of the blood itself. The large quantity of the artist's blood would have been medically removed over an extended period: in fact, the quantity of blood in each of these busts, ten pints, is about the same as the total quantity of blood present in an adult human body at a given time. Blood thus removed is often used for therapeutic purposes, though not in this case. Blood gives life, but its loss through injury and wound can result in death. Blood records what we've been up to, as Quinn, who has a history of alcoholism, would be acutely aware.[22] Blood can prompt squeamishness, unease, and for good reason: contact with others' blood can be dangerous. Of course, by freezing the blood and putting it in a vitrine, Quinn confuses these reactions a bit: the blood is, for the moment, safely contained.

There is thus a duality to our encounter with this work: even while it is present to us as a solid object, clinically frozen, its potential for decay and loss animates our emotions and our bodily imaginings. It matters that the work is actually made of blood and actually has the potential to melt, which we feel and experience as potential not only for it but for us.

The power of Quinn's work rests partly in the fact that it exemplifies both the very substance of the artist's person and exquisitely fine detail of the topography of his skin and features. It evokes a particular moment in Quinn's life by virtue of what it shows about his age but even more by revealing aspects of the casting process, such as the pressure placed on the skin and material clumped in the eyelashes. We see the variability of color and surface texture of frozen blood, but also the signs of a body being impinged upon in very specific ways.

Goodman says that "all expression is exemplification." How does what this work exemplifies, both literally and metaphorically, connect to its expressive power? Quinn refers to these works as being "on life support": by literally exemplifying the potential to melt and a technological support to resist this potential, they metaphorically exemplify the precarity of human life and our heroic efforts to sustain it.[23] The metaphor has its power not merely by showing us the fine features of something, a function that could be served as well by a realistic simulacrum. Our knowledge of the presence of the artist's real blood, tenuously suspended in a solid state, activates a whole sequence of bodily imaginings and emotional responses having to do with bodily injury, disease, and mortality. The object invites a sort of squeamish fascination, helping us quiet any disgust response through the presentation of a frozen object clinically sealed off. These bodily and emotional responses are what activate the expressive energy of the metaphor of a temporary marvel of self-preservation and its inevitable failure.

We may turn now to Kara Walker's 2014 work *A Subtlety or the Marvelous Sugar Baby, an Homage to the unpaid and overworked Artisans who have refined our Sweet tastes from the cane fields to the Kitchens of the New World on the Occasion of the demolition of the Domino Sugar Refining Plant.* This work, like Quinn's, has a pronounced duality: it was simultaneously present as a magnificent monument and as an impending ruin. The fact that the building in which it was displayed was soon to be demolished is signaled in the title. The grand sphinx figure was made with a surface of white sugar, which is soluble, and the object is described as a subtlety, which is traditionally a small sugar sculpture designed to be eaten. But most markedly, the child worker figures, some cast in brown sugar and others cast in resin and coated in molasses, were collapsing before our eyes.[24] The visible disintegration, along with the associated smell, heightened the sense of ephemerality, of impending loss. And this activated a variety of bodily, emotional, and imaginative responses.

The most widely publicized reactions were disrespectful and exploitative, objectifying the Black female body in ways that hark back to the sexual violence against Black women that was used, in the sugar industry as elsewhere, to produce a labor force to create wealth for whites.[25] But other forms of longing and bodily engagement were in play as well. Children were seen playing with and even licking the child laborer figures.[26] Walker had video shot of audience engagement with the work, and this showed that some treated the sphinx figure with gentle, reverent caresses.[27] While there were signs indicating that touching of the work is prohibited, the knowledge and visible signs of the work's degradation and eventual destruction may have helped to disable this prohibition and allow people to feel more fully justified in indulging their longings and yielding to the work's physical allure.[28] Why not engage with the work fully, given that there was no chance of the objects' long-term preservation and that there would be no opportunity to return for further encounters in later years?

Most of us, of course, did not lick the objects or dip our hands into the syrupy goo puddled on the floor. But the signs of loss and disintegration nonetheless heighten our sensory imaginings of texture, viscosity, and sweetness. Our actual visual perception of the objects is, as Langer suggested, "molded by the imagination," connecting to both embodied desire and emotional passion. The literal exemplification of specific materials yielding visual and olfactory signs of their deterioration, paired with the literal exemplification of impending destruction, activates a metaphor of the fragility and preciousness of our beings, the power we have over each other, the tragedy of disrespect and disregard.

And the knowledge that the site—which previously housed part of an industry that enacted racialized and gendered forms of violence upon workers—will be literally bulldozed may activate the wistful metaphoric imagining of a prospect of transforming the entire system of white supremacist patriarchy into a ruin.

A simulacrum, of course, also engages imagination and emotion, but not in the same way. The potentialities that we know it to offer for us are different. Consider Ai Weiwei's 2010 work *Sunflower Seeds*, which was displayed by placing one hundred million individually hand-painted porcelain sunflower seeds in the Turbine Hall at Tate Modern in 2010–11.[29] Ceramic seeds function differently than real seeds in our experience because of their very different potentialities for us: real sunflower seeds are edible, while ceramic seeds will not nourish and might break the teeth. The work really does show us what sunflower seeds are like—it exemplifies their external color and morphology, though not their other material qualities. Precisely through this careful simulation, which is a form of selective exemplification, the work activates our bodily imaginings associated with an experience of chewing and eating while placing them in collision with the other qualities it exemplifies.

Another instructive contrast is between items directly included in a work and items photographically depicted. Sam Taylor-Johnson's *Still Life* (2001; collection Museum of Fine Arts Boston) is a remarkable time-lapse film, lasting nearly four minutes, of a bowl of decomposing fruit.[30] For almost the first minute, little appears to happen; but then some of the fruit visibly shifts position almost as if alive, and not long afterward a dramatic bloom of mold begins, other fruits move, and rotten spots begin to emerge and spread. After first expanding outward in a greenish fuzz, the mold and the objects beneath it begin to collapse, gradually shifting toward a darkening gray invaded by spots of opaque white, as a haze of mold creeps onto the bowl rim and surrounding table. Things seem stable for a while, until the scene is animated by the movement of flies and other insects as parts of the fruit darken to black.

The objects shown, though not the film itself, exemplify florid states of decay. As with the tailor's swatch, part of the value of exemplification here is epistemic: the work shows us the unfolding of processes we may know little about, since we go to considerable pains to prevent anything like this happening in our living spaces. In addition, the work certainly evokes emotions and bodily imaginings: to my mind, it resembles a naturalistic short horror film, pinning me in place with fascination even as I experience a welling up of physical disgust. I even imagine the smell of the rotting fruit, until things have progressed to the point where I no longer know what to imagine.

At the same time, this decay has no real bodily potential for me. The objects and events that fascinate me are not present in my space; the features of interest are not exemplified by the work itself. The smell remains purely hypothetical. I can watch the film as many times as I like, and things always unfold exactly the same way; the remains of those fruits are long discarded. What I have access to is only a record, not the things themselves. While the blood of Marc Quinn's self-portraits is clinically separated from us by freezing and containment, it is still *right there*. There is an immediacy to how it encroaches on my sensibilities, in contrast to the detachment fostered by technological reproduction.

Artists use unconventional materials expressively, and they set in motion states of disintegration and decay for expressive purposes. Goodman and Elgin give us some resources to account for this when they discuss exemplification as a process whereby a work can render salient some of its own characteristics, thereby signaling that we are to take them seriously. These characteristics, once highlighted, can then be deployed to metaphorical ends.

But to complete the explanation of why literal exemplification—the actual inclusion and salience of decaying materials—has expressive significance that differs from that of depiction or simulation, we must look to further experiential phenomena. Sculptural materials have potential that we feel as potential for ourselves; we have embodied and emotional reactions to them, sometimes only felt and sometimes acted upon. Actual states of decay may alter our sense of the prohibitions associated with contact, as they undermine any illusion of an untouchable object preserved for posterity. And our awareness of fragility, even in an object not yet exhibiting decay, creates a duality in our experience, with the object present to our senses even as its degraded state lurks in our imaginations.

Let us return briefly to Leonard's *Strange Fruit*. The literal exemplification of decay activates a powerful metaphorical exemplification of mortality; the literal exemplification of hollow shells that have been futilely stitched back together, only to continue their process of disintegration, metaphorically exemplifies emptiness, loss, and the persistence of grief. The activation of these metaphors is emotional and embodied: our knowledge that Leonard and her friends ate all this fruit and collected the peels over a period of years, and that she painstakingly sewed them back up again, activates our embodied awareness of eating, handling, and acts of care. The decay of the objects constitutes real loss, not merely loss alluded to. We know and feel that we are encountering the transitory: the transitory genuinely present to us, genuinely exemplified. The presence of real decaying objects activates our bodily, imaginative, and emotional responses, and this explains the metaphor's expressive power.

Conservation practices, either applied or withheld, are key to maintaining the salience of features that is required for exemplification and expression. Leonard's work expresses what it does because of the withholding of traditional processes to arrest decay. The force of Walker's work depends on the active erosion of objects combined with the fact that the entire site is slated for destruction. Quinn's work, on the other hand, depends critically on conservation of the refrigeration apparatus to prevent its rapid destruction. The unfolding of processes governing active matter may either promote or threaten the presence and salience of the work's expressive features. Conservation practices designed to foster those features, either by permitting change to progress unimpeded or by arresting it, are essential to the ability of these works to realize their full expressive and aesthetic power.

NOTES

I am grateful to Bard Graduate Center for the invitation to present an earlier version of this work in 2019 and to A. W. Eaton and Ivan Gaskell for helpful comments and discussion.

1. Goodman, *Languages*, 53.
2. Ibid., 52.
3. Elgin, "Making Manifest," 400.
4. Elgin, "Understanding," 15.
5. Ibid., 17.
6. Goodman, *Languages*, 50–51.
7. Elgin, "Making Manifest," 401.
8. Wharton, "Challenges of Conserving."
9. Goodman, *Languages*, 53.
10. Elgin, "Understanding," 16.
11. Ibid., 24.
12. Temkin, "*Strange Fruit,*" 47.
13. I regard establishing rules governing the conservation of their works as part of the artist's prerogative in determining the work's nature and

boundaries. However, the extent to which such rules are binding on the institution is controversial, and there are circumstances in which violating them is justified. I discuss related issues in "Authenticity, Misunderstanding" and *Immaterial.*

14. Langer, *Feeling*, 89. Emphasis in original.
15. Ibid., 90.
16. Herder, *Sculpture.* For discussion, see Zuckert, "Sculpture and Touch."
17. For example, Fadiga et al., "Motor Facilitation."
18. Langer, *Feeling*, 372. Emphasis in original.
19. "Marc Quinn ('Self')." Quinn made works in the series every five years from 1991 to 2011 but does not appear to have made one in 2016.
20. Yu, "Blood Writing."
21. "Blood Sculpture 'Melted.'" Quinn has reportedly refused to comment on whether the work, *Self 1991*, was subsequently remade. Akbar, "National Portrait Gallery." Nelson, *Freezing People.*
22. "Ephemeral Works."
23. Ibid.
24. Carpio, "On the Whiteness," 551–52.
25. See, for example, the accounts by Elizabeth Chin and Dána-Ain Davis in Reeser, "Reflections."
26. Watts, "Audacity."
27. "Kara Walker—an Audience."
28. It is worth noting, however, that audience members violating the prohibition on touching artworks in museum collections is a frequent occurrence, as conservation records attest.
29. "Unilever Series." Further information about the work's production and exhibition history is available at the Faurschou Foundation's website Ai Weiwei Sunflower Seeds, accessed October 8, 2020, https://www.aiweiweiseeds.com/.
30. "Still Life, Sam Taylor-Johnson." A video of the work is available to watch on the artist's website, accessed October 8, 2020, https://samtaylorjohnson.com/. I am grateful to Ivan Gaskell for suggesting discussion of this work.

BIBLIOGRAPHY

Akbar, Arifa. "National Portrait Gallery Acquires Marc Quinn's Bloody Head." *Independent*, September 10, 2009. https://www.independent.co.uk/arts-entertainment/art/news/national-portrait-gallery-acquires-marc-quinns-bloody-head-1785133.html.

"Blood Sculpture 'Melted.'" *BBC News.* July 4, 2002. http://news.bbc.co.uk/2/hi/entertainment/2093053.stm.

Carpio, Glenda. "On the Whiteness of Kara Walker's *Marvelous Sugar Baby.*" *ASAP/Journal* 2, no. 3 (2017): 551–78.

Elgin, Catherine Z. "Making Manifest: The Role of Exemplification in the Sciences and the Arts." *Principia: An International Journal of Epistemology* 15, no. 3 (December 2011): 399–413.

———. "Understanding: Art and Science." *Synthese* 95, no. 1 (April 1993): 13–28.

"Ephemeral Works (on Life Support)." Marc Quinn (website). Accessed September 29, 2020. http://marcquinn.com/artworks/ephemeral-works-on-life-support.

Fadiga, Luciano, Leonardo Fogassi, Giovanni Pavesi, and Giacomo Rizzolatti. "Motor Facilitation during Action Observation: A Magnetic Stimulation Study." *Journal of Neurophysiology* 73, no. 6 (1995): 2608–11.

Goodman, Nelson. *Languages of Art: An Approach to a Theory of Symbols.* Indianapolis: Hackett, 1976.

Herder, Johann Gottfried. *Sculpture: Some Observations on Shape and Form from Pygmalion's Creative Dream.* Translated by Jason Gaiger. Chicago: University of Chicago Press, 2002.

Irvin, Sherri. "Authenticity, Misunderstanding, and Institutional Responsibility in Contemporary Art." *British Journal of Aesthetics* 59, no. 3 (2019): 273–88.

———. *Immaterial: Rules in Contemporary Art.* Oxford: Oxford University Press, in press.

"Kara Walker—an Audience (trailer)." Sikkema Jenkins & Co. Accessed June 8, 2019. https://vimeo.com/112396045.

Langer, Susanne. *Feeling and Form: A Theory of Art.* New York: Charles Scribner's Sons, 1953.

"Marc Quinn ('Self')." National Portrait Gallery (website). Accessed June 7, 2019. https://www.npg.org.uk/collections/search/portrait/mw138260/Marc-Quinn-Self.

Nelson, Bob. *Freezing People Is (Not) Easy.* Guilford, CT: Lyons Press, 2014.

Reeser, Doug. "Reflections on Kara Walker's 'A Subtlety, or the Marvelous Sugar Baby.'" *AnthroNOW.* July 24, 2014. http://anthronow.com/reach/reflections-on-kara-walkers-a-subtlety-or-the-marvelous-sugar-baby.

"Still Life, Sam Taylor-Johnson." Museum of Fine Arts Boston (website). Accessed October 8, 2020. https://collections.mfa.org/objects/518595.

Temkin, Ann. "*Strange Fruit.*" In *Mortality, Immortality? The Legacy of 20th Century Art,* edited by Miguel Angel Corzo, 45–50. Los Angeles: Getty Conservation Institute, 1999.

"The Unilever Series: Ai Weiwei; Sunflower Seeds." Tate (website). Last modified October 22, 2010. https://www.tate.org.uk/whats-on/tate-modern/exhibition/unilever-series/unilever-series-ai-weiwei-sunflower-seeds.

Watts, Stephanye. "The Audacity of No Chill: Kara Walker in the Instagram Capital." *Gawker*, June 4, 2014. https://gawker.com/the-audacity-of-no-chill-kara-walker-in-the-instagram-1585944103.

Wharton, Glenn. "The Challenges of Conserving Contemporary Art." In *Collecting the New: Museums and Contemporary Art*, edited by Bruce Altshuler, 163–78. Princeton, NJ: Princeton University Press, 2005.

Yu, Jimmy. "Blood Writing as Extraordinary Artifact and Agent for Socioreligious Change." *Humanities and Social Sciences Communications* 7, no. 1 (2020): 1–9.

Zuckert, Rachel. "Sculpture and Touch: Herder's Aesthetics of Sculpture." *Journal of Aesthetics and Art Criticism* 67, no. 3 (Summer 2009): 285–99.

THREE

The Look of Age: Appearance and Reality

Carolyn Korsmeyer

Abstract

Gradual degradation is one of the slower processes of active matter, and it can create aesthetic features that mark the passage of time. Worn stone, faded paint, damaged details on antiques, and ruins are appreciated for the very look of age that they manifest. So admired can such features be that objects are often produced to mimic that look. Sometimes they are intended to stand in for real old things, such as the sham ruins that were popular in Europe in the nineteenth century; other times the look of age is employed as a contemporary style of decoration. This chapter argues that aesthetic encounters with sham and authentic are in principle different, no matter how convincing and evocative a replica or fake might be. This is because perceiving degradation entails recognizing its cause, implicitly noting the events that produced the appearance. By employing Nelson Goodman's notion of exemplification and its contrast with other referential types such as imitation and illusion, I conclude that an object that really is old invites a different kind of aesthetic apprehension from an object that only looks old.

The "look of age" refers to perceivable deterioration that seems to have accrued over time or to have been inflicted at some recognizable point in the past. To be sure, it is not always an appearance that is valued. An object that is worn out, damaged, or otherwise degraded usually invites discard and replacement, especially if its function has been compromised. Positive appreciation of damaged appearance is the last response one is likely to have when confronting crumbling piles precariously holding up a bridge, a house wrecked by a tornado, or a tooth eroded by decay. It is unlikely that such things exert much appeal.

However, there are damaged artifacts that actually have an aesthetic pedigree, foremost in the case of ruins, where the marks of age, accident, and destruction become a focus of attention. Although the progressive deterioration of ruins is often too slow to be perceived by any one person, they count as venerable "active matter" among objects of appreciation. The ravages of time can produce outlines and surfaces that are beautiful or picturesque, such as the soft edges of eroded stone, the pale lichen growing on old headstones, or a graceful line of lonely arches that once buttressed a roof. Encounters with ruins and with other objects of great age often engage moral and existential attitudes—awe at the power of time and nature. In cases such as these, appreciation includes feelings of loss and melancholy, signaling awareness of bygone times and the impermanence of human endeavors.

In his widely read study of ruins, Paul Zucker invites us to savor what he calls the "fascination of decay" when we contemplate "the patient, deliberate, and often beautiful wearing away of stone by time."[1] This perspective not only fosters appreciation of the look of age; it suggests that removing the appearance of degradation can produce aesthetic damage. Such, for instance, was Walter Kaufmann's objection to the cleaning of a frieze of Buddhas at Borobudur in 1971. He lamented that "the symbiosis of art and nature is broken, the ruins that are more than a thousand years old have a new, artificial, lifeless look that brings to mind plastic. The spirit has been driven out of the stone, and what is left behind is dead."[2]

While ruin appreciation of this sort is now sometimes dismissed as a symptom of obsolete romanticism, it is worth noting that admiration of the changes wrought by time has deep roots that are not only aesthetic but also ethical. Marcus Aurelius, for one, observes that the capacity to find beauty in decay is a sign of wisdom regarding the nature of things.

> For instance, when a loaf of bread is baked there are cracks in the surface, and these breaks, which are contrary to the purpose of the baker, are beautiful in their way, and stimulate the appetite. Again, figs when they are quite ripe gape open; and ripe olives when they are near to rotting are particularly good to look at. . . . Thus if a man has a feeling for and deep insight into the things produced in the universe, there is hardly one of their characteristics that will not seem to him of a sort to give him pleasure.[3]

In other words, finding beauty in deterioration is a mark of aesthetic sensibility that has roots in moral wisdom. In this case, the wisdom is Stoic, although other perspectives yield similar attitudes. As Diderot put it many centuries later, "Everything dissolves, everything perishes, everything passes, only time goes on. . . . How old the world is. I walk between two eternities. . . . What is my existence in comparison with this crumbling stone?"[4]

Ruin appreciation of a romantic, reflective sort was especially popular in the eighteenth and nineteenth centuries. During that period, European wealthier classes even built sham or artificial ruins in their parks and gardens—follies and crumbling walls that appear to the eye as having once been whole but over time to have fallen into fragments. Shams remind us that the appearance of degradation evoking a sense of the distant past need not be the result of any actual process of degradation over a proportionately long time. Shams are not real remnants of the past but rather represent a taste for a certain style of architecture that mimics such remnants. A liking for the faux antique is enduring, for it extends into the twentieth century, as with the "medieval" structure known as Gillette Castle, which was built in Connecticut in 1919 (fig. 1).

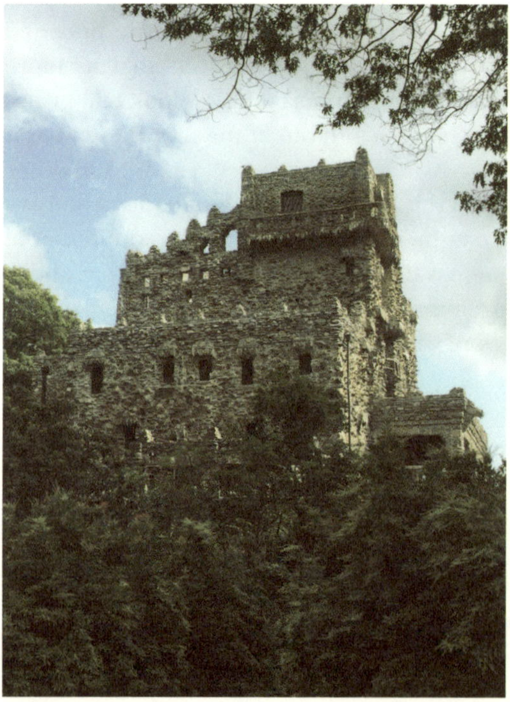

Fig. 1. Gillette Castle, 1914–19, East Haddam, CT. Photo by author.

It may also be found with opportunistic repurposing of architectural fragments to appear like the remnants of antiquity fallen into ruin over time rather than recent victims of urban clearance, as with the arrangement of columns illustrated here that were once part of a bank but that now grace my university campus (fittingly, beside an artificial lake; fig. 2).[5]

A taste for the degraded also includes the popular category of the "shabby chic" or the "distressed" artifact, where either a new object is made to look worn or an old one is left in its faded condition. Some whole interiors display the current taste for faux degradation, such as the restaurants from the Cheesecake Factory chain, with their apparently crumbling plaster (fig. 3).[6] Less over the top but in a similar mode are the exposed brick and broken beams displayed at another restaurant in my local mall (fig. 4). It is clear that the look of age is popular even with an artifact that has no pretense for being actually aged. With these examples, we can see that the taste for apparent deterioration extends from ancient ruins to shams that expertly mimic the look of antique architectural remnants to interiors that fool no one but are decorative or even amusing.

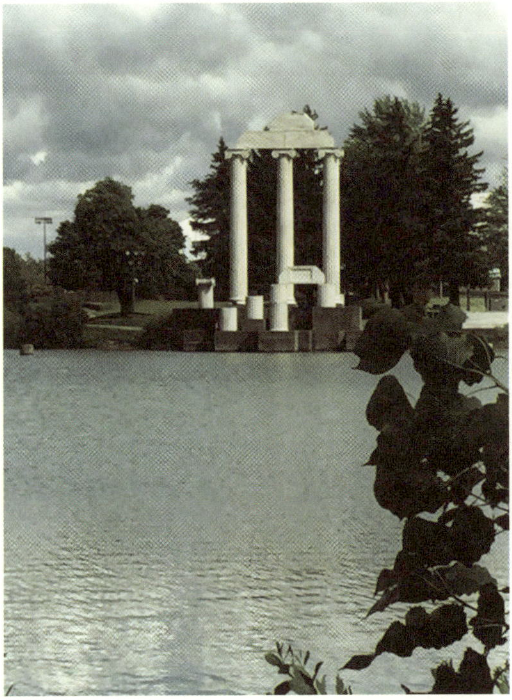

Fig. 2. Baird Point, University at Buffalo. Photo by author.

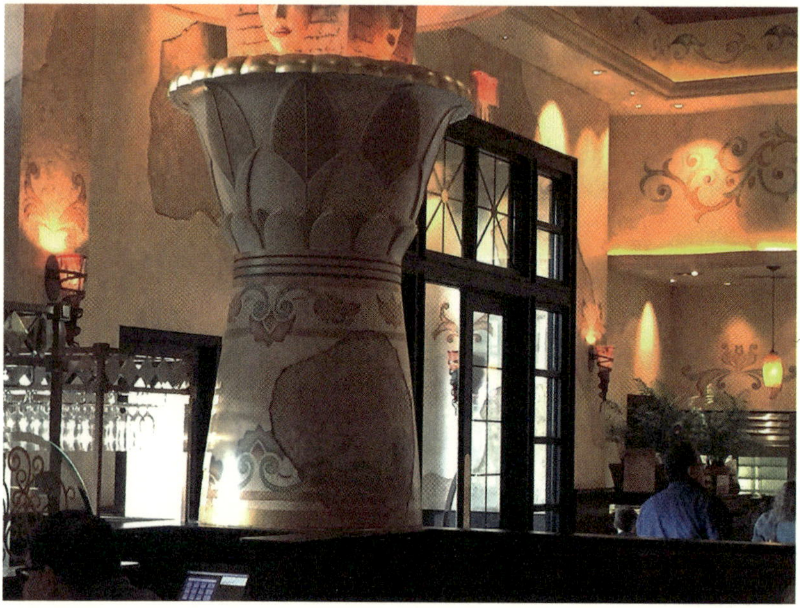

Fig. 3. Cheesecake Factory, Walden Galleria Mall, Cheektowaga, NY. Photo by author.

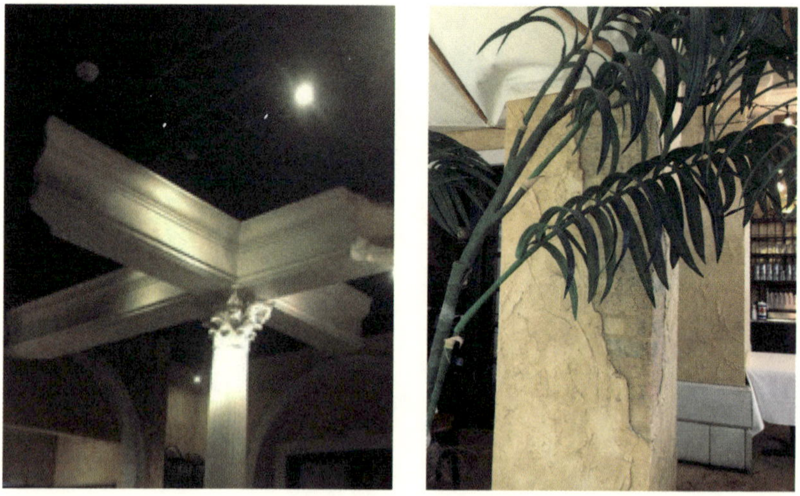

Fig. 4. Bravo! Italian Kitchen, Walden Galleria Mall, Cheektowaga, NY. Photos by author.

I am interested in the relationship between the mere *appearance* of wear and decay and *actual, "authentic"* wear and decay; that is, between the intentional degradation and arrangement of materials to look older than they really are and the gradual wearing away of surfaces over time. Objects in both categories can be beautiful, picturesque, attention-grabbing, grotesque, curious, and decorative. Is their aesthetic appeal the same, regardless of their histories and the way their appearances have come about? While at first glance one may be tempted to say that the real and the sham share aesthetic properties and differ only in their historical and cultural standing, I believe that this is not the case, and the purpose of the rest of this chapter is to examine the look of age—including degradation and damage—as an aesthetic property.

Appearance, Reality, and the Aesthetic: What Is in the Frame?

The term "aesthetic property" refers to a quality of an object that arouses appreciation for its presentation to the senses and the imagination, that is, for the way it appears. The distinction between "appearance" and "reality" is one of the most ancient in philosophy, arguably an original guiding question of the field.[7] Plato set the stage for the strong preference for reality over appearance, a view that persists for rather obvious reasons. In contrast, and somewhat ominously, aesthetic qualities were traditionally—and still often are—construed chiefly in terms of immediate perceptual effect, especially the way something looks or sounds. In other words, as matters pertaining to appearance.

According to a strong (though waning) tradition in philosophy, aesthetic qualities are always matters of appearance, of the *perceptual surface,* so to speak. As philosopher J. O. Urmson put it when this view prevailed, "What makes the appreciation aesthetic is that it is concerned with a thing's looking somehow without concern for whether it really is like that."[8] A real ruin and an artful sham, a true antique and an appropriately distressed artifact of an older or retro style, can be difficult to tell apart just by looking. Therefore, locating their appreciable properties on the perceptual surface of objects equalizes their aesthetic standing.

Although still in circulation both theoretically and popularly, this is a view that many philosophers now reject, for it depends upon a dubious concept of perception that severely restricts what count as aesthetic features, ignoring nonperceptual attributes that contribute to appreciation. It is more common today to conceive of aesthetic attributes in a robust fashion that includes reference to historical context and to meanings that are partly determined by that context. Neither is conveyed by immediate perceptual presentation alone. To discern something's aesthetic qualities competently, one must have at least some degree of understanding of the object of attention, including apposite artistic category and cultural position, both of which exceed simple appearance.

Recognizing the contextual embeddedness of aesthetic properties is especially imperative for artifacts where some condition of degradation is perceivable, such as visible wear and tear, fading, decay, and damage—the sorts of conditions that yield a look of age. Such qualities evoke what we might call a narrative of the object: the history, the story behind what brought it to the condition before us. Degradation is different from other sorts of changes that beset active matter, such as ripening or curing, changes of appearance that artists often anticipate—the desired development of verdigris on copper or of rust on iron sculptures, for example. When one perceives an artifact as damaged or degraded, one implicitly recognizes that there was some prior, more complete or pristine condition; and perhaps one also identifies the cause of the damage or decay that has brought about the present state. A mistake about a cause engenders a mistake about how its properties should be perceived.

I offer a progression of claims about degradation as an aesthetic property, all of which add to the evidence that both perception of aesthetic properties and competent appreciation are rarely just surface matters. First, *to perceive something as damaged or degraded entails noting that something caused it to come to its present condition.* It is not necessary to know just what that cause was, but missing walls, exposed plumbing, breakage, worn and chipped paint, command recognition that the object used to be different and that some event or events brought about degradation. This implicit recognition is part of the "contents" of perception with objects such as these.[9]

This fact is especially obvious when damage was caused not by gradual deterioration over time, but by a specific, dramatic event.[10] For example, the Genbaku Dome in Hiroshima is an object where the precise cause of the damaged condition is readily recognized: the atomic bomb dropped on the city on August 6, 1945 (fig. 5). Indeed, that cause is the very reason for its preservation and display, and if one regards this structure as just a worn-out industrial building, then one misperceives it badly. Similarly, if you were to encounter the large metal object illustrated here out of the context of its display, you might adopt

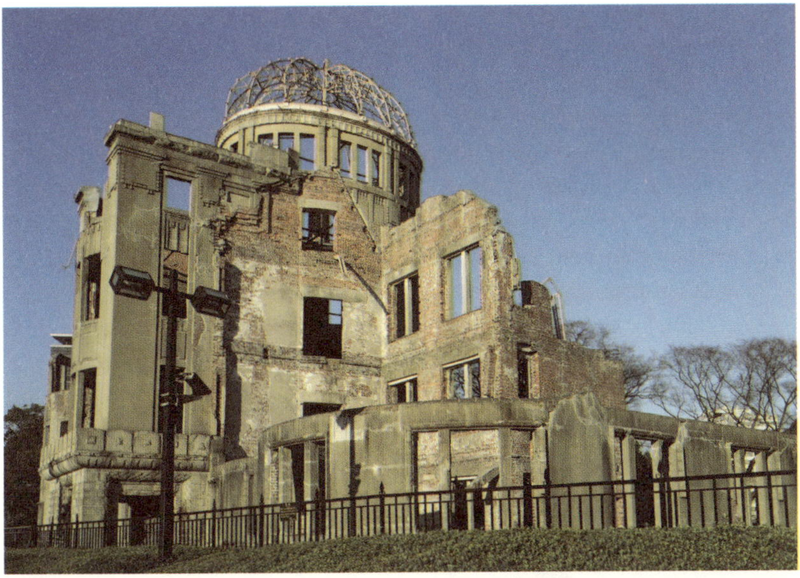

Fig. 5. Hiroshima Peace Memorial/Genbaku Dome. Photo: Wikimedia Commons/Oilstreet, 2013. https://commons.wikimedia.org/wiki/File:Genbaku_Dome04-r.JPG.

a modernist, formalist perspective and find it dramatically twisted, perhaps whimsically grotesque (fig. 6). The presumption at work is that the shape is a matter of artistic design. In fact, this is a gigantic fragment of the North Tower of the World Trade Center, whose distorted shape was caused by the terrible impact of the 9/11 attacks. A twist in an abstract sculpture is rightly perceived as a deliberate effect for visual interest; in this case, the same shape is the effect of malignant damage. This example dramatically demonstrates that once one recognizes the identity of an object, the way it appears can manifestly shift.

Even when one is not aware of the exact sequence of events that brought about an injured condition, one must be at least vaguely aware that there was some such cause. The historical distance between now and the time when the abbeys of Britain flourished has erased (at least among tourists) once-common knowledge that their first major destruction came about when Henry VIII dissolved the monasteries in the mid-sixteenth century. But anyone visiting landmarks such as Tintern Abbey, surrendered to the crown in 1536, will note missing roofs, foundations without walls to support, and eroded stone, thereby noticing that something or other brought these buildings to their present state.

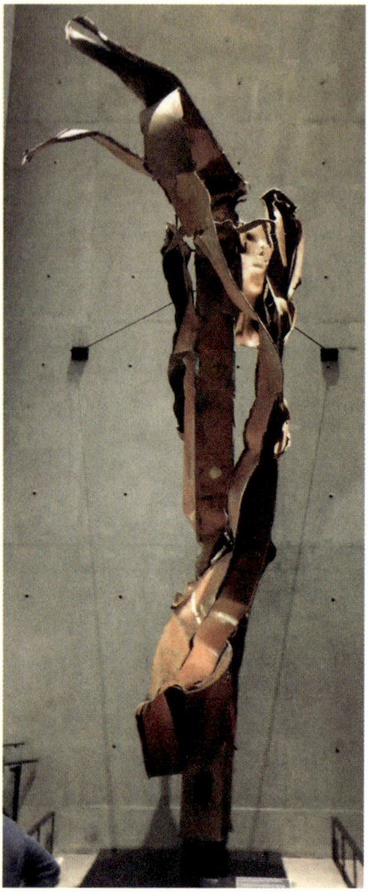

Fig. 6. Fragment of the North Tower of the World Trade Center, 9/11 Memorial & Museum, New York City. Photo by author.

In short, cause is embedded in the very concepts of damage, deterioration, and degradation, which situate objects in relation to a set of external events. The mere notice of damage or degradation is pro tanto notice of some injurious cause. Thus, degradation as an aesthetic property commands what we might term a particularly deep *narrative aura*, for it summons a story accounting for the damaged condition.

An additional claim that separates real from sham follows upon this one: *When degradation is a perceptual property of an object, aesthetic encounters between real and sham are not the same.* By "genuine" and "real" I refer to the enduring material thing—continuity of the same physical object over time—where perceptible damage records and displays its history. Only with the persisting physical object does the appearance of degradation match its cause. (Potential difficulties with this claim will be addressed at the end.) With shams and retro objects, damaged appearance is manufactured rather than "earned"; their faux narrative aura emanates from counterfactual imagination.

Exactly why do different histories amount to aesthetic differences in such cases? What is the aesthetic difference between something that is degraded because it is *really old* and something that only *appears old*? After all, they look the same, and the possibilities for error about what caused their appearance are manifold (a complicating factor that must not be gainsaid). How different, really, can the marks of age be that appear on sham objects, or for that matter, newly manufactured replicas of real old things? Many would argue that my devotion to the genuine thing is misdirected because sham and real can achieve the same aesthetic ends, and also because imitation can, in practice, assume the aesthetic standing of a real thing. Before advancing my additional claims about degradation, therefore, I address this possibility by considering two especially sophisticated versions of those positions.

Champions of Sham, Fake, and Substitute

Saul Fisher asserts that "designers of shams aim at the same aesthetic effects offered by ruins."[11] He equalizes the aesthetic status of ruins and shams by considering both as abstract architectural objects—the former actually built, the latter fantasy—whose formal perceptual properties invite the same kind of imaginative extrapolation to a complete structure, either one that formerly existed or one that might have been.[12] Fisher endorses a formalist approach, aligning himself with those who regard aesthetic features as matters of appearance.

> What we can rely on across actual and sham ruins is their formal similarities, in appearing to the spectator as parts of an original, now corrupted whole, built structure. That a whole structure never existed as built, in the case of sham ruins, or even that no such built structure was ever intended, is immaterial to their common features with actual ruins, of having that sort of *appearance* to the spectator.[13]

The fact that often we cannot tell a ruin from a sham is evidence that tagging aesthetic properties to the history of an object is irrelevant and even, in Fisher's view, prejudicial.[14] To his way of thinking, the history of an artifact is outside the aesthetic frame.

A contrasting approach that reaches a similar conclusion regarding aesthetic equivalence comes from archaeologist and anthropologist Shannon Lee Dawdy. Although she considers historical and cultural context to be crucial in justifying substitution of fake (her term) for real, under some conditions, real thing, replica, and fake can serve the same aesthetic purposes. She makes this point in her study of New Orleans as it recovered from the devastation wreaked by Hurricane Katrina in 2005, and thus her position is bolstered by empirical evidence of practices among the residents of that damaged city. Dawdy uses the term "patina" to refer generally to the look of age, including the deterioration and damage that count as degradation. She claims, "One way . . . to understand patina is as a medium of aesthetic value *perceived* to have accumulated through time."[15] (The emphasis on *perceived* is hers.) But that perception need not be veridical, for patina "can be artificially created, or faked."[16] This view is sustained by what occurred as the city rebuilt and reclaimed its historic structures. Dawdy found that fake items that purport to be artifacts from old New Orleans gained in popularity—not only for tourists seeking souvenirs, but also for longtime residents remembering their city before the storm. Of these she says,

> New Orleans overflows with small objects that reference the past, some merely with a vague aura of another time, some with a quite specific historical narrative spun around and through them. . . . To do so, they need not be "real"; they only need reference a relevant narrative or evoke an experience of pastness. . . . That fake patina works just as well in some cases highlights the fact that it is the qualia of the material appearance of age that is the relevant signifier in the present.[17]

Thus, we have two different approaches that push back against my allegiance to the real material thing. Fisher argues that, formally speaking, shams and ruins can be indistinguishable and hence achieve aesthetic equivalence; Dawdy observes that, in practice, fakes work just as well as real things and occupy the same aesthetic role. Both stress the appearance of the object, no matter how it came into being.

However, when we examine the different symbolic standing of real thing, sham, fake, or even replica, we can see that their aesthetic properties of necessity differ. Although I disagree that fake and real are aesthetically equivalent, I agree with Dawdy that *reference* is salient for understanding both. While genuine things and shams both evoke the past by means of their aesthetic properties, their modes of reference are importantly different.

Exemplification, Imitation, Allusion

When a physical object persists over years, its look of age is achieved by having endured through time; it embodies its past and carries a narrative aura from that past. Similarly, when artifacts are damaged by a significant historical event, their appearance manifests the cause of that damage, for they have become, in the words of artist Eric Fischl, objects that "bear the scars of their survival."[18]

A particular analytical term describes this well, and that is "exemplification." "Exemplification" is a term taken from Nelson Goodman's *Languages of Art,* later elaborated by Catherine Z. Elgin in *With Reference to Reference.*[19] I do not adopt the more elaborate aspects of their complicated system, but I find their methods useful to distinguish the different referential functions that similar-appearing objects possess, thereby furnishing insight as to the particular power of damaged objects.

"Exemplification" is defined as possession of a property and reference to that same property, a reference that calls attention to a particular feature of an object. Goodman's own examples are paint chips and fabric swatches. A paint chip both possesses the property of blue and refers to that property. It also possesses the property of being three inches square but does not exemplify that feature because color is the only property in the frame of relevance for a paint chip. Swatches and chips are actual samples of the things they exemplify. If you select upholstery fabric from a swatch, then you can be sure that what you have ordered is the same thing (allowing for contingencies such as dye lot variations). To put it in simpler terms, exemplification presents one with what an object *is*; it refers to its own salient properties.

Exemplification is an apt way to think about the look of age or degradation, but only when the object in question *really is* old or degraded. Only the real thing both possesses the properties *old* and *degraded* and foregrounds those properties for attention. A replica ruin or a sham has the appearance of being old and degraded but imitates rather than exemplifies age or deterioration; for although the marks of destruction can be replicated, the symbolic function of exemplification cannot. Hence my third claim about degradation: *a real old thing exemplifies degradation, which amplifies aesthetic apprehension of the look of age.*

Recognition of the identity of an object and what it exemplifies delivers particular perceptual impact. This fact underlies the affective power of real objects that bear the scars of their survival, such as the Genbaku Dome and other damaged objects that have become memorials to the events that caused their destruction.[20] The role of recognition in the perception of damage underscores the cognitive foundation of degradation as an aesthetic property, for the appearance of an object rides on some understanding of its reality. The real thing delivers a singular thrill—marvel at being in the presence of something of historical or artistic value or of great rarity that has survived the ravages of time. One can replicate the visible appearance of a new thing so that it appears old, but veridical encounters with old things presume proximity to a real old thing. Calling attention to the referential relation of exemplification is an efficient way to distinguish apprehension of the real thing from a replica, a sham, or a fake.

If it is correct that perceiving degradation involves recognizing its cause, then the real thing and the replica are not perceptually—aesthetically—indistinguishable. They may look the same with regard to their formal properties, but knowledge that different processes account for their appearance affects their impact. This claim often invites the rejoinder: "You're just supplying the difference in your imagination." However, it is not imaginary that the real thing exemplifies; the imitation refers but does not exemplify. Recognizing "this *is* the ancient object and its damage accrued over time" is different from recognizing "this *looks just like* the ancient object and the damaged look has been replicated." Referential function identifies the object of attention, the kind of thing it is, thereby setting in motion the aesthetic encounters.

Although visual appearance is the usual way to approach aesthetic properties, my loyalty to the real thing is also partly grounded on the sense of touch, for not only do old things have an evocative appearance to the eye; touch affords tangible contact with objects from bygone times, whether artifacts with a venerable history, of great rarity, or from the hand of an acclaimed artist.[21] The possibility of contact makes touch an especially intimate sense, for bodily engagement is invited. On a battlefield, one walks on the very ground where war was fought and soldiers bled. When visiting a ruin, one can enter its precincts, touch its walls, feel its age. Zoltán Somhegyi refers to the "tangible perception of temporality" in the presence of ruins.[22] The physical proximity required by touch can bring the past before one; exemplification excels at summoning presence.

Imitation and Varieties of Reference

Shams, replicas, and vintage styles also impart an impression of the past, but they do so by means of imitation. Imitation is a familiar feature of art and artifacts, and it takes its place alongside exemplification as a mode of reference.

There are many types of imitation, and the roles they play are numerous and varied. Replicas, for instance, refer to their originals in virtue of looking just like them, or sufficiently like them to be easily recognized. In certain cases, a replica stands in for an original in practice. For example, the reproduction Paleolithic caves at Lascaux and Chauvet that were constructed for visitors substitute for the real ones because to permit tourism to the actual caves would damage them irreparably. The experience for some is thrilling because it shows what the real caves are like; for others, it is disappointing because it does not acquaint them with the caves themselves.[23] Employing the terminology suggested here, one could say that some appreciate the imitation whereas others demand exemplification. A replica might also take the place of an original that was destroyed or is no longer safe or possible to use, such as the reconstructed bridge of Mostar, Bosnia. Not only does the new bridge reproduce the look of the sixteenth-century Ottoman original, obliterated by Croatian artillery fire in 1993, it stands in the same place and permits continued use.

Many reproductions make no pretense of substitution, as with the small figurines that are sold in museum shops. Both their reduced size and the fact that they may number in the hundreds forestall any confusion as to whether they are original or not, and they are never taken for the real thing. Some replicas are so precise, however, that they could be misidentified as a genuine original.

Replicas can be informative, beautiful, historically important, and technological wonders in themselves. And when they are all that is left, they convey a narrative aura borrowed from the original thing, keeping alive the memory of artifacts now irretrievably lost. Replicas can be profoundly significant—such as the resurrected Palmyra Arch, the real one having been deliberately destroyed by the Islamic State (Daesh) in 2015. They can also be whimsical or perplexing, like the Sphinx I once encountered guarding the entrance to a movie theater in a Baltimore shopping mall.

Artificial ruins are not replicas of anything specific but represent an aesthetic kind in themselves.[24] With shams, imitation of style makes reference to a period in the past when such things might have been created, generally evoking "pastness" in Dawdy's sense. Her own examples have a particular poignance, helping to salvage the spirit as well as vestiges of the material heritage of a severely damaged community. However, perceiving the look of age with a sham article is not understanding "patina as a medium of aesthetic value perceived to have accumulated through time," as she puts it. Rather, fake patina manifests an aesthetic style perceived to have the characteristics of something that is old enough to have accumulated a degree of degradation.

In the case of shams, the style itself refers either to an older style or to a real thing from which that older style is derived.[25] The reference may be meticulous, as with William Hooker Gillette's careful research that informed the building of his castle, or little more than a vague allusion to bygone times, as with the eclectic mélange of Cheesecake Factories.[26] (Elgin calls allusion "a form of referential action at a distance."[27])

The appearance of something that manifestly is not what it is designed to look like engages multiple layers of recognition. Consider again the arrangement of columns of the sham ruin at Baird Point (fig. 2). In this case, the allusion invited is not to the original whole building, which was a bank, but to what that original itself imitated by virtue of its Greek revival style. To a viewer even slightly familiar with European history and art history, lone columns standing beneath open sky are more likely to summon an image of classical antiquity than nineteenth-century commercial buildings.[28] At the same time, the impossibility of there being a classical ruin in upstate New York automatically contributes realization of the fact that one appreciates a sham rather than a vestige of antiquity. The ease with which informed perception transposes Ionic columns from bank building to imagined ancient ruin indicates the potent, near automatic, role of recognition in aesthetic perception.

Conclusions and Complexities

I hope that these examples demonstrate that the look of age is appreciated in different ways with real thing, replica, or sham. Perceiving degradation, whether damage, wear and tear, or fading, implicitly recognizes that the object got that way somehow. Recognition identifies an object and takes in its narrative aura. This is why it is mistaken to assume that what makes appreciation aesthetic is a matter of appearance detached from reality. Quite apart from the scope of properties in the aesthetic frame, a thing does not have the same affective impact when it *is* what it looks like and when it merely is *like* what it looks like. The error is twofold: in restricting the source of aesthetic impact to surface, usually formal properties; and in assuming that perception itself is limited to the visible surface. Replicas and shams imitate the look of the past in ways that help our imagination to explore bygone times. But with a truly old artifact whose degraded appearance testifies to its endurance over time, the past stands before us.

However, in practice, exemplification and imitation are not always mutually exclusive. The appearance of age gives rise to complexities that become especially tangled in the case of restorations. My allegiance to the enduring physical artifact leads me to value conservation of old things—preservation and repair of original materials—over restoration—replacement of damaged segments with replica parts. (Certainly, either is preferable to renovation, which is apt to destroy much of the original altogether.) Such a scruple, however, cannot be maintained steadfastly. When artifacts are a part of a living environment, they are quite likely to be altered in the course of their lives. Moreover, in order to preserve certain kinds of very old things, one must repair them regularly, for example, stabilizing precarious ruins or buildings to comply with safety standards. In the case of artifacts admired for their historical or artistic significance, those alterations might be designed to match the style of the original, replicating earlier aesthetic properties, including the look of age that time has wrought. Just as a heavily retouched painting might be as much by the hand of the restorer as of the original artist, so a heavily restored artifact might imitate as much as it exemplifies.

In short, although I have distinguished appearance from reality by means of the notions of imitation and exemplification, considerable muddling of reference occurs when only a part of an old thing is replicated in order to repair something that is no longer intact. Objects that are composed partly of original material and partly of substitute repairs confuse any easy typology of reference types. Not to mention the ready confusion for those who admire something they think persisted whole for centuries, unaware that portions of it only look old since they were made to fill in missing pieces. Unless one is an expert, it is unlikely that one will notice the difference between original whole object and nicely patched restoration.

Complexities such as these might tempt one to concede that, after all, there is in practice no aesthetic difference between a real old thing in a state of decay and a good replica (or a restored artifact or a sham) that mimics the characteristics of decay. If one focuses only on immediate appearance, this may be the case, but only in the moment. It is a shallow kind of aesthetic savor that does not withstand deeper understanding, nor does it characterize the likely operation of perception. Aesthetic judgments can be mistaken, just as epistemic judgments can; indeed, because complex aesthetic responses have an epistemic content, the possibility of error is always present. But errors are correctable with understanding that includes proper identification of the object of attention and a fuller grasp of the story of its past.

NOTES

1. Zucker, *Fascination of Decay*, 249.
2. Walter Kaufmann, *Time Is an Artist* (1978), 74, quoted in Moore, "On the Road to Ruin," 101.
3. Marcus Aurelius, *Meditations*, 3:2, p. 26.
4. Denis Diderot, quoted in Macaulay, *Pleasure of Ruins*, 23.
5. Ronald Mercer called my attention to Ottawa's Gatineau Park and its "ruins" established by Prime Minister Mackenzie King.
6. The owner of the chain, David Overton, and the designer of these interiors, Rich McCormack, selected an eclectic décor that represents the taste of nowhere and everywhere: rococo-ish ceilings, bulging columns that appear to be losing their finishing plaster, and art deco–style stencils. "We have these French limestone floors, then we throw in some Egyptian columns, Victorian beadboard wood paneling—a really eclectic mix—which most people wouldn't be brave enough to try, but, fortunately, the way we assembled them, it worked pretty well," Overton says. Quoted in Cills, "Cheesecake Factory Designer."
7. An extrapolation from Danto, *Transfiguration of the Commonplace*, esp. chap. 3.
8. Urmson, "What Makes a Situation," 562.
9. Sedivy, "Aesthetic Properties." There are a number of other theories of perception that also serve to support my claims concerning the perception of aesthetic properties, including Stokes, "Cognitive Penetration."
10. As Susan Stewart notes, "Ruination happens at two speeds: furious and slow—that is, sudden and unbidden or inevitable and imperceptible." *The Ruins Lesson*, 5.
11. Fisher, "Ruins and Sham Ruins," 111.
12. Ibid., 109.
13. Ibid., 114.

14. Ibid., 117.
15. Dawdy, *Patina*, 5.
16. Ibid., 11.
17. Ibid., 117.
18. Fischl, "Trauma of 9/11," 197.
19. Goodman, *Languages of Art*; Elgin, *With Reference to Reference.* Before Goodman examined the mode of reference he labeled exemplification, it was more or less overlooked. Foreword to Elgin, *With Reference to Reference*, 1.
20. I have developed these ideas about exemplification with regard to war memorials in "When Damage Becomes Memorial."
21. I make this case in *Things.*
22. Somhegyi, *Reviewing the Past*, 18.
23. See, for example, these contrary assessments: Jones, "Don't Fall for a Fake"; Hammer, "Finally, the Beauty."
24. I take "aesthetic kind" from Lamarque, "Values of Ruins," 91.
25. Elgin, *With Reference to Reference*, 145–46. For an analysis of the importance of authenticity for style, see Giombini, "Originalism and Anti-Originalism."
26. Stewart notes that several artificial ruins also borrowed from many different antique architectural traditions. *The Ruins Lesson*, 220–25.
27. Elgin, *With Reference to Reference*, 142.
28. In fact, that was the stated goal of those who salvaged the columns. These fragments were originally part of the Federal Reserve Bank in downtown Buffalo, New York. When that building was slated for demolition in 1959, its columns were procured for what was then the University of Buffalo. They sat unused for some years and in 1978 were erected at the new North Campus of the University at Buffalo, State University of New York, at a site now named Baird Point in honor of an early Buffalo philanthropist. The current arrangement was designed by architect Peter G. Castle. For a history of the site, see "Baird Point."

BIBLIOGRAPHY

"Baird Point." UB History, University Archives, University at Buffalo (website). Accessed July 31, 2020. https://library.buffalo.edu/archives/campuses/detail.html?ID=39.

Cills, Hazel. "Cheesecake Factory Designer Knows You Think the Restaurants Look Horrible." *Jezebel*, June 21, 2017. https://jezebel.com/cheesecake-factory-designer-knows-you-think-the-restaur-1796305525.

Danto, Arthur. *The Transfiguration of the Commonplace*. Cambridge, MA: Harvard University Press, 1981.

Dawdy, Shannon Lee. *Patina: A Profane Archaeology*. Chicago: University of Chicago Press, 2016.

Elgin, Catherine Z. *With Reference to Reference*. Indianapolis: Hackett, 1983.

Fischl, Eric. "The Trauma of 9/11 and Its Impact on Artists." In *Ethics and the Visual Arts*, edited by Elaine A. King and Gail Levin, 195–97. New York: Allworth Press, 2006.

Fisher, Saul. "Ruins and Sham Ruins." In *Philosophical Perspectives on Ruins, Monuments, and Memorials*, edited by Jeanette Bicknell, Jennifer Judkins, and Carolyn Korsmeyer, 108–20. New York: Routledge, 2020.

Giombini, Lisa. "Originalism and Anti-Originalism: Style and Authenticity in Aesthetic Appreciation." *Debates in Aesthetics* 15, no. 1 (2020): http://debatesinaesthetics.org/vol-15-no-1/#GIOMBINI.

Goodman, Nelson. *Languages of Art: An Approach to a Theory of Symbols*. 2nd ed. Indianapolis: Hackett, 1976.

Hammer, Joshua. "Finally, the Beauty of France's Chauvet Cave Makes Its Grand Public Debut." *Smithsonian Magazine*, April 2015. https://www.smithsonianmag.com/history/france-chauvet-cave-makes-grand-debut-180954582/.

Jones, Jonathan. "Don't Fall for a Fake: The Chauvet Cave Art Replica Is Nonsense." *Guardian*, April 15, 2015. https://www.theguardian.com/artanddesign/jonathanjonesblog/2015/apr/15/chauvet-cave-art-replica-is-nonsense.

Korsmeyer, Carolyn. *Things: In Touch with the Past*. New York: Oxford University Press, 2019.

———. "When Damage Becomes Memorial." In *Heritage in War: Philosophical Analyses of Ethical Issues*, edited by William Bülow, Helen Frowe, Derek Matravers, and Joshua Lewis Thomas. Oxford: Oxford University Press, forthcoming 2022.

Lamarque, Peter. "The Values of Ruins and Depictions of Ruins." In *Philosophical Perspectives on Ruins, Monuments, and Memorials*, edited by Jeanette Bicknell, Jennifer Judkins, and Carolyn Korsmeyer, 83–94. New York: Routledge, 2020.

Macaulay, Rose. *Pleasure of Ruins*. London: Weidenfeld and Nicolson, 1953.

Marcus Aurelius. *Meditations*. Translated by George Long. New York: Walter J. Black, 1945.

Moore, Ronald. "On the Road to Ruin: Anticipating and Appreciating the Natural Degradation of Human Constructions." In *Philosophical Perspectives on Ruins, Monuments, and Memorials*, edited by Jeanette Bicknell, Jennifer Judkins, and Carolyn Korsmeyer, 95–107. New York: Routledge, 2020.

Sedivy, Sonia. "Aesthetic Properties, History and Perception." *British Journal of Aesthetics* 58, no. 4 (October 2018): 345–62.

Somhegyi, Zoltán. *Reviewing the Past: The Presence of Ruins.* London: Rowman and Littlefield, 2020.

Stewart, Susan. *The Ruins Lesson: Meaning and Material in Western Culture.* Chicago: University of Chicago Press, 2020.

Stokes, Dustin. "Cognitive Penetration and the Perception of Art." *Dialectica* 68 (2014): 1–34.

Urmson, J. O. "What Makes a Situation Aesthetic?" In *Art and Philosophy*, edited by W. E. Kennick, 552–64. New York: St. Martin's, 1964. First published 1957.

Zucker, Paul. *Fascination of Decay: Ruins; Relic—Symbol—Ornament.* Ridgewood, NJ: Gregg Press, 1968.

FOUR

The Aesthetics of Repair

Yuriko Saito

Abstract

The material world is never static. Once created, objects' appearances inevitably undergo transformations through aging processes, wear and tear from use, and accidental damage. These changes are generally regarded as degradation, deterioration, and decay. Such negative aesthetic judgments lead to the premature disposal of perfectly functional objects and their replacement by brand-new products in perfect condition, exacerbating the mounting environmental problems. The practice of repair, once a way of life in many parts of the world, has become overtaken by this throwaway culture fueled by rampant consumerism. This chapter takes repair as a site of cultivating a care relationship with the material world. By responding to and working with the specific condition of an object, we not only encourage its longevity but also collaboratively create a narrative of growth and vicissitude of our intertwined lives. In particular, the mode of visible repair embodies this attitude, and it has provided inspiration to many contemporary art projects. However, the promotion of repair must be accompanied by a practical wisdom to avoid indiscriminate repair of everything and vigilance against creating another fashion trend.

When I was growing up in Japan, I used to fix runs in stockings because cheap pantyhose were not available then. I also remember a handyman coming around the neighborhood regularly to fix holes in pots and pans, as well as broken umbrellas. I believe that repairing and making do with what is available is still a way of life in many parts of the world. However, repairing everyday objects ceased to be a way of life in developed nations, particularly in the United States, where rampant consumerism and fast fashion have given rise to a throwaway lifestyle. This situation is further exacerbated by the notorious irreparability of many electronic products.

Today, however, repair seems to be making a comeback. In the public sphere, deferred maintenance of infrastructure has prompted calls for urgent repair work. Right-to-repair legislation has been passed in twenty US states to date, to fight against industry's strategy of planned obsolescence and its monopoly of repair manuals, parts, tools, and designated repair shops. In the private sector, there is growing concern over excessive consumerism that is responsible for environmental devastation, as well as human rights violations of factory workers and garbage pickers in developing nations. These concerns are generating grassroots initiatives to promote repair. For example, a Dutch group, Platform 21, issued the "Repair Manifesto" in 2009, soon followed by another "Repair Manifesto" issued by Ifixit, a California company. The same year saw the opening of the first Repair Café in Amsterdam, a model that since then has spread to 1,500 venues globally, where experts share their knowledge and skills with people to help them repair broken objects.[1]

Repair wouldn't be necessary, as philosopher Elizabeth Spelman observes, "if things never broke, never frayed, never splintered or fell to pieces." Such a world would be "filled with unchanging unbreakable eternal objects," a material version of Platonic forms.[2] With the rapid advancement of material science, it is conceivable that manufactured objects could soon self-clean, self-heal, and remain unchanged. In fact, Daniela Rosner, a human-centered design and engineering scholar who has "always loved cracks," entertained this possibility when a reporter asked her to comment on such a scientific development taking place in the United Kingdom.[3] This possibility may be welcome for things like airplanes and bridges. We may also appreciate the freedom from tedious and often time-consuming and labor-intensive chores of cleaning and fixing things.

Aesthetically, too, there is a preponderant preference for material objects in "perfect" condition that do not show signs of aging and accidental damage. This penchant for the new, the unblemished, and the undamaged seems to transcend historical and cultural borders. David Lowenthal, a cultural geographer, observes that "the balance of evidence . . . shows general dislike of age and decay. We prefer youth, not only in living creatures but in our surroundings, including our own creation." While he is fully aware that "distaste for the marks of age is far from universal," he admits that appreciation of an aged appearance "is the exception."[4]

Even the advocates of Japanese *wabi* aesthetics, one of the few aesthetic movements extolling aged and damaged appearances, present their view as a contrast to a more common taste toward opulence and perfection. For example, a fourteenth-century Buddhist priest, Yoshida Kenkō, whose writing anticipates later wabi aesthetics, argues for the aesthetic value of cherry blossoms when falling and the moon hidden behind a cloud by observing that "people *commonly* regret that the cherry blossoms scatter or that the moon sinks in the sky, and this is *natural*."[5] Similarly, the sixteenth-century founder of the wabi tea ceremony, Sen no Rikyū, instructs his disciples to use tea utensils that "fall . . . short of perfection," while acknowledging that "there are people who find it repugnant to have a tiniest defect in them."[6]

Privileging the unblemished and undamaged appearance is also indicated by the method favored by conventional mending practice: invisible repair. At the 2018–19 exhibit *Repair and Design Futures* at the Rhode Island School of Design Museum, many objects, predominantly fabric, were covered by a glass top with a marking indicating where the mending occurred.[7] Even then, I had to look closely to find the repaired parts, and the resulting amusing experience was similar to finding natural creatures in their camouflaged appearance. In addition, I could not help but admire the considerable skills involved in such meticulous and exquisite results.

In our daily lives, too, we normally try to make the signs of repair as inconspicuous and, better still, invisible as possible. We use the same-colored thread and fabric to patch up a rip and yarn to darn socks.[8] When putting broken pottery pieces back together, we apply a transparent glue. The paint used for touching up the scratches on a car needs to match the rest of the body. There is an implicit aesthetic judgment that the original appearance of an object is superior to its later changed appearance, which is almost always characterized negatively as damage, dilapidation, defect, or degradation. Such a fall from grace needs to be made as invisible as possible.

What accounts for this aesthetic judgment to privilege the new, the unblemished, and, in short, the perfect? To regard signs of aging and breakage, an inevitable aspect of empirical reality, as degradation and deterioration implies a yearning for permanence of material objects in their perfect condition. In the Western tradition, favoring timelessness over vicissitude goes back to Plato's identification of reality as forms or ideas that are immutable and exist beyond the empirical world.

A more recent force behind the penchant for permanence of the material world is the Enlightenment agenda, which equates the good life with a life free from constraints of nature. Although he did not specifically take up the issue of material changes, Francis Bacon presents a utopia in *The New Atlantis* (published posthumously in 1627) that is filled with technological wonders to help humanity overcome nature's limitations. Most of the projects that he conceptualized—airplanes, submarines, genetic engineering, and holography, to name a few—have already been accomplished.[9] I have no doubt that Bacon would have embraced self-cleaning, self-healing, and self-repairing materials.

Buddhism, on the other hand, identifies reality as impermanence and teaches followers to accept and affirm it by renouncing clinging. While one could claim that Buddhism confronts the reality of this world squarely, its teaching can be interpreted as a response to the common human yearning for permanence of youth, power, possessions, and ultimately life. In other words, the emphasis on the importance of overcoming such clinging underscores our desire for unchanging reality.

This stance of privileging permanence is manifested in architectural practice. As Rumiko Handa, herself an architect, documents, architects tend to exert what she calls an "authorial authority" that prohibits any modifications once a building is completed.[10] At the time of completion, the architect deems the structure perfect, and any change, whether intentionally inflicted or naturally developed, counts as deterioration. Some architects even dictate the activities and objects used within their designed space, regardless of the lifestyle and wishes of its users. Frank Lloyd Wright, for instance, was notorious for designing everything within the space, including furniture and even plates, so as to create an overall harmony. In Philip Johnson's Glass House, "even tabletop bric-a-brac are discreetly marked with indications of their correct location," making the space "unlivable as a domestic space," according to one critic.[11] Architects' desires and attempts to freeze buildings and things in them at the time of completion reflects an unrealistic expectation that defies the reality of the lived world.[12] Those who privilege permanence over impermanence, such as these architects, will find the imaginary world of unchanging material objects a true utopia.

There is another, and more recent, factor to explain the penchant for permanence: today's industrial production system. This system identifies a manufactured object as the end state of the production process when the product is in "mint" condition. Any deviation from this condition is considered defective, deficient, or degraded. Information and technology scholar Steven J. Jackson terms this privileging of the end result of a production process a "productionist bias" or "production-centered ethos," which dictates the design process to create a finished product that is perfect and unalterable.[13] We the consumers are encouraged to think that the object stops its life after the point of purchase, instead of continuing its history through its own aging process and our interaction with it, each stage exhibiting a unique characteristic rather than compromising or damaging the original integrity of the object.

In addition, today's fast-fashion goods are deceptively cheap, hiding their true cost. Considered disposables, they are thrown away at the first sign of aging or damage.[14] This problem is exacerbated by the industry strategy of planned obsolescence, which is commonly regarded as first targeting functionality and only recently focusing on fashionableness, creating so-called aesthetic obsolescence or perceived obsolescence. However, it is noteworthy that this strategy was first proposed to stimulate the economy during the age of the New Deal in the United States. In an influential 1932 essay, "Ending the Depression through Planned Obsolescence," Bernard London, a real estate broker, argued for increasing industry production by encouraging the replacement of "old articles with new *for reasons of fashion and up-to-dateness.*" According to him, "furniture and clothing and other commodities should have a span of life. . . . When used for their allotted time, they should be retired, and replaced by fresh merchandise."[15] Thus, privileging perfection and the resultant negativity associated with deterioration and defects has an almost century-old lineage in the United States.

This aesthetic ethos has led our relationship with the material world to become rather dysfunctional. It neither encourages an enduring relationship with objects as they proceed with their inevitable changes nor promotes an active interaction with them through maintenance, care, and repair. I believe this situation impoverishes our life in several ways. Let me explore what we lose if we do not appreciate and actively engage with the changing nature of the material world.

Recall the imaginary world of permanently unchanging objects. Such a world compromises our aesthetic life by severely reducing the diversity of sensuous surfaces brought about by ever-changing materials. We will be deprived of the opportunities to gain an aesthetic appreciation of things such as ruins, patina, and various forms of so-called imperfection, including stains, scratches, cracks, frays, faded colors, peeling paint, accidental damage, and the like. These things have often provided inspiration for contemporary artists. To take only one example, consider the *Stain* project by Bethan Laura Wood, which features teacups' interiors that gradually reveal elegant patterns, such as birds in flight, that have been pretreated to stain with tea drinking.[16]

There are other aesthetic values provided by these "imperfections." In order to explore them, let me take the Japanese wabi sensibility associated with the tea ceremony, one of the most prominent aesthetic movements to celebrate the appearance of aging and imperfection.[17] Although tea gatherings originally started in Japan among the powerful and the wealthy as social entertainment and an occasion to display expensive wares from China, the tea ceremony as an exemplar for wabi aesthetics became established in the sixteenth century. Initially proposed as an alternative to the prevailing taste for opulence and luxury, wabi aesthetics celebrates irregularity, rough surfaces, asymmetry, and defects in tea bowls, other implements, and tea huts. These qualities result from aging processes, happenstance, or intentional acts by tea masters, such as breaking a vase's handle.[18]

In addition to diverse sensuous qualities, wabi appeal is activated by exercising the imagination. To his own rhetorical question, "Are we to look at the moon and the cherry blossoms with our eyes alone?" Kenkō answers, "[H]ow much more evocative and pleasing it is to think about the spring without stirring from the house, to dream of the moonlit night though we remain in our room!"[19] For the same reason, Rikyū criticizes those who lament a landscape devoid of flowers and colorful foliage because they "are merely capable of taking pleasure in the colorful sights which appear to their physical eyes alone."[20] The imagination is stretched even further when Kenkō invokes a poignant appreciation of a festival's aftermath—when various objects used in the procession are taken apart and a sense of desolation sets in—by drawing a parallel to human life: "you realize with a pang of grief that life is like this."[21]

The aesthetic loss incurred by living in the imaginary world of permanence thus has an existential dimension. Without objects that show aging effects and accidental damage, we will lose an effective medium through which to reflect upon the parallel transience of our own life. Such objects provide a kind of consolation by reminding us that nothing existing is exempt from the law of nature. If we remain mortal in this imaginary world, the contrast between us and the permanently unchanging world will be too glaring and painful to bear. Thus, material objects' inevitable vicissitude provides both aesthetic and existential benefits.

However, accepting and appreciating the changing nature of the material world does not exclude our active engagement with its care through maintenance and repair. Recall Jackson's notion of productionist bias.

He proposes an alternative model, "broken world thinking," which foregrounds "erosion, breakdown, and decay, rather than novelty, growth, and progress, as our starting points."[22] While erosion, breakdown, and decay all have a negative connotation, these processes *are* actually the world's movers and shakers. Without these processes carried out by largely invisible actors, such as bacteria and insects in the natural world, and the various maintenance workers, cleaning crews, and repairers in the human world, more visible, and presumably more productive and creative work, cannot take place.[23] Instead of relegating these processes of breaking down, restoring, and repairing to the background, broken world thinking illuminates "the remarkable resilience, creativity, and sheer magnitude of the work represented in the ongoing maintenance and reproduction of established order" and encourages developing "a deep wonder and appreciation for the ongoing activities by which stability . . . is maintained."[24] The productionist bias promotes creativity, originality, and imagination as features of an ex nihilo conception of design and production when in fact they also characterize the tasks involved in care, maintenance, and repair.[25]

An American artist, Mierle Laderman Ukeles, makes such invisible work visible in a series of installations and performances that highlights the maintenance work performed by New York City's sanitation workers. In conjunction with her 1969 *Manifesto for Maintenance Art: Proposal for an Exhibition "Care,"* she asks, "[A]fter the revolution, who's going to pick up the garbage on Monday morning?"[26] Similarly, in a series of paintings and installations, Ramiro Gomez features those laborers, such as domestic help, gardeners, and garbage collectors, whose largely invisible work is essential in keeping society functioning. Regarding one such painting, *On Tenth Avenue (Chelsea)* (2018), he comments, "the labor and repair performed by the workers on 10th Avenue also play an absolutely necessary, vital role to [*sic*] the survival of our society. My paintings . . . make visible labor that otherwise is not recorded."[27] Even outside the artistic context, Jackson asks us to "think . . . of the differential visibility of faculty and nighttime cleaning staff on American university campuses and its relation to the highly skewed distributions of income that follow."[28]

As Ukeles's project title indicates, there is an important connection between maintenance and repair work and the ethics of care. Care ethics was proposed primarily by feminist philosophers who were dissatisfied with the justice-centric ethics that emphasizes rationality, objectivity, universality, fairness, and impartiality. They point out that our everyday life is inseparable from specific relationships we develop toward certain people, such as family and friends, as well as nonhuman animals and, I would argue, certain objects. Repair must attend to the *specifics* of the object, the material, and the nature of damage. Katrina Rodabaugh characterizes her fabric mending practice as "respond[ing] to each individual repair as the garment demands. It means that every patch, stitch, darn, or other combinations of mending techniques can be in response to that particular damage."[29] Spelman also reports that her acquaintance, a successful car repairer who enjoys his work, "comes up with nifty case-specific solutions to the constant stream of unique challenges" and draws an analogy with care ethics. "[T]he ethics of care highlights the intimacy of the knowledge of the moral agent as problem solver: intimate both in the sense of having or seeing specific and nuanced and contextualized knowledge of the people involved and the situation they are in, and in the sense of acknowledging or creating a close relationship to the people involved."[30] What is important for my purpose is that a close attention to and relationship with the particularity of an object thus required in care work also characterizes aesthetic experience.[31] I shall explore this point by examining a practice that gives maintenance and care work a prominent voice: visible repair.

Contemporary repair projects challenge the conventional invisible repair and all its assumptions by accentuating the damage and its repair. Probably the best-known example of visible repair is proudly displayed patches and stitches on clothing, sometimes literally worn on the sleeves.[32] Some contemporary art projects also feature visible repair. Let me list several: Daniel Eatock's *Visible Vehicle Repairs* (2017–), which results in two-colored cars; Rachel Sussman's *Sidewalk Kintsukuroi* (2016–), which appropriates the Japanese *kintsugi* method in repairing sidewalk cracks by filling them with gold-colored paint; Jan Vormann's *Dispatchwork* (2007–), in which, across different parts of the world and often together with area residents, he repairs crumbling brick structures with Lego pieces; Charlotte Bailey's repaired vases, wrapped in fabric with the same pattern and sewn with gold threads (2016); Tomomi Kamoshita's patchwork chopstick rests and mismatched earrings that piece together ceramic shards (2014–); Yee Sookyung's *Translated Vases* (2002–) that assemble shards from multiple vases using the kintsugi method; Elisa Sheehan's *Kintsugi Eggshells*, an ongoing project presenting "repaired" eggshells with the kintsugi method, described by the artist herself as "a visual representation of imperfection as a true value and where flaws are celebrated and viewed as beautiful"; and Tatiane Freitas's *My Old New Series* (2010–), broken wooden furniture "fixed" with parts made with acrylic resin.[33]

As is evident from this list, many artists are inspired by the Japanese repair practice called kintsugi (gold joinery) or *kintsukuroi* (repair by gold). Kintsugi originated as a mending method for tea bowls used in the tea ceremony. According to a legend, Shōgun Ashikaga Yoshimasa (1436–90) had a prized celadon porcelain tea bowl from China that became cracked. He sent it back to China for repair. It came back repaired with metal staples, which did not satisfy his aesthetic taste, so he ordered a craftsman to come up with a more attractive repair. The craftsman repaired the bowl with the kintsugi method. This repair method was subsequently popularized with the sixteenth-century establishment of the wabi tea ceremony, as indicated by Rikyū's observation: "We might naturally find it awkward to use . . . a cracked tea bowl of present-day porcelain. On the other hand, however, we are accustomed to make use of or even very ready to use, despite defects, the antique tea-caddy imported from China for instance, sparing no pain of having it mended with lacquer."[34] As this passage indicates, despite the impression given by the term "gold joinery" or "repair by gold," lacquer is used as an adhesive and gold, sometimes silver or tin, flakes are applied as an ornamental touch.

Although cracks may be the best-known example, there are different kinds of "damage" repaired by this method: incomplete cracks called *nyū* that do not compromise functionality, chips called *kake*, peeling away of the material surface called *hotsure*, and a complete breakage called *ware*. While most cases of kintsugi put broken pieces of an object back together, sometimes a piece from a different object is used for filling, the technique called *yobitsugi* or *yosetsugi*, meaning "repairing by calling forth" or "joint call." This is the most extreme form of kintsugi, which, as I will explain later, has been effectively used in an art project.[35]

Kintsugi suggests several important implications that are applicable beyond its specific cultural and historical origin. First, it regards various forms of so-called damage as having their own integrity for what they are and providing an opportunity for exercising imagination and creativity. In keeping with Zen Buddhism, the philosophy underlying the tea ceremony, which advocates respecting and appreciating the Buddha nature of everything whatsoever, wabi aesthetics celebrates, rather than denigrates, the damage or imperfection. It is noteworthy that the new pattern created by those imperfections is considered to evoke a landscape, *keshiki*, and it becomes further articulated through visible repair. For example, one of the best-known tea bowls repaired with kintsugi by a noted craftsman, Hon'ami Kōetsu (1558–1637), is named Seppō, a snow-clad mountain peak.[36]

Second, accepting impermanence and transience of the material world does not necessarily exclude *working with* this condition. This is in contrast with working *in spite of* or *regardless of* the specific ways in which individual objects go through vicissitudes. The kintsugi method requires close observation to devise the intervention most appropriate for and respectful of the particular damage. For example, if a deep crack needs repair, it is recommended that the break be completed by splitting the object open. In order to determine whether the crack is deep enough, we put our ear against the object while squeezing the object to see whether a clicking sound is heard. In other words, although mending intervenes with the object's status quo to fix damage, it requires us to *work with* the object's specific state. Similarly, in mending fabric, one practitioner states that "every stitch requires listening and responding to what the fabric, and the hole, might need."[37] This way of mending is echoed by a denim repairer: "Rather than having a predetermined vision of the finished garment, we let the contours of the damage dictate the repair."[38] In short, our repair work expresses our respect and care for the object with its specific condition, bringing us back to the care ethics referenced above.

Thus, the aesthetic paradigm of visible repair shuns uniformity and accentuates the individuality of a particular object and its unique history by presenting a tangible record of repair, as well as the care and affection for the object. The owner, the user, or the repairer cared enough about the object to take the time and effort to give it a renewed life. For the spectator, the visible mark of repair stimulates the imagination to speculate on its unique (hi)story and its prior condition, the cause of breakage, and its future life, while at the same time encouraging the appreciation of the skill and care involved in the repair.

With the increasing popularity of repair among amateurs today, its practice offers an aesthetic experience not only to spectators but also to those who engage in this activity.[39] While aesthetics discourse has traditionally been concerned with a spectator's experience, there is growing attention to the experience of executing activities, such as cooking, eating, cleaning, gardening, and playing sports. Like these activities, mending practice engenders a bodily experience of handling and working with the object's materiality. Mending fabric, for example, occurs where "this active space of at/tending—assessing, touching, thinking, and intuiting—entwines into an embodied knowledge, a soft technique, during which the ameliorative thread is sewn this way and that."[40] Insofar as the activity of mending requires an embodied knowledge, skills, and working according to the material's dictate, the process can be considered an apt example of the reciprocal, cumulative, and continuous relationship of "doing" and "undergoing" proposed by John Dewey when characterizing an aesthetic experience.[41] In addition to providing a possible occasion for "an" experience, such active and care-full engagement with the object cannot but nurture one's affection for the object, rendering it not only an object of aesthetic appreciation but also an object graced with longevity.[42]

So far, I have presented aesthetic benefits associated with repair. However, I want to offer a couple of caveats. First, the aesthetic value of repair may depend upon a judicious balance between repaired and undamaged objects. If our everyday environment is surrounded by too many repaired objects, particularly with visible repair, its aesthetic potency may get diluted. Imagine what it would be like if every crumbled built structure were repaired by Legos, or if kintsugi-inspired repair appeared on every cracked pavement. Certainly, in light of serious environmental problems, we should strive to facilitate the longevity of what we have through caring for it by maintenance and repair. But an object may reach a point where it makes more sense to either discard it or turn it into something else through recycling or repurposing. As Spelman warns, "a voracious appetite for fixing can lead to poor judgment about what is and is not desirable or even possible to repair. Pride in our repairing abilities may push us into believing that whatever has been broken can be and ought to be fixed."[43] We may need what the cultural geographer Caitlin DeSilvey calls "curated decay," rather than indiscriminate repair of everything.[44]

In addition, some broken objects with a special historical significance should not be repaired but rather preserved as they are. For example, some, but perhaps not all, ruined structures that result from a war or an evil human deed should be preserved in their ruined state as historical witnesses and cautionary reminders of "never again" for future generations.[45] Similarly, Shannon Lee Dawdy, an anthropologist, argues against completely eradicating the traces of destruction caused by Hurricane Katrina in New Orleans and replacing them with a sanitized version of what things used to be or, worse, their Disneyfication for consumption by tourists. She instead argues for a judicious preservation of patina, which she defines as "a medium of aesthetic value *perceived* to have accumulated through time that represents the social palimpsest" and supports "a process of both accumulation and decay that must be kept in a cared-for balance."[46]

Second, the recent popularity of visible repair and repair cafés may be culturally and economically situated in affluent societies with material abundance that can afford to take pleasure and amusement in the signs of repair. What about those who are living in circumstances marked by poverty and scarcity and are forced to make do with what little they have? After all, the stunning beauty of quilts in the American South reflects the poor living condition of the makers. The same is true for the Japanese fishermen's kimonos mended with *boro*, tattered rags.[47] These repeatedly mended clothes resulted from the extreme scarcity of resources in the northern part of the Japanese mainland. In fact, wabi aesthetics, tea ceremony, and kintsugi supported by the wealthy and powerful were criticized by a nineteenth-century Japanese Confucian scholar: "Today's tea men take filthy and damaged old bowls, whose ages they cannot know, repair them with lacquer and other materials, and then use them. It is an unspeakably disgusting custom," because "whatever tea dilettantes do is a copy of the poor and humble. It may be that the rich and noble have a reason to find pleasure in copying the poor and humble. But why should those who are, from the outset, poor and humble find pleasure in further copying the poor and humble? . . . All that tea dilettante does is to copy everything which looks poor and shabby."[48]

In a similar vein, contemporary advocates of visible repair warn against "the exploitative chic-ing of the shabby," "the idealization of repair," and "the romanticization of strategies of survival," so as not to be complicit with capitalism and colonialism.[49] Fetishization and commercialization of visible repair can endanger this practice by rendering it another fashion trend within a capitalist framework that can easily be replaced by another trend. This is the same concern with today's so-called ruin porn and poverty tourism in cities like Detroit and New Orleans as well as impoverished nations, where physical manifestations of social ills and suffering become an object of the gaze primarily for (more affluent) nonresidents.[50]

Regardless of these questions and possible pitfalls, however, I want to conclude by pointing out the potentials of kintsugi that go beyond material repair. Nishiko, a Japanese artist practicing in the Netherlands, has been working on the *Repairing Earthquake Project* since the earthquake and tsunami hit the northeastern coast of Japan in March of 2011. In addition to documenting its aftermath through eyewitness accounts and photographing various objects, her project includes collecting fragments and debris from this catastrophe and putting them back together. The repaired item, cased in a box specifically created for it, is either returned to the owner, if found, or placed with a foster family for safekeeping. The objects and processes are presented through photography, video, installation, performance, publication, and blogging. The broken pieces symbolize the scars left by this natural disaster and the act of repair, a process of healing.[51]

Earthquakes, frequent occurrences in Japan, provide inspiration for another kintsugi project. Kunio Nakamura, a Japanese kintsugi artist, volunteers his time and expertise to put broken pottery pieces back together for residents of Kumamoto prefecture, which was hit by a massive earthquake in 2016.[52] He also articulates his hope for a peaceful world by combining plate pieces from two nations caught in a problematic relationship, such as South Korea and North Korea. Utilizing the technique of *yobitsugi* (repair by calling forth), mentioned above, this art project called *Kintsugi Pieces in Harmony* presents the possibility and hope for a harmonious coexistence of feuding nations by symbolically attaching two disparate pieces.[53]

Finally, this method of material repair is sometimes taken as a metaphorical expression of repairing a fractured psyche and fragility turned into resilience in the field of psychology. Scott Barry Kaufman features a piece of kintsugi-repaired pottery in his article "Post-Traumatic Growth: Finding Meaning and Creativity in Adversity," published with the tagline "Resilience and strength can often be attained through unexpected routes."[54] Reflecting on her practice of mending fabric, Lisa Morgan offers a similar observation: "Tending to the wound in the garment facilitates a tending of ourselves, and through mending a hole there is the sense of stitching oneself almost whole, a reflecting on what has gone before and a bringing together and uniting of what remains. The scar may never disappear, but it indicates in detail how it was healed. The darn becomes the celebration of a story."[55]

The practice of repair thus offers a surprisingly rich reservoir for exploring the aesthetics of transience, care ethics, today's consumerism, human relationships with the ever-changing material world, and the efficacy of aesthetic power for repairing the psyches and lives of humans shattered by various traumas, which unfortunately are very much a part of this world.

NOTES

I thank Ivan Gaskell for his comments and edits on the first draft. I also thank Florence Grant for thorough and meticulous copyediting, including the creation of the bibliography.

1. Platform 21, "Repair Manifesto"; Ifixit, "Repair Manifesto"; Repair Café (see website for each in Bibliography).
2. Spelman, *Repair*, 5.
3. Rosner, "Beautiful Oops," 77.
4. Lowenthal, *The Past*, 127.
5. Kenkō, *Essays*, 115. Emphasis added. I follow the Japanese custom of referring to historical figures by the given name.
6. Sōkei, "Record of Nanbō," 146.
7. For details of the exhibition, see the RISD Museum's website, accessed June 7, 2020, https://risdmuseum.org/exhibitions-events/exhibitions/repair-and-design-futures.
8. For discussion and examples of invisible fabric mending, see Gwilt, "Producing Sustainable Fashion."
9. Bacon, *New Atlantis.*
10. Handa, *Allure of the Incomplete.*
11. Melchionne, "Living in Glass Houses," 191.
12. Rem Koolhaas acknowledges this tendency of his own profession in the interview following the film *Koolhaas Houselife*, which features a cleaning lady's tasks in a rather unusual house he built for a French family (directed and produced by Ila Bêka and Louise Lemoine, 2013).
13. Jackson, "Rethinking."
14. I explore these issues in "Consumer Aesthetics." Also see *The True Cost*, directed by Andrew Morgan (Bullfrog Films, 2013).
15. London, "Ending the Depression," 2 (emphasis added), 5. As a real estate developer, London must have been familiar with the notion of architectural obsolescence as an economic strategy to justify tax advantage based upon the buildings' depreciation over time. Tax deductibility for building depreciation was incorporated in the federal income tax code, introduced in 1909. Daniel Abramson discusses how various writings at the time made "obsolescence appear to be an inevitable and incontrovertible law of development and change—something that just happens to buildings—rather than the result of specific business practices and historical actors." *Obsolescence*, 34. I thank Ivan Gaskell for pointing out this historical context for London's writing and the reference to Abramson's book.

16. For visual images, see the project's page on the artist's website, accessed June 7, 2020, http://www.bethanlaurawood.com/work/stain/. For other examples, see Slemmons, "On Imperfection"; Weiss, "Toilet Bowl"; Weiss, "Cracks"; Martin, "Patina of Cloth."
17. The other prominent aesthetics in this regard is the eighteenth-century British picturesque, which gave rise to the cult of ruins. There are remarkable similarities between wabi aesthetics and the picturesque, but here I will limit my discussion to wabi because of the kintsugi repair I discuss later. I give a detailed account of both in chapter 4 of *Everyday Aesthetics*, 149–204.
18. Saito, "Japanese Aesthetics."
19. Kenkō, *Essays*, 115, 118.
20. Sōkei, "Record of Nanbō," 156–57.
21. Kenkō, *Essays*, 120. The eighteenth-century British picturesque advocates' obsession with ruins is also partly motivated by their parallel to human existence.
22. Jackson, "Rethinking," 221.
23. Fujiwara Tatsushi, a Japanese scholar of agricultural history, points out that soil formation is largely due to decay rather than production, but the former rarely garners proper recognition. With industrial production, if it weren't for the work of taking things apart for recycling and reusing, the earth would be buried in garbage. *Bunkai no Tetsugaku*, 46, 48. The same point regarding industrial production is made by Jackson with the example of a shipbreaking operation in Bangladesh that is portrayed in Edward Burtynsky's photographs. "Rethinking," 224–26.
24. Jackson, "Rethinking," 222.
25. For further exploration of this point, see Jackson and Kang, "Breakdown, Obsolescence."
26. Cited in Lippard, "Never Done," 17. See this book for visual images of Ukeles's oeuvre, including her Manifesto.
27. Gomez, "*On Tenth Avenue*," 35.
28. Jackson, "Rethinking," 229. This hierarchy of visibility and accompanying structure of prestige and reward also applies to the celebration of "essential workers" during the COVID-19 pandemic of 2020–21. Those featured as "heroes" are primarily the medical professionals but rarely the cleaning crew whose work is as essential and dangerous as those heroes'. Hospitals cannot operate if things are not cleaned and sanitized or if their garbage does not get picked up.
29. Rodabaugh, *Mending*, 141–42.
30. Spelman, *Repair*, 47.
31. I explore the importance of attention to the singularity and particularity

of the object of an aesthetic experience developed in Ronald W. Hepburn's aesthetic theory in "Aesthetic Experience." I also explore the comparison between aesthetic experience and care ethics in "Aesthetics of Care."

32. See Gwilt, "Producing Sustainable Fashion," and Rodabaugh, *Mending*, for innovative examples of visible repair.
33. See examples at the following websites of each artist: https://eatock.com/books/visible-vehicle-repairs/, http://www.rachelsussman.com/portfolio#/sidewalk-kintsukuroi, https://www.janvormann.com/testbild/dispatchwork/, https://hangingbyathreadembroidery.wordpress.com/, https://rittau.jimdofree.com/kin-tsugi/, https://www.yeesookyung.com/translated-vase-, https://elisasheehan.com/collections/eggshells, and https://tatianefreitas.com/My-Old-New-Series, all accessed on June 7, 2020. More images and their discussion can be found in Pellegrini, "Broken Vases," and Richman-Abdou, "Kintsugi." I thank Sanna Lehtinen for the Vormann reference.
34. Sōkei, "Record of Nanbō," 146.
35. The information about kintsugi in this and subsequent paragraphs is culled from the following sources: Irahara and Makamura, *Zero kara no Kintsugi Nyūmon*; Nakamura, *Kintsugi Techō*; Ozawa, *Kintsugi no Susume*.
36. See the image on the website of Hatakeyama Kinenkan Museum, which holds this piece, accessed August 5, 2021, https://www.ebara.co.jp/foundation/hatakeyama/information/collection.html. (Click on the rectangle at top right, which opens the collection of tea implements. The *Seppō* piece is the third one from the top.)
37. Morgan, "Kate Kittredge's Stockings," 70.
38. Matt Rho, cited in Rodabaugh, *Mending*, 119.
39. Fujiwara points out that today transparent epoxy can repair the breakage much faster and easier without leaving any trace of repair, which indicates a particular attraction of kintsugi beyond actual repair. *Bunkai no Tetsugaku*, 289–90.
40. Morgan, "Kate Kittredge's Stockings," 70.
41. "An experience has pattern and structure, because it is not just doing and undergoing in alternation, but consists of them in relationship." Dewey, *Art as Experience*, 44. I thank Ivan Gaskell for calling attention to this passage.
42. Fujiwara states, "Clothes, houses, bicycles, and cars are all repaired, used, and repaired again when broken again. They are cleaned, washed, polished, and maintained. After repeating these activities, we develop attachment to these objects and want to keep them with us as long as possible. Attachment does not mean not hurting. Rather, it is to

appreciate thoroughly the scars and frays as much as possible and as long as functionality is not lost." *Bunkai no Tetsugaku,* 285. My translation. I also explore the seeming paradox of facilitating the longevity of consumer goods through accepting and appreciating their transience in "Consumer Aesthetics."

43. Spelman, *Repair,* 6.
44. DeSilvey, *Curated Decay.*
45. Saito, "Reflections."
46. Dawdy, *Patina,* 5, 16.
47. For *boro,* see Rissanen, "Designing Endurance," 130; Wallinger, "Mottainai." For quilts, see Arnett et al., *Quilts of Gee's Bend*; Beardsley et al., *Gee's Bend.*
48. Dazai Shundai, *Dokugo* (1816), cited in Minami, *Psychology,* 90.
49. "The exploitative chic-ing" is from Rosner, "Beautiful Oops," 78–79. The other two phrases are from Mattern, "Maintenance and Care."
50. Conroy, "Rust Belt Ruins"; Wells, "Detroit"; Gullen, "Gods and Monsters."
51. The description and visual images of this project can be found in the booklet *Repairing Earthquake Project* and on the exhibition's page of the Stroom Den Haag website, accessed November 1, 2019, https://www.stroom.nl/activiteiten/tentoonstelling.php?t_id=408800. I thank Joanna van der Zanden for this example.
52. It is described in Mitchell, "Craftsman." Fujiwara observes how people's sympathy and compassion toward the victims of the earthquake have also become directed toward broken wares. *Bunkai no Tetsugaku,* 293.
53. You can see *Kintsugi Pieces in Harmony* in a YouTube video uploaded April 12, 2017, https://www.youtube.com/watch?v=i8m_AeTGcmk.
54. Kaufman, "Post-Traumatic Growth."
55. Morgan, "Kate Kittredge's Stockings," 70.

BIBLIOGRAPHY

Abramson, Daniel M. *Obsolescence: An Architectural History.* Chicago: University of Chicago Press, 2016.

Arnett, William, Alvia Wardlaw, Jane Livingston, and John Beardsley. *The Quilts of Gee's Bend.* Atlanta: Tinwood Books, 2002.

Bacon, Francis. *The New Atlantis.* Project Gutenberg. Posted October 23, 2008. https://www.gutenberg.org/files/2434/2434-h/2434-h.htm. First published 1627.

Beardsley, John, William Arnett, Alvia Wardlaw, and Jane Livingston, eds. *Gee's Bend: Women and Their Quilts.* Atlanta: Tinwood Books, 2002.

Bicknell, Jeanette, Jennifer Judkins, and Carolyn Korsmeyer, eds. *Philosophical Perspectives on Ruins, Monuments, and Memorials.* New York: Routledge, 2020.

Conroy, Renee M. "Rust Belt Ruins." In Bicknell et al., *Philosophical Perspectives,* 121–32.

Dawdy, Shannon Lee. *Patina: A Profane Archaeology.* Chicago: University of Chicago Press, 2016.

DeSilvey, Caitlin. *Curated Decay: Heritage beyond Saving.* Minneapolis: University of Minnesota Press, 2017.

Dewey, John. *Art as Experience.* New York: Capricorn Books, 1958. First published 1934.

Eatock, Daniel. *Daniel Eatock.* Series Thirty-Two. New Haven: Draw Down, 2017. https://eatock.com/books/visible-vehicle-repairs/.

Fujiwara Tatsushi. *Bunkai no Tetsugaku: Huhai to Hakkō o meguru Shikō* [Philosophy of decomposition: Thoughts on decay and fermentation]. Tokyo: Seidosha, 2019.

Gomez, Ramiro. "*On Tenth Avenue.*" *Manual: A Journal about Art and Its Making* 11 (Fall 2018): 35–37.

Gullen, Christopher T. "Gods and Monsters: A Solastalgic Examination of Detroit's Ruins and Representation." In Lyons, *Ruin Porn,* 31–44.

Gwilt, Alison. "Producing Sustainable Fashion." In Gwilt and Rissanen, *Shaping Sustainable Fashion,* 59–73. London: Earthscan, 2011.

Gwilt, Alison, and Timo Rissanen, eds. *Shaping Sustainable Fashion: Changing the Way We Make and Use Clothes.* London: Earthscan, 2011.

Handa, Rumiko. *Allure of the Incomplete, Imperfect, and Impermanent: Designing and Appreciating Architecture as Nature.* Abingdon, UK: Routledge, 2015.

Ifixit. "Repair Manifesto." Accessed June 7, 2020. https://www.ifixit.com/Manifesto.

Irahara Mitsumi and Makamura Makoto. *Zero kara no Kintsugi Nyūmon: Utsuwa o Yomigaeraseru, Urushi no Tsukoroi* [Introduction to kintsugi from zero: Restoring containers by mending with lacquer]. Tokyo: Seibundō Shinkōsha, 2017.

Jackson, Steven J. "Rethinking Repair." In *Media Technologies: Essays on Communication, Materiality, and Society,* edited by Tarleton Gillespie, Pablo J. Boczkowski, and Kristen A. Foot, 221–39. Cambridge, MA: MIT Press, 2014.

Jackson, Steven J., and Laewoo Kang. "Breakdown, Obsolescence and Reuse: HCI and the Art of Repair." In *CHI 214: Proceedings of the SIGCHI Conference on Human Factors in Computing Systems,* 449–58. New York: Association for Computing Machinery, April 2014.

Kaufman, Scott Barry. "Post-Traumatic Growth: Finding Meaning and Creativity in Adversity." *Scientific American* (blog), April 20, 2020. https://blogs.scientificamerican.com/beautiful-minds/post-traumatic-growth-finding-meaning-and-creativity-in-adversity/.

Lippard, Lucy R. "Never Done." In *Mierle Laderman Ukeles: Maintenance Art,* edited by Patricia C. Phillips, 14–21. New York: Queens Museum, 2016.

London, Bernard. "Ending the Depression through Planned Obsolescence." Wikimedia Commons. First published 1932. Accessed June 7, 2020. https://upload.wikimedia.org/wikipedia/commons/2/27/London_(1932)_Ending_the_depression_through_planned_obsolescence.pdf.

Lowenthal, David. *The Past Is a Foreign Country.* Cambridge: Cambridge University Press, 1990.

Lyons, Siobhan, ed. *Ruin Porn and the Obsession with Decay.* Cham: Palgrave Macmillan, 2018.

Martin, Lois. "Patina of Cloth." *Surface Design Journal* 28, no. 4 (Summer 2004): 16–21.

Mattern, Shannon. "Maintenance and Care." *Places Journal* (November 2018): https://placesjournal.org/article/maintenance-and-care/.

Melchionne, Kevin. "Living in Glass Houses: Domesticity, Interior Decoration, and Environmental Aesthetics." *Journal of Aesthetics and Art Criticism* 56, no. 2 (Spring 1998): 191–200.

Minami, Hiroshi. *Psychology of the Japanese People.* Translated by Albert R. Ikoma. Toronto: University of Toronto Press, 1971.

Mitchell, Kristine. "Craftsman Voluntarily Repairs Any Family Heirloom Ceramics Damaged by Earthquakes in Japan." *My Modern Met,* April 23, 2016. https://mymodernmet.com/kunio-nakamura-kintsugi-pottery-repair/.

Morgan, Lisa Z. "Kate Kittredge's Stockings." *Manual: A Journal about Art and Its Making* 11 (Fall 2018): 70–71.

Nakamura Kunio. *Kintsugi Techō: Hajimete no Tsukuroi* [Kintsugi notebook: First repair]. Tokyo: Genkōsha, 2017.

Nanbō Sōkei. "Record of Nanbō." In *The Theory of Beauty in the Classical Aesthetics of Japan.* Edited and translated by Toshihiko and Toyo Izutsu. The Hague: Martinus Nijhoff, 1981. Sen no Rikyū's teaching recorded by his disciple, Sōkei Nanbō.

Ozawa Noriyo. *Kintsugi no Susume: Mono o Taisetsuni suru Kokoro* [Recommendation on kintsugi: The mindset to cherish things]. Tokyo: Seibundō Shinkōsha, 2014.

Pellegrini, Leah. "Broken Vases Are Restored by Sewing Them Back Together with Gold Thread." *My Modern Met,* May 5, 2016. https://mymodernmet.com/charlotte-bailey-kintsugi-patchwork-porcelain/.

Platform 21. "Repair Manifesto." 2009. http://www.platform21.nl/page/4315/en.

Repair Café (website). Accessed June 7, 2020. https://repaircafe.org/en/.

Repairing Earthquake Project. Den Haag: Stroom Den Haag, 2019.

Richman-Abdou, Kelly. "Kintsugi: The Centuries-Old Art of Repairing Broken Pottery with Gold." *My Modern Met*, November 5, 2019. https://mymodernmet.com/kintsugi-kintsukuroi/2/.

Rissanen, Timo. "Designing Endurance." In Gwilt and Rissanen, *Shaping Sustainable Fashion*, 127–38.

Rodabaugh, Katrina. *Mending Matters.* New York: Abrams, 2018.

Rosner, Daniela K. "A Beautiful Oops." *Continent* 6, no. 1 (2017): 77–80.

Saito, Yuriko. "Aesthetic Experience as an Educational Journey." In *Aesthetics, Nature and Religion: Ronald W. Hepburn and His Legacy*, edited by Endre Szécsényi, 215–33. Aberdeen: Aberdeen University Press, 2020.

———. "Aesthetics of Care." In *Aesthetics in Dialogue: Applying Philosophy of Art in a Global World*, edited by Zoltán Somhegyi and Max Ryynänen, 187–202. Berlin: Peter Lang, 2020.

———. "Consumer Aesthetics and Environmental Ethics: Problems and Possibilities." *Journal of Aesthetics and Art Criticism* 76, no. 4 (Fall 2018): 429–39.

———. *Everyday Aesthetics.* Oxford: Oxford University Press, 2007.

———. "The Japanese Aesthetics of Imperfection and Insufficiency." *Journal of Aesthetics and Art Criticism* 55, no. 4 (Fall 1997): 377–85.

———. "Reflections on the Atomic Bomb Ruin in Hiroshima." In Bicknell et al., *Philosophical Perspectives*, 201–14.

Slemmons, Kiff. "On Imperfection." *Metalsmith* 28, no. 1 (2008): 26–29.

Spelman, Elizabeth V. *Repair: The Impulse to Restore in a Fragile World.* Boston: Beacon Press, 2002.

Wallinger, Sasha Rabin. "Mottainai: The Fabric of Life, Lessons in Frugality from Traditional Japan." *Textile: The Journal of Cloth and Culture* 10, no. 3 (November 2012): 336–45.

Weiss, Allen S. "Cracks." *ArtUS* 29 (2010): 62–67.

———. "The Toilet Bowl and the Tea Bowl." *ArtUS* 27 (2009): 80–85.

Wells, Kate. "Detroit Was Always Made of Wheels: Confronting Ruin Porn in Its Hometown." In Lyons, *Ruin Porn*, 13–29.

Yoshida, Kenkō. *Essays in Idleness: The Tsurezuregusa of Kenkō.* Translated by Donald Keene. New York: Columbia University Press, 1967.

FIVE

Death and Entanglement: Some Thoughts about Life, Love, and the Aims of Art Conservation

Alva Noë

Abstract

What is an artwork such that we can care for it and preserve it even in the absence of the world in which the artwork found its original meaning? This question is in some ways like, and in others unlike, questions faced when working to protect endangered species. This chapter looks at these dual questions in relation to each other as well as in the light cast by competing conceptions of loss, death, and life.

Back in the early 1980s, when the California condor was on the verge of extinction, the pioneer environmentalist and conservationist David Brower, who was one of the founders of the Sierra Club, opposed efforts to save the species by breeding it in captivity for later release in the wild. His objection was principled. A condor is at most 5 percent blood, brain, bones, and feathers. The rest, he argued, is environment. A condor bred in captivity, in isolation from its environment, is no condor. Perhaps, he bitterly suggested, it is a flying pig.[1]

Brower appreciated that an organism, in this case a bird, is no mere body. It is a locus of activity, and this activity is world involving. The animal, to borrow Susan Hurley's phrase, is a "dynamic singularity."[2] Where do we find it? In its nexus with the wider world.

In this brief essay I explore Brower's insight about living beings but in relation to questions of art and conservation. But to do so I would like, first, to warn against a way of misunderstanding the basic phenomenon he appreciated.

To bring this, the misunderstanding, into focus, let us consider Patricia Westerford, the fictional botanist in Richard Powers's novel *The Overstory*. At a crucial juncture in the story, Westerford

> describes how a rotting log is home to orders of magnitude more living tissue than the living tree. . . . She tells . . . about the ambrosia beetle. The alcohol of the rotting tree summons it. It moves into the log and excavates. Through its tunnel system, it plants bits of fungus that it brought in with it, on a special formation on its head. The fungus eats the wood; the beetle eats of the fungus. . . . She tells . . . : everything depends on everything else. There's a kind of vole that needs old forest. It eats mushrooms that grow on rotting logs and excretes spores somewhere else. No rotting logs, no mushrooms; no mushrooms, no vole; no vole, no spreading of fungus; no spreading of fungus, no new trees.[3]

On the basis of reflections such as these, she finds herself entertaining the hypothesis that "a tree's real task on Earth is . . . to bulk itself up in preparation to lying dead on the forest floor for a long time."[4] Life, she considers, is for death.

Westerford conceives of the living old-growth forest as "reciprocal nations tied together in life," but her conception has everything to do with death.[5] Earlier in the book Westerford discovers this. Here is the crucial passage:

> She presses on fissures of bark and her fingers sink in knuckle-deep. A bit of bushwhacking reveals the extent of the prodigious rot. Crumbling, creature ridden boles, decaying for centuries. Snags gothic and twisted, silvery as inverted icicles. She has never inhaled such fecund putrefaction. The sheer mass of ever-dying life packed into each single cubic foot, woven together with fungal filaments and dew-betrayed spiderweb leaves her woozy. Mushrooms ladder up the sides of trunks in terraced ledges. Dead salmon feed the trees. Soaked by fog all winter long, spongy green stuff she can't name covers every wooden pillar in thick baize reaching higher than her head.
>
> Death is everywhere, oppressive and beautiful.[6]

Powers here gives effective voice to ideas that are growing in popularity—think of Peter Wohlleben's bestseller *The Hidden Life of Trees* and

Michael Pollan's article a few years back in the *New Yorker* about the new plant "neurobiology."[7] I teach about plant intelligence, so called, in my introductory classes, precisely in order to get students to think in fresh ways about what is required for an organism to have a mind, about what mind is or might be. The case for plant minds is all the more exciting when we notice that "plant neurobiology" is a misnomer, for plants have no brains and no nervous system. Maybe brains and nervous systems are not required for consciousness or thought, after all?

But this is all an aside. What I'd like us to pay attention to here now is this idea of "ever-dying life" that Westerford, Powers's character, encounters in the old-growth forest. Life and death, the idea seems to be, are joined symmetrically in a kind of processual, circular unity. Death is the source of life, which is itself thought of as a condition of steady dying.

From this point of view, life and death are an unending process of metamorphosis and transformation in relation to which, very importantly, trees and other large-scale biological individuals—you, me, the California condor—are but moments or phases; we are living bridges from one state of death or dying to another. Authentic death is a kind of eternal life, a beautiful oneness with the biomass from which we come and to which we will return.

This impersonal, and perhaps consoling, picture of living death or dying life, of an authentic death, oppressive and beautiful, is implicitly contrasted with something else, a different kind of end-event, something much more frightening and total and absolute than the fertile death of the death-life circle, namely, the very cessation of the death-life circle itself. This cessation—brought about by the devastation of biodiversity and the destruction of ecosystems, by human interventions such as clear-cutting—is something new and terrible and *wicked*; from the standpoint of the ecological naturalism I have been describing, this megadeath, this sterilization and undoing of four billion years of organic evolution, is uniquely human and very much the aftermath of something like *sin*: greed, rapaciousness, ignorance, hungry indifference.

But the essential piece in all this, for my present purposes at least, is not the politics, not the very genuine threat, not the moral challenge, but the focus shift, implied here, from the individual, as a unit of significance, to the biomass collective. Ecological naturalism, here articulated by the fictional scientist Patricia Westerford, posits life and death as mutually enabling processes and, crucially, it views living agencies—trees, voles, people—as, in a way, mere epiphenomena: we come and go as the active microbial and fungal matter, which is our real ground, blooms and decays.

Now the mistake comes into view. Yes, we are dynamic singularities. We are spread out and world involving.[8] We are 95 percent environment; Brower was right about that. But we are not epiphenomena, and our relation to death, and life, is not at all like what the imagined ecological naturalist seems to suggest.

We do not enter into life and then depart, preserving a great continuity. Rather, we are, each of us, every single one of us, born to our human mothers; we are pushed or squeezed, or pulled, sucked, or even cut out of her body. We do not emerge from a living death, a vast preconscious unknowing; we emerge from *life*, from the very specific, personal, female lives of our specific individual mothers. And our lives start there, not exactly in medias res, but certainly *already in relation* to another person. It is the fact of this inescapable intimacy with our mothers that gives life its basic asymmetry, its unidirectionality. However you tell the story, ours is the tale of a breakup, and its final chapter is always the same. We die. But death for us now is not a sacrifice to the microorganisms alive within us and around us, nor is it a spiritual eternal rebirth. Our deaths, in this way like our births, can only ever be for us great traumas, for they are forced separations and unwilled goodbyes. Death is a terrible and undeniable loss of everything we have ever loved and everything we have ever cared about.

In this admittedly somewhat melodramatic telling, we do not stand in a symmetrical relationship to life and death. And while we are food for the fishes and fodder for the worms, no doubt, we are more than that. We are persons, we are agencies, we are protagonists.

This is not melodrama, however; it is what we might look to biology to teach us.

Consider: The simplest living thing, the unicellular organism, makes itself by intaking energy from the surrounding field and using it to build a wall, the membrane that literally defines the cell, or the organism itself, and marks where the rest of the world begins. Life is this self-making, autopoietic, but fundamentally xenophobic activity of producing, preserving, conserving, and insisting on the integrity of oneself.

This process also delivers something like meaning and value. For the cell's universe spontaneously bifurcates into that which is good for the cell and that which it cannot abide.[9] The cell has interests, needs, goals, and in a way, problems, even if, of course, it cannot frame these or think about these for itself. The simplest organism is already a protagonist, and a protagonist is not merely a locus of biomechanical activity. The cell is not a process or a dynamic but an *it*—as opposed to a he, she, or they—and it is that in virtue of which it is an *it*—a kind of unity or integrity—that makes it different from any nonliving things that there are.

Now, an amoeba is no person, to be sure. But this much is shared in common. To be a person, no less than to be a bacterium, is to be organized. We are organisms, after all. Moreover, neither the amoeba nor we ourselves are mere chemistry; we are characters, and our stories, whether told by Charles Darwin, by the journalist or novelist, or by ourselves, reveal an order and meaning that have no reflection, none at all, in mere physics.

I pause here, to recapitulate and to set our sights more clearly on what I'm trying to bring into focus here. Human life and consciousness is not something that happens inside us; it is something we achieve; like everything else we achieve, it depends not only on our brains, bones, feathers, and blood but on our dynamic involvement with a surrounding environment, including, in our own case at least, the cultural environment. This was Brower's insight, and it was the basic argument of my book *Out of Our Heads.*

But to insist on this world-coupled collective dynamic distributedness is not, or should not be, to eliminate or reduce or demote the existential and the ontological significance of the organism or the person. No living being is merely active matter.

And this has everything to do, first, with the fact that life not only self-organizes but always does so in the interest of homeostasis. We make ourselves, we self-organize, and we do this by denying that mere process, that is to say, entropy, can rule us and reduce us to mere biomass. We make ourselves up. You lose sight of this if you give too much attention to the fragility of our hold on ourselves and of our vulnerability to, as well as our productive exploitation of, the fungi and the microbiome.

But there is a second point. To be an organism such as we are is always to be a kind of servant. In the first instance, a servant to the ceaseless demands of life making, self-making, self-preserving, self-conservation that is our first order of business. But things are more poignant still: We didn't sign up for this, did we? To be an organism is to be *organized,* organized by the demands of this fundamental, existential, and, in a way, terribly constraining task: the conservation of ourselves. But none of us chooses this. In fact, for the most part we do not even understand the ways we are organized and made up by the ways we are organized.

Ludwig Wittgenstein wrote that a philosophical problem has the form: I don't know my way about.[10] And the aim of philosophy is, then, to enable us to recover our bearings or to obtain an orientation. We need, or we feel the need anyway, to gain an overview of, basically, where we find ourselves.

This is art's basic work too.

Art aims at ecstasy. I don't mean passion, emotion, or pleasure. Its push is for release, and its point is to undo the stasis that is our ordinary, our normal, and our most familiar condition, namely, the organization that governs us at so many levels, in the large and in the small.

The key thing is to appreciate that the conditions from which we seek release—for example, in the human case, the basic habits, intelligent automatisms, and unthinking skills that constrain us—are precisely that which make us what we are. Release means something more than enlightenment—Wittgenstein's coming to "know our way around"—it holds forth the possibility of reorganization, something kin to liberation.

Two examples:

Choreography. People dance. Dancing is an organized activity. It is an activity built out of habits. Choreography, the art of dance, however, is not just more dancing. Choreography puts dancing on the stage. It puts the fact that we are dancers, that we are organized by dancing, on the stage. It puts us on the stage. It offers a representation of ourselves dancing. Think of this as writing ourselves down, or as the introduction of a perspicuous representation, like a new notation. And this representation of ourselves changes us, for it gives us new resources for thinking about what we are doing when we dance, and so finally new ways of dancing. You dance differently in a world in which there is choreography. Choreography loops down and changes the dancing of which it is the representation.[11]

Second, painting. People make pictures; they use pictures. We have been doing so for millennia, much as we have been talking and thinking, and dancing, for millennia. We are participants in a picture-based economy of exchange. Pictures—sketches, drawings, photos, you name it—are tools for showing, a technology of display. We use them for all manner of purposes. For example, we use them to show something we want to sell, or to record what happened, and so on. Crucially, artists who make pictures are not merely participating in this economy of show and display. They are not really making pictures at all; they are making art *out of pictures*, that is, out of the fact that we are organized by pictures, that pictures occupy an existentially organizing place in our lives. Painting as an art puts all this on display and does so in ways that give us resources, finally, for making new kinds of pictures and for thinking about the place of pictures in our lives in new ways. Painting starts with habits, with the ways we find ourselves organized by pictures, and then it reorganizes us.

Crucially, again, this is art's job, its work. This is, to borrow Martin Heidegger's pun, the work of art.

If I am right that art serves in this way to unveil us to ourselves and to reorganize us, that art is, fundamentally, about *ek-stase,* a breaking up of the stasis that holds us captive and makes us what we are, then it follows that we must resist identifying the art *object* or the art *thing* with the art work. John Dewey warned us about this. To do so is about as wrong-headed as thinking that the California condor is simply a structure of blood, brains, bones, and feathers. To save the condor you need to save the condor's world, for the condor is a way of being in the world. And to save the artwork, or to keep it alive, as it were, you need to preserve the conditions in which it can provide us the needed opportunity to unveil and reorganize.

It is useful to reframe our question in relation to choreography. Useful because a choreographic work is an artwork, fully fledged. But also because it is an artwork where, in a way, the ratio of thinghood or materiality to true significance is as low as can be, and so it serves as a reminder that artwork is not thing, but that, rather, the thing is at most a prop. If our job is to conserve or preserve Merce Cunningham's choreographies—for example—what else is there for us to do but, as it were, stage his works anew? Or perhaps our job is in some other less specific way to carry on an art-making practice inaugurated by him.

But isn't that what we need to do, too, if we wish to preserve paintings, sculptures, films or videos, architectures, or whatever? It isn't just that we need to resist the microbes and the fungi and so keep things from falling apart; we need to preserve the very environment, or our access to the very environment, in which alone the works are actually even the works they are in the first place.

And that isn't something we can do only conditionally, as it were. We can't pretend art objects can serve to open up new possibilities for us or give us new resources for understanding if they no longer really do. Art preservation, art conservation, isn't just historical excavation, bracketing, or place holding. Thought of as artifacts, artworks are just shards or pictures or traces. Real conservation must be therefore a kind of dedication or commitment to the whole system of relationships, which is the artwork. This sort of commitment can't be undertaken indiscriminately or promiscuously. And so conversation will always be tied up not only with politics but also with the most basic questions about what we value.

But there is another lesson choreography teaches. Every living work of art, of whatever kind you care to select, is, in the sense once articulated by William Forsythe, a *choreographic object.*[12] It choreographs us. Recall Barnett Newman's note on the wall of the Betty Parsons Gallery urging visitors to step in close to the works. Newman's note brings out that every painting sets the viewer in motion, bodily, but also cognitively and emotionally. It is in this setting into motion that the work of art happens.

Of course, it is we who set the work into motion by caring about it. So that's the main job of the conservator. To make it possible for us to care about the work of art.

Perhaps we should say that every work of art, then, is a kind of choreographic score; and our question is, How to play it? or, Do we want to play it? or, Why should we play it? Art conservation, then, is about keeping it real, about restaging, about maintaining the environment, the whole contexture, in which the work can afford what it affords, can score what it can score.

Now this *may* and very often will require that we attend to the object itself and its needs. That is, it is important that the Old Master paintings and other treasures of that sort receive the kind of careful maintenance that lets them remain physically intact. For their material survival is essential to what they are and to their continued existence, in any form, for us.

But this is not always the case and this need not always be the case. Which is the truer artwork: Paolo Veronese's painting in the Louvre or Factum Arte's copy, restored to the original Italian church setting from which it was stolen by Napoleon's troops?

I end with this:

There was an engraving on Roman pre-Christian headstones that was once as common as RIP is today on ours: *NF F NS NC—non fui / fui / non sum / non curo.* Translation: "I was not / I was / I am not / I do not care."

Perhaps it is sometimes good and right that we let artworks die their natural death. It is terrible and scary to let things we love go, but there is no alternative.

NOTES

1. As recounted in Montgomery and Strombeck, *Condor Comeback.*
2. Hurley, *Consciousness in Action.*
3. Powers, *Overstory*, 282.
4. Ibid.
5. Ibid.
6. Ibid., 134.
7. Wohlleben, *Hidden Life of Trees*; Pollan, "Intelligent Plant."
8. This I something I have been arguing for the last fifteen years; it is the central theme of my books *Action in Perception* and *Out of Our Heads.*
9. This idea, as well as the conception of life as autopoiesis in the previous paragraph, are due to Maturana and Varela, *Autopoiesis and Cognition.* For a later development, see Thompson, *Mind and Life.*
10. Wittgenstein, *Philosophical Investigations*, § 123.
11. This conception of art as looping down and reorganizing us is the main topic of my *Strange Tools* and is developed further in *Entanglement.* This paragraph and the next borrow text from this latter book. The idea of "looping" itself is due to Ian Hacking. See, for example, *Social Construction of What?*
12. Forsythe, "Choreographic Objects."

BIBLIOGRAPHY

Forsythe, William. "Choreographic Objects." Accessed August 14, 2021. https://www.williamforsythe.com/essay.html.

Hacking, Ian. *The Social Construction of What?* Cambridge, MA: Harvard University Press, 1999.

Hurley, Susan L. *Consciousness in Action.* Cambridge, MA: Harvard University Press, 1998.

Maturana, Humberto R., and Francisco G. Varela. *Autopoiesis and Cognition: The Realization of the Living.* Boston: Reidel, 1980.

Montgomery, Sy, and Tianne Strombeck. *Condor Comeback.* New York: Clarion Books, 2020.

Noë, Alva. *Action in Perception.* Cambridge, MA: MIT Press, 2004.

———. *The Entanglement: Art before Nature.* Princeton, NJ: Princeton University Press, 2022.

———. *Out of Our Heads: Why You Are Not Your Brain, and Other Lessons from the Biology of Consciousness.* New York: Hill and Wang, 2009.

———. *Strange Tools: Art and Human Nature.* New York: Hill and Wang, 2015.

Pollan, Michael. "The Intelligent Plant." *New Yorker*, December 15, 2013.

Powers, Richard. *The Overstory*. New York: Norton, 2018

Thompson, Evan. *Mind in Life: Biology, Phenomenology, and the Sciences of Mind.* Cambridge, MA: Belknap Press of Harvard University Press, 2007.

Wittgenstein, Ludwig. *Philosophical Investigations.* Edited by G. E. M. Anscombe and Rush Rhees. Translated by G. E. M. Anscombe. New York: Macmillan, 1953.

Wohlleben, Peter. *The Hidden Life of Trees.* Vancouver: Greystone Books, 2016. Published in German in 2015 by Ludwig Buchverlag.

PART 2

History

SIX

Introduction: Conserving Active Matter and the Art Historian's Craft

Ittai Weinryb

Abstract

This short essay introduces the Bard Graduate Center working group on historical approaches to active matter. The so-called ontological and agentive turns in the humanities have encouraged rethinking the framework of dealing with agency and ontology in historical perspective. The essays in the section offer various ways to deal with historical objects and materials as they manifested themselves in different historical instances and explore the implications of such moments in history for studying the past.

Around mid-March 2020, the world came to a halt. The COVID-19 pandemic paralyzed the economy, forcing much of the populace to quarantine at home, able to continue to work only with the aid of digital platforms. One immediate effect of the closure was the shortage of fresh produce combined with the plentitude of free time in lockdown.[1] The combination of the two made people start baking at home, preparing bread as a form of returning to the "basics."[2]

One type of bread that became immensely popular was sourdough, in which the fermentation of the dough was achieved through naturally occurring lactobacilli and yeast. The lactic acid produced by the lactobacilli gives the bread a slightly sour taste and longer shelf life. Sourdough is made using a starter (or leaven): a mix of living bacteria, colonies of microorganisms, and naturally occurring enzymes. A small portion of the starter is taken in order to bake each fresh batch of bread. Starters are living organisms, brought to life in various ways, fed on a regular basis, and at the right moment, put to work as leavening agents in the making of bread. Some old strains have been thriving for hundreds of years. Bakers can become attached, celebrating their starters' birthdays and even naming them.[3]

Maybe the starter is not the same kind of active matter as a pinecone. The yeast, after all, was always alive in ways that the dead cellulose cells of a pinecone are not. On the other hand, this is not activation through a collaborative process of discussion, conservation, and curation. But sourdough bread is the introduction that millions of people have just received to what we could call the "active matter spectrum." Beyond that, active matter as seen through sourdough starter takes us directly to the frontier of the human and nonhuman.[4]

In many instances we, art historians and conservators both, think that looking for the ur-moment of the entity, the moment when it was created, provides clues to the nature of the living matter and thus the key to any subsequent campaign to conserve it. Let us look, for example, at the world and the ur-moment of its creation. In one of the frontispieces of a *Bible moralisée* (Vienna, Österreichische Nationalbibliothek, MS 1179; fig. 1) made, most likely, for the education of King Louis VIII of France (1187–1226), Christ holds the world on his lap in the moment of creation. The compass in his right hand indicates that he is measuring the shapes that will constitute the world, both human and nonhuman, from unformed matter. The inscription on the top of the page reads, *hic orbis figulus disponit singulus solus* (Here, the potter of the universe arranges each separate [element]).

The four Aristotelian elements, earth, wind, fire, and water, are presented as painted forms that resemble marble, undulating within the circle that makes up the world. The elements never become one, and so they seem to be always moving and changing position, constantly pushing against one another, making a world that is itself active matter.[5]

Christ is here named a *figulus*—a potter. The potter, who creates shapes and useful forms out of clay or earth, becomes a metaphor for God the Creator. Adam, after all, was created by God from the "dust of the ground" and was enlivened by God breathing the breath of life through his nostrils. Since 200 CE God has been referred to as the potter of Adam, and Jewish and Muslim legends tell of the boy Jesus who enlivened clay birds by breathing the breath of life through their beaks.[6] The material that makes the clay alive is the creator's breath, and it is the invisible thing that is the hardest to discern. Unless the living entity—the world, Adam, or Christ's birds—tells of the breath of life that animated it during that ur-moment, knowledge of its creation

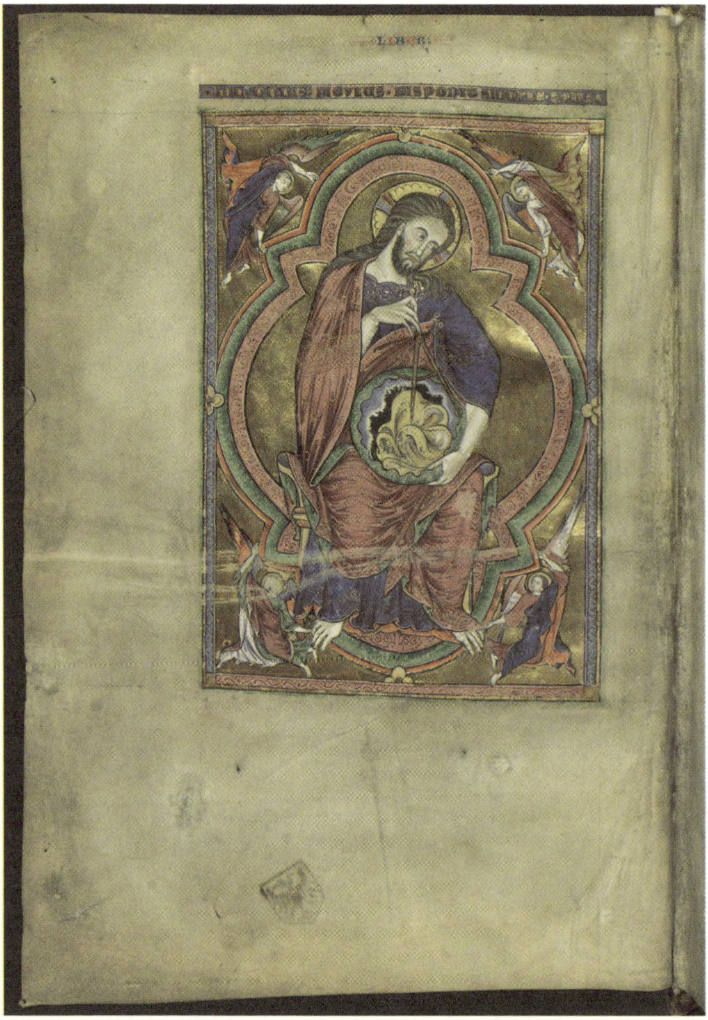

Fig. 1. Austrian National Library, Cod. 1179, "Bible moralisee." ANL/Vienna Cod. 1179, fol.1v.

is impossible. A telling moment in the book of Isaiah describes the folly of denying the creator: "as if the clay should meditate against the potter, and the work would say to its maker, 'you did not make me.'"[7] The verse indicates that if the clay entity does not speak up, so to speak, no one can know what gave it life. From the ur-moment of creation, the nonhuman entity has a life of its own, eager, so to speak, to cut itself loose from its creator.

In the context of the image of Christ creating the world, we must think of knowledge paradigms, paradigm shifts, and their potential influence on the interpretation of active matter, both for historical figures and for ourselves. In the image of Christ as potter, the border of the mandorla surrounding Christ is adorned with Arabicizing markings that appear like Arabic script but denote nothing; this inclusion may give further clues as to the reception of ideas surrounding active matter.[8] While early theories regarding creation in Western Europe were chiefly Platonic and relied heavily on the *Timaeus,* by the thirteenth century, a surge of translations from Arabic introduced new Aristotelian concepts found in texts such as *De anima* and *De animalibus,* challenging the understanding of the boundaries between humans and nonhumans and changing the understanding of what could be defined as active matter.[9] The four elements, those painted in the Vienna frontispiece, not only were thought to represent an ur-moment of creation but, through Aristotelian science, were linked closely to the bodily humors—the chemical system that governed the human system. Therefore, the four elements, which took part in the story of creation of the world, became a living entity that affects humanity on a daily basis. Intellectual paradigm shifts brought about by the embrace of Aristotelian science influenced thirteenth-century makers and illuminators and how they understood notions of living matter, particularly the four elements.[10] The frontispiece image is generated through the conflation of biblical and scientific traditions, responding to the new translation of knowledge from Greek and Arabic into Latin; its active potency resides in the textual sources that generated the pictorial representation.[11]

Potential alterities play a central role when thinking of conceptualizations of active matter and its conservation. While the ur-moment of creation suggested in the Vienna manuscript envisioned a configuration of its core values through the importation of new theories on the formation of matter, in that same period travelers began narrating ethnographic information regarding the activity of things in other Indigenous cultures. In fact, Louis IX, the son of Louis VIII, for whom the frontispiece was made, had sent such a traveler to the easternmost part of the known world. William of Rubruck, a Flemish Franciscan monk, was sent by the king on a mission to convert the Tatars to Christianity.[12] He wrote a report of his journey that includes lengthy descriptions of the local customs of the Tatars and Mongols, including the practice of making deities out of felt:

> Over the master's head there is always an effigy like a doll or statuette, made of felt, which they call the master's brother, and a similar one over the mistress's head, which they call the mistress's brother, [both] fastened to the wall. Higher up between these two is a small thin one, which is the guardian, as it were, of the whole dwelling. The mistress of the house places on her right, in a prominent position at the foot of the couch, a goat's skin stuffed with wool or some other material, and near it a tiny statuette facing her attendants and her women. Close to the entrance on the women's side is yet another effigy with a cow's udder for the women who milk the cows, since milking the cows is one of the women's tasks: on the other side of the entrance, facing the men, is a second statuette with a mare's udder, for the men who milk the mares.[13]

Here, William describes the worship of guardian spirits associated with material images crafted out of an associated substance. Felt is a material made by saturating pieces of animal hair in hot water in order to create a fabric, a time-consuming process in which the material, so to speak, "grows" from the mechanical agitation of animal hair in water. This process, while not biological, is facilitated by the alkaline environment that is incidentally produced during the cleaning and processing of animal hair by suint (the sweat of the animal) fermentation. Felt is thus associated with biological processes in its making, albeit indirectly as compared to the case of the bread starter: it does not continue to grow and requires no maintenance. Animated Mongol fetishes remain ritual objects in Mongolia to this very day. A set of felt deities from early twentieth-century Mongolia now in the National Museum in Copenhagen appear consistent with William's account (fig. 2), as they had similar ritual function to the deities in the description.[14] Removed from their ritual context, the question becomes to what extent the textual description and the material artifacts are identifiable with each other, even though they are separated by more than eight hundred years. To what extent do the qualities described in medieval words correlate to the actual felt objects in the museum today? Is there a ritual consistency that manages to link the textual entity to the material one? To what extent is the "active" quality of the object preserved in the museum case, where it now lacks all ritual maintenance? What can we say about objects that are stripped of their functionality, that have lost their active nature, and that belong to cultures that may not even exist any longer? How can we extract meaning regarding the nature of active matter in bygone cultures from their texts describing the nature and function of such active entities? Our dissonance here lies in the gaps between past and present, knowing and believing, object and word, self and alterity.

The contributions in the following section all relate to the three main examples I discuss here. Spike Bucklow's essay deals with the rood screen, which, like our bread starter, involves material that grows and develops through environmental means. André Laks's and Guido Giglioni's contributions contemplate the basic questions underlying the illustration of Christ the Creator in the Vienna frontispiece. What are the historical developments, and what underlies these developments regarding our understanding of living matter? For Giglioni, the answer lies in histories of oppositions—living versus dead and human versus nonhuman. For Laks, the answer lies in how, in the end, we can talk of active matter in Presocratic thought. The difference between "living" for Giglioni and "active" for Laks is crucial and becomes a central theme in Lee Palmer Wandel's essay: the Eucharist wafer becomes active through rituals of consecration and prayer, but it becomes living, so to speak, during the Mass, when the holy wafer becomes the actual liv-

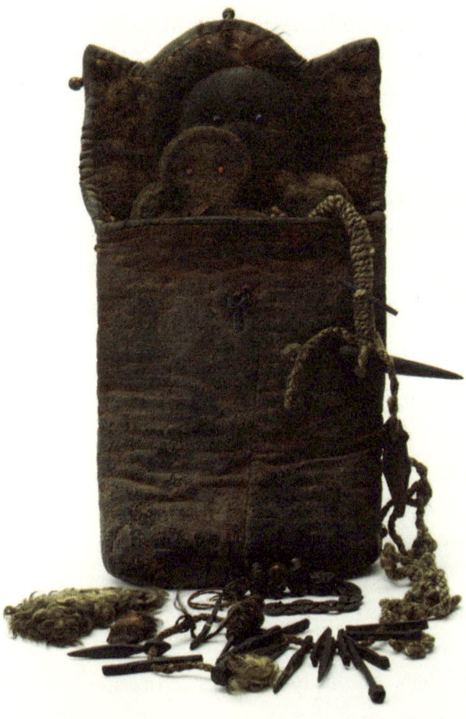

Fig. 2. Felt deities from Inner Mongolia, early twentieth century. Felt, leather, and animal hair. National Museum of Denmark, Copenhagen. Photo: Katrine Vintov.

ing body of Christ. Similar issues surrounding the Mongolian felt deities are crucial in the discourse regarding the nature of the Eucharist: how material artifacts can possess divine attributes and how these animated qualities are treated in internal and cross-cultural discourse over time.

Making a bread starter, representing the creation of the world, and discovering the material deities of another culture are all encounters with alterities in which the subject needs to place some faith in the activity and even animation of the entity in its own context, even if that world is far removed. We may owe it to contemporary materials scientists and contemporary artists to have refocused attention on the activity of matter. But once we start to pay attention to the category of activity, we find that it opens onto avenues that connect the work of conservators, anthropologists, philosophers, and historians, insofar as all of them are engaged in the project of understanding. Historians might argue that their remit extends from the moment in which matter becomes active through to all the later moments of activity—another way of saying "biography of the object." My argument here, and perhaps this is me speaking specifically as an art historian, is that the ur-moment is crucial to understanding subsequent relations between humans, their active materials, and their desire to conserve whatever form of activity those materials are assumed to possess. Mechanisms for generation and maintenance are found in theories of creation and of the magical efficacy of manmade objects. The challenge for the historian is to make sense of endless complexities of cultural circumstance, communal intentionality, and ritualistic consecration that unfold from that originating moment, without denying artistic license—or the ability of actors to make new meaning at any moment. The essays in this volume, each in its own way, grapple with the different ways in which active nonhuman entities are generated, harnessed, and preserved.

NOTES

1. For more, see the essays in *Gastronomica* 20, no. 3 (2020).
2. Ocklenburg, "Distractibaking."
3. A bible for those who really want to dig into the details is Gobbetti and Gänzle, *Handbook on Sourdough.*
4. On living and the organic, and their close relations, see, from a growing body of literature, Bennett, *Vibrant Matter*, 1–38; De Landa, *Thousand Years*, 11–103. See also Cohen, *Animal, Vegetable, Mineral*; Weinryb, "Beyond Representation"; Baader and Weinryb, "Images at Work."

5. On the frontispiece and its inscription, see Lowden, *Making of the Bibles Moralisées*, 87–88.
6. On concepts regarding the animation of objects and their relations to God's creation of Adam, see Cole, *Cellini and the Principles*, 43–78; Guérin, "Saisir le sens"; Panofsky, "Neoplatonic Movement"; Weinryb, *Bronze Object*, 33–54.
7. Isiah 29:16 (Douay–Rheims Bible).
8. On arabicizing script, see Nagel, "Twenty-Five Notes."
9. On the Timaeus and attitudes toward active matter, see Weinryb, "Living Matter."
10. Green, "De genecia"; Lawn, *Prose Salernitan Questions*; Baader and Keil, *Medizin im mittelalterlichen Abendland*; Hunt, *Anglo-Norman Medicine*; Green, *Trotula*; Delogu and Peduto, *Salerno nel XII secolo*; Jones, "Image, Word, and Medicine"; Jacquart and Paravicini Bagliani, *Scuola Medica Salernitana*; Arikha, *Passions and Tempers*. For an overview of these issues, see David C. Lindberg, *Beginnings of Western Science*, 321–56.
11. On the translation movement in the Middle Ages, see Alverny, "Translations and Translators"; Burnett, *Introduction of Arabic Learning; Burnett*, "Coherence." See also Lindberg, *Beginnings of Western Science*, 321–56.
12. For William of Rubruck and his travels, see Jackson, *Mission of Friar William*, 1–56.
13. Ibid., 75.
14. Braae and Haslund-Christensen, *Among Herders*, 159.

BIBLIOGRAPHY

Alverny, Marie-Thérèse d'. "Translations and Translators." In *Renaissance and Renewal in the Twelfth Century*, edited by Robert L. Benson and Giles Constable, 421–62. Cambridge, MA: Harvard University Press, 1982.

Arikha, Noga. *Passions and Tempers: A History of the Humours*. New York: Ecco, 2007.

Baader, Gerhard, and Gundolf Keil. *Medizin im mittelalterlichen Abendland*. Darmstadt: Wissenschaftliche Buchgesellschaft, 1982.

Baader, Hannah, and Ittai Weinryb. "Images at Work: On Efficacy and Historical Interpretation." In "Images at Work." Special issue, *Representations* 133 (2016): 1–19.

Bennett, Jane. *Vibrant Matter: A Political Ecology of Things*. Durham, NC: Duke University Press, 2010.

Braae, Christel, and Henning Haslund-Christensen. *Among Herders of Inner Mongolia: The Haslund-Christensen Collection at the National Museum of Denmark.* Aarhus: Aarhus University Press, 2017.

Burnett, Charles. "The Coherence of the Arabic-Latin Translation Program in Toledo in the Twelfth Century." *Science in Context* 14 (2001): 249–88.

———. *The Introduction of Arabic Learning into England: The Panizzi Lectures, 1996.* London: British Library, 1997.

Cohen, Jeffrey Jerome, ed. *Animal, Vegetable, Mineral: Ethics and Objects.* Washington, DC: Punctum/Oliphaunt, 2012.

Cole, Michael. *Cellini and the Principles of Sculpture.* Cambridge: Cambridge University Press, 2002.

De Landa, Manuel. *A Thousand Years of Nonlinear History.* New York: Zone Books, 1997.

Delogu, Paolo, and Paolo Peduto, eds. *Salerno nel XII secolo: Istitutzioni, società, cultura; Atti del convegno internazionale, Raito di Vietri sul Mare Auditorium di Villa Guariglia 16/20 giugno 1999.* Salerno: Centro Studi salernitani "Raffaele Guariglia," 2004.

Gobbetti, Marco, and Michael Gänzle, eds. *Handbook on Sourdough Biotechnology.* Netherlands: Springer US, 2012.

Green, Monica H. "The *De genecia* Attributed to Constantine the African." *Speculum* 62, no. 2 (1987): 299–323.

———. *The Trotula: A Medieval Compendium of Women's Medicine.* Philadelphia: University of Pennsylvania Press, 2001.

Guérin, Sarah M. "Saisir le sens: Les ivoires gothiques et le toucher." In *Les cinq sens au Moyen Âge,* edited by Éric Palazzo, 589–622. Paris: Le Cerf-Alpha, 2016.

Hunt, Tony. *Anglo-Norman Medicine I: Roger Frugard's "Chirurgia," the "Practica Brevis" of Platearius.* Woodbridge: Boydell Brewer, 1994.

Jackson, Peter. *The Mission of Friar William of Rubruck: His Journey to the Court of the Great Khan Möngke, 1253–1255.* Farnham: Ashgate, 2011.

Jacquart, Danielle, and Agostino Paravicini Bagliani, eds. *La Scuola Medica Salernitana. Gli autori e i testi. Convegno internazionale Università degli Studi di Salerno 3–5 novembre 2004.* Florence: Sismel-Edizioni del Galluzzo, 2007.

Jones, Peter Murray. "Image, Word, and Medicine in the Middle Ages." In *Visualizing Medieval Medicine and Natural History, 1200–1550,* edited by Jean A. Givens, Karen M. Reeds, and Alain Touwaide, 1–24. Aldershot, UK: Ashgate, 2006.

Lawn, Brian. *The Prose Salernitan Questions: An Anonymous Collection Dealing with Science and Medicine Written by an Englishman c. 1200, with an Appendix of Ten Related Collections.* Oxford: Oxford University Press, 1979.

Lindberg, David C. *The Beginnings of Western Science.* Chicago: University of Chicago Press, 1992.

Lowden, John. *The Making of the Bibles Moralisées.* University Park: Pennsylvania State University Press, 2002.

Nagel, Alexander. "Twenty-Five Notes on Pseudoscript in Italian Art." *RES: Anthropology and Aesthetics,* no. 59/60 (2011): 228–48.

Ocklenburg, Sebastian. "Distractibaking: A Note on the Psychology of Baking in Times of the COVID-19 Crisis." *Gastronomica* 20, no. 3 (2020): 5–6.

Panofsky, Erwin. "The Neoplatonic Movement and Michelangelo." In *Studies in Iconology: Humanistic Themes in the Art of the Renaissance,* 171–230. New York: Icon, 1962.

Weinryb, Ittai. "Beyond Representation: Things—Human and Nonhuman." In *Cultural Histories of the Material World,* edited by Peter N. Miller, 172–86. Ann Arbor: University of Michigan Press, 2013.

———. *The Bronze Object in the Middle Ages.* Cambridge: Cambridge University Press, 2016.

———. "Living Matter: Materiality, Maker, and Ornament in the Middle Ages." *Gesta* 52, no. 2 (2013): 113–32.

SEVEN

Active Matter in Presocratic Thought?

André Laks

Abstract

In this contribution written with an eye to a nonspecialized readership, I (provocatively, I hope) challenge a hypothesis arguably derived from Nietzsche and Heidegger, according to which Presocratic authors, who thought and wrote before the appearance and development of Plato's dualism of mind and matter and body and soul, provide intellectual resources, notions, or worldviews that would help us to think differently about matter being an active rather than a passive entity. In order to map the territory of the issue at stake, I point to some conceptual and terminological presuppositions that are encapsulated in the hypothesis in question and try to give a sense of the fragmentary and often biased material that historians of early Greek philosophy have to deal with. The upshot of my presentation, which draws examples from Thales, Parmenides, Empedocles, Leucippus, and Anaxagoras, is that early Greek philosophers, while they did not have at their disposal the Aristotelian terminology that has shaped the history of Western philosophy, developed various means to discuss the issues that this terminology implies.

> Someday the demonstration will be achieved that feeling or touching is a sense that is common to all beings. There are already phenomena that point in that direction; in that case matter in general will have five or six essential properties, dead force or live force, length, breadth, depth, impenetrability, and feeling; I would have added attraction if it were not perhaps a consequence of movement or of force.
>
> —Denis Diderot

Can Presocratic authors such as Thales, Anaximander, Heraclitus, Parmenides, Empedocles, and Anaxagoras (to name a few), who all wrote before the appearance and development of Plato's dualism of mind and matter and body and soul, provide intellectual resources, notions, or worldviews, to think or to think with about matter being an active rather than a passive entity? In order to map the territory of possible answers to this question, I shall first point to some conceptual presuppositions that are encapsulated in the question itself and, second, give a sense of the material (an active material of sorts too) that historians of ancient philosophy have to tackle when dealing with works that are no more extant, as is the case with early Greek thinkers. My own construal of the question is sketched in the third section, before a conclusion in the form of a short apologue.

"Presocratics"

There are quite legitimate reasons to draw a line between Presocratic thinkers and later philosophical reflection. For Socrates undoubtedly represents a turning point in the development of Greek thought, although different versions of what this turning point consists of are possible—the two main ones being that Socrates "brought back," as Cicero said, "philosophy to earth" (meaning that he directed attention to ethics and politics rather than to the natural world) or that he raised for the first time the question of essence and definition (the "What is x?" question)—a question that opened the way to Plato's doctrine of forms.[1] The key text in this respect is Plato's *Phaedo.* There, Socrates relates that he had been interested in his youth by a trend of inquiry that looks for the causes of natural phenomena (the so-called inquiry into nature); puzzled by conceptual questions raised by this kind of investigation (such as those concerning growth and decay), he had also been disappointed not to find in Anaxagoras's book, in spite of promising declarations about "intelligence" (*nous*) being the ultimate cause of everything, a teleological explanation of the world.[2] Having thus "failed in [his] inquiries into reality [*ta onta,* the things that are]" Plato's Socrates takes a new turn—a "second voyage [. . .] to discover the cause":[3]

> I thought I should take great care not to suffer what people do who study and observe the eclipse of the sun. [. . .] I was afraid I would be altogether blinded in my soul by looking at

> these matters with my eyes and each of my senses in my attempt to seize hold of them. So it seemed to me that I should take refuge in arguments [*logoi*] and consider the truth of the realities in these.[4]

Since Nietzsche's celebration of the "tragic period" in Greek culture before it declined under the assaults of Socratic epistemological and theological optimism (opening the way first to Christianity and then to socialism) and Heidegger's reformulation of the contrast—Plato's assignment of truth (*alêtheia*) to his eternal forms now puts an end to the reign of phenomenological, preconceptual disclosure and "nonoblivion" (*a-lêtheia*, according to Heidegger's analysis of the Greek word)—there has been a strong tendency in postmodernist thinking to interpret the divide in terms of a decadence, with Socrates and Plato initiating the history of a certain Fall after the paradisiac life of early Greek philosophy.[5]

Now, it is perfectly true that Presocratic authors did not speak the language of later philosophers (which has become part of our current languages and conceptual framework of intelligibility—"substance," "substrate," "genre," "essence," "energy," and, last but not least, "matter" and "form" are all Aristotle's terms, most of the time with Platonic antecedents). To take an example that is both simple and directly relevant to the present discussion, Empedocles considers that earth, water, fire, and air, which were to become from Aristotle onward the four classical elements, are divinities, and he refers to them not as "elements" (*stoikheia* in Greek), which carries the notion of a constitutive part of any given whole, but as "roots" (*rhizômata*), which is also a part, but one that is specifically responsible for the nourishment of plants and trees.[6] This is certainly suggestive of a definite worldview, oriented toward the model of vegetal growth. But it should immediately be added that the alternative model, based on the notion of a constitutive element without anything divine about it, is Presocratic too: Leucippus, a contemporary of Empedocles, and his better-known follower Democritus compared their atoms, out of which bodily compounds (including soul) are made, to the letters of the alphabet, which is usually considered to be the original sense of the word *stoikheia*.[7] One can generalize: it is not easy to find a global characterization of the Presocratic language or of its philosophical orientation that would contrast with "ours."

The depiction of Socrates giving up first-order questions about the natural world and engaging in second-order considerations implying hypothetical reasoning and conceptual coherence in Plato's *Phaedo* (quoted above), while extremely powerful, also obfuscates the fact that there is more second-order reflection in early Greek philosophy than the separation between the two levels suggests. An alternative, more accurate, and less Platonizing characterization of the situation would recognize that second-order considerations and implicit arguments are in fact *embedded* in the first-order explanations that Presocratic thinkers have to offer—a specific brand of "intellectual hylomorphism," to risk a phrase, in virtue of which form/*morphê* cannot be isolated from matter/*hulê* ("hylomorphism" is the term that is traditionally applied to Aristotle's theory of material substance). The emergence of a type of inquiry based on an autonomized conceptual argumentation did not put an end to embeddedness (far from it), although it certainly reconfigured the general argumentative landscape.

The notion of an embedded (hylomorphic) reflection helps approaching early Greek thought from the perspective of its diversity: not only did early Greek philosophers come up with different answers to the same questions, but they also worked out new questions and conflicting paradigms, such as the one I evoked through the opposition between roots and components to refer to the elements. We should firmly resist the irenic depiction that Gadamer, in the wake of Heidegger, gives of the beginnings of Greek philosophy, when, having talked about "[the] astonishing uniformity in its basic motif for all the Presocratic philosophy of the first period," he adds,

> the Milesians [i.e., Thales, Anaximander, Anaximenes], Parmenides, and Heraclitus express the same basic view of the unity of difference. [. . . W]e must learn to free ourselves not only from the Aristotelian idea, but also from the Hegelian and modern one, that these thinkers form a connected sequence. They do not ever philosophize against one another but instead as philosophers against the non-philosophy of mortals.[8]

This is a prejudiced misdescription. Early Greek philosophy took from the start the form of a fierce intellectual fight by means of embedded argumentation. It is worthwhile trying to reconstruct the stakes, and recover sequences, both historical or conceptual, which is not to say that the outcome will not differ from the pictures we find in Aristotle, Hegel, or, for that matter, Cassirer.[9]

The Material

Any reconstructive task is confronted with difficult hermeneutical problems, but in the case of Presocratic thought, as in other sectors of ancient Greek philosophy, it is made even more thorny by the kind of material that is at our disposal. When we talk about Presocratic philosophers, we must always keep in mind that their works, when they are not simply lost, have been disintegrated. The original wording of early Greek thinkers is known only through a small number of fragments, in some cases only one sentence (as the famous "sentence of Anaximander"[10]); these fragments, even when there are more than a few of them for a given author (which happens), are in the rule short, ranging from a couple of words (sometimes a single word) to a dozen lines. To complete these debris, we must heavily rely on reports, paraphrases, interpretations, and criticisms of later authors, which are themselves dependent to a very large extent on Platonico-Aristotelian views and terminology. This makes the hermeneutical situation both terribly complex and fascinating, because our own assumptions and prejudices about what beginnings can or cannot, or for that matter should, look like inevitably play a fundamental role in the assessment of the available evidence. Conservation of the intellectual substance, as is the case for material objects, has its flipside too.

The situation can be illustrated with a piece of information that is appropriate in the present context, since it connects the early history of the "soul" (*psukhê* in Greek) with a paradigmatic example of an "active matter."

Here is what Aristotle writes about Thales, whom he knew of only by hearsay (Thales did not write anything), in the first book of his treatise *On the Soul*:

"Thales too seems, from what is reported, to have thought that the soul is something that moves, for he says that the stone [referring here to the magnet] has a soul, given that it moves iron."[11]

Thanks to Diogenes Laertius (who, some five centuries later, relies on several intermediary sources), we are able to identify the author of the report Aristotle relies on, namely, the sophist Hippias of Elis, a contemporary of Socrates who had undertaken to collect the opinions of ancient poets and prose writers (Greek and non-Greek alike) in order to show their commonalities:

"Aristotle and Hippias say that he [i.e., Thales] attributed a soul to inanimate beings too, judging from the evidence of the stone from Magnesia [i.e., the magnet] and of amber."[12]

This piece of information, in spite of its brevity and the absence of any indication about the context in which Thales may have formulated the view, prompts some elementary observations and interesting questions.

First, we can be fairly confident that Hippias was familiar with the distinction between "animate" (*empsukhon*; lit., "with a soul") and "inanimate" (*apsukhon*; "without a soul"). As a matter of fact, the contrast between living human beings and the soul that departs to Hades at the moment of their death is already fully present in Homer (as the first lines of the *Iliad* show), and an explicit distinction between body and soul is attested in eschatological traditions, where Orphics cross Pythagoreans.[13] What is really at stake is less the distinction itself than the manner in which it is construed. There is no doubt that Aristotle's considered characterization of the soul as "the first actuality [*entelekheia*: "the coming to its end"] of a natural body having life potentially in it," to mention this epochal formulation (which, by the way, is deeply anti-Platonic and not dualistic in any straightforward sense), is not appropriate to speculate about what a Thaletian soul might have been.[14] But "life," which is a perfectly legitimate rendering of the original notion of the Greek word *psukhê*, and one that is still very much active in Plato and Aristotle, does capture the point.

Second, the fact that Hippias attributed to Thales the idea that inanimate items have a soul (i.e., are living beings) can mean more than one thing. It depends, first, on whether we attribute the word "inanimate" to Thales, sometime at the hinge of the seventh and sixth centuries BCE, or to Hippias, in the second half of the fifth century BCE, and, second, on whether the term "inanimate beings," in the phrase "Aristotle and Hippias say that he [i.e., Thales] attributed a soul to inanimate beings," refers to inanimate beings in general or only to some subset among them. (1) Thales may have ignored the distinction and considered that all that would later be classified as inanimate possesses a soul. Or (2) he may have been familiar with the distinction (although not necessarily expressed through the words *empsukhon* and *apsukhon*, which feature in Aristotle's and arguably already in Hippias's catalogue of parallels), and thought that what people normally consider as inanimate is in fact animate, taking the magnet as a privileged instance of a general situation. Or (3) he may have known the distinction but thought that some items that were usually thought to be devoid of soul (i.e., inanimate), as the magnet is, were in fact animate (plants, i.e.), on the basis of an implicit argument (later fully shared by Plato and Aristotle) in virtue of which a soul is an item that has the capacity to move (whether itself or other things).[15]

To decide between these different options is a challenging task and may be in the end impossible, although a series of reports lends some plausibility to the option according to which the magnet is a visible instance of a general but usually unnoticed state of affairs (option 2). These reports concern what is probably the most famous of the many sayings attributed to Thales:

"All is full of gods."

The following sequence of four testimonies, presented in inverse chronological order—from the most recent source to the older ones—is a clear example of the interpretive process to which archaic sayings such as this one were submitted in the course of the centuries, a clear instance of conservation by destruction. I have italicized the words that must have triggered in the first place the interpretive process (namely, the reference to Thales's saying); the terms or formula that I take to be the upshot of this interpretive process I italicize.[16]

T1 Diogenes Laertius 1.27 (third century CE)
[He, Thales, thought] that the world is animated [*empsukhon*; has a soul] and *full of divinities.*

T2 Cicero, *On the Nature of the Gods,* 1.10.25 (first century BCE)
For Thales of Miletus, who was the first to investigate these matters, said that water is the beginning of things but that god is the intelligence [*mens*] capable of making all things out of water.

T3 Aëtius 1.7.11 "On God" (going back to the third century BCE)
Thales: god is the intelligence [*nous*] of the world, but the whole is ensouled [*empsukhon*] and at the same time *full of divinities*; and a divine power passes also through the elementary moisture and moves it.

T4 Aristotle, *On the Soul,* 1.5, 411a7–8 (fourth century BCE)
Some people say that it [i.e., the soul] is mixed in the whole, which is perhaps also the reason why Thales thought that *all things are full of gods.*

The phrase "all is full of gods" is the most famous of the (many) sayings attributed to Thales, next to the idea that water is the origin of all that is. It is uncertain whether Aristotle was the first to link Thales's gods with souls (he might have relied, as in the preceding case, on Hippias, but evidence is not available in this case). Anyway, the move he makes in his report (T4), which is the oldest attributed evidence we have about the saying,[17] consists of suggesting a physical interpretation of Thales's dictum by reference to an anonymous doctrine according to which soul (apparently taken as a mass noun) is mixed "in the whole"—a typical example of an allegorical procedure that Aristotle embraces here with a characteristically Aristotelian caution ("perhaps"). We can't be sure about the implications of this anonymous doctrine, possibly of Orphico-Pythagorean origin, but the most probable is that it was linked to the fact that the word *psukhê*, taken in the sense of "life," was easily associated with breath and so with air.[18] In Aristotle's report, the phrase "in the whole," which paraphrases in a cosmological perspective Thales's encompassing "all things" (in the formula "all things are full of . . ."), is combined with the idea that Thales's "divinities" (*daimones*), which Aristotle calls "gods" (*theoi*), are souls. This double move opens avenues to stronger (and less cautious) interpretations, for it allows Thales's statement to be read in the typically Platonic terms of a cosmos animated by a world-soul, as happens in T1–T3. To be sure, this is not yet the view mentioned by Aristotle, since to be mixed in something does not boil down to an ensoulment of the thing in question. Still, we are making our way toward the formulas found in T1 ("ensouled world") and in T2 ("the whole is ensouled"). This last testimony (which is Cicero's) contains one further interpretative step, whereby "soul" stands for "intelligence" or "mind" (*mens*, translating the Greek *nous*). This identification points to an even stronger Platonizing reading of Thales's formula, with the integration into the world-soul of the Platonic Demiurge (who represents in Plato's *Timaeus* the organizing intelligence that Socrates was looking for in the *Phaedo*). Thales's first principle, water, now appears as a living, organizing intelligence, akin to the Stoic rational *pneuma* (another word that refers to breath).[19]

This series of interpretative steps, which are clearly identifiable, thus amounts to a Platonic-Stoicizing reading of Thales's phrase.[20] But what can we make of the phrase itself, once we strip it from its later receptions? Once again, to settle an original scope and meaning is probably beyond reach; pithy statements of this sort are by nature open-ended. To give it a sense, we would need a context, intellectual or more broadly cultural, which is basically lacking. We have to speculate: Maybe

Thales was endorsing, at a very general level, current polytheism, either by reference to the gods of his own city (Miletus) or, in a more Panhellenic and hence universalizing vein by reference to Hesiod's *Theogony*, where gods are so numerous that the world, in the most encompassing sense of the word, can be said to be full of them.[21] The perspective might be, in a more restrictive way, to a specific passage of Hesiod's *Works and Days* that presents the deceased members of the now-extinct golden race (the first of Hesiod's five races) as "fine divinities" now taking care of humans "everywhere upon the earth."[22] The point is that to the extent that something got lost with later appropriations, which is surely the case given that by definition what is appropriated ceases to be what it was before it was appropriated, it does not look as if what was lost in the process could offer a plausible alternative to the new determination, except under the general heading, perhaps, of a re-enchantment of the world—in whatever way we might want to flesh out this improbable but surely attractive program.

Terminological Issues

I now turn to the pair "matter"/"activity" itself. "Matter" (*hulê*), as opposed to "form" (*eidos*), is one of these numerous terminological innovations, still unknown to Plato, that were introduced by Aristotle and that have penetrated our everyday language far beyond technical vocabulary. When Plato speaks in the *Timaeus* about the ultimate "substrate" of sensible bodies, of which it has a clear notion although no word for it, he calls it *khôra*, which means "place"—a strange kind of place that is moving chaotically before the divine craftsman geometrizes it with its "consent," as Plato claims.[23] No wonder that ancient commentators debated about whether Plato's *khôra*/place was or was not animated, given that soul is for him the principle of movement. Aristotle successfully replaced this chaotic but eventually obedient "place" with the stiller *hulê*, a word whose original meaning, "wood," accords more immediately with the notion of a material to which a craftsman gives a certain form. The case of the abstract concept "activity," which strictly speaking corresponds to Aristotle's *energeia*—another neologism, ever more linguistically audacious than the metaphorical use of *hulê*—is somewhat more complicated.[24] For the present purpose, it will be enough to keep in mind that its root, *ergon* (the common word on which Aristotle builds *energeia*), primarily refers to the work shaped by a craftsman when he imposes a form on his material.

The fact that early Greek philosophers did not have the words "matter" and "activity" at their disposal does not mean that they did not have functional equivalents for them. Water, air, fire, earth, the four so-called elements that Plato and Aristotle took up from Empedocles, as mentioned before, are all over the place in Presocratic thought (especially in medical writings, many of which are extant, contrary to philosophical works). They are active in the unproblematic sense that they have active properties. The hot and fire heat, the watery and water evaporate. More interestingly, life initially emerges out of an elemental rotting—a scheme that will remain, even though in a marginalized form, in Aristotle's worldview (this is the doctrine of spontaneous generation). These active features of matter are often called "powers," *dunameis*, a fascinating term in its own right, since it is the one that Aristotle will choose in order to refer to matter itself (the marble), as being "in potentiality" (*en dunamei*) the item (the statue) of which the form is the shape. To this extent, activity does not constitute a strong dividing line between Preplatonic thinkers and Plato, apart from its being an abstract, second-order concept.

However, whereas it is true that there is no concept of passive matter in Presocratic thought insofar as elemental qualities and bodies are endowed with active properties, it is also the case that several thinkers make a distinction between a still-indeterminate material and a determining active principle that is active in another way than the active powers they possess. Thus, Parmenides, Empedocles, and Anaxagoras, to take three clear cases, operate with an equivalent of a Platonic Demiurge, that of a primary craftsman who organizes available materials and gives shape to the world.

Parmenides, as the majority of other Preplatonic thinkers, extensively develops, in the second part of his doctrinal exposition, an account of how the world came about and of its actual state. He does that on the assumption that there are two basic constituents of the world, which he calls *morphai*, one of the Greek words that means "form." These are fire and night, with their respective qualities and powers:

> For they [i.e., the mortals] have established two forms
> [*morphai*] [. . .]
> And they have divided their body into opposites and posited
> signs
> separate from each other: for the one, the ethereal fire of
> flame,
> Being mild, very light in weight, the same as itself everywhere,
> And not the same as the other one; and that one too, in itself,

> The opposite, night without without light [or: without knowledge], a dense and heavy body [*demas*: build][25]

And, later in the poem:

> But since all things have been named light and night
> And what belongs to their powers [*dunameis*] is assigned to these and to those,
> The whole is altogether full of light and of ungleaming night,
> Both of them equal, since nothing is amidst either of them.[26]

Thus, the world and its population, from stars to living beings, result from the activity of a goddess whose function is to trigger the mixture of these constituents, which in this sense are passive:

> And in the middle of these, the divinity who steers all things.
> For she begins the hateful birth and mingling of all things,
> Leading the female to mingle with the male and again, in the opposite direction,
> The male with the female.[27]

It is possible, although not certain, that Parmenides named his demiurgic divinity Aphrodite, the goddess of love, but its working is in any case akin to hers. Empedocles's physics is easily construed as an extension and complexification of Parmenides's scheme. Beyond the four roots, fire, ether (air), water, earth, which Empedocles substitutes, not without implicit criticism, for Parmenides's pair of bright fire and obscure night (not only are they four in number rather than two, allowing a fuller range of explanations, but they are roots endowed with a life of their own, rather than simple forms), there are two demiurgic forces, instead of a just one—strife next to love:

> For as I already said, when I was clearly indicating the boundaries of my words,
> Two-fold is what I shall say: for at one time they grew to be only one
> Out of many, at another time again they separate to be many out of one,
> Fire, water, earth, and the immense height of ether;
> And baleful Strife is separate from them, equivalent everywhere,
> And Love in them, equal in length and in breadth.[28]

The work of love is explicitly compared to that of specific craftsmen:

> As when painters color many-hued sacrificial offerings,
> And, by reason of their skill, very expert in their art,
> They grasp many-colored pigments in their hands,
> Then, having mixed them in harmony, the ones more,
> the others less,
> Out of these they compose forms similar to all things,
> Creating trees, men, and women,
> Wild beasts and birds, water-nourished fish,
> And long-lived gods, the greatest in honors:
> In this way may your mind not succumb to the error that it is
> from elsewhere [sc., than from the four elementary roots]
> That comes the source of all the innumerable mortal things
> whose existence is evident.[29]

To be sure, the two demiurgic powers active in the world are material, since Empedocles thinks of them as being extended and acting locally.[30] On the other hand, Parmenides's and Empedocles's basic constituents—fire and night for Parmenides; air, fire, earth, and water for Empedocles—are endowed with a cognitive power. Aristotle's disciple Theophrastus tells us that much about Parmenides in his doxography *On the Sensations*:

> The fact that he [i.e., Parmenides] explains sensation by one of the contraries [i.e., the cold and the hot] taken by itself is clear from the lines in which he says that a corpse does not perceive light, heat, and sound, because fire has withdrawn, but does perceive cold, silence, and the contraries, and in general that everything that exists possesses some knowledge.[31]

As for Empedocles's four "roots," they have feelings and thoughts. Here is what was arguably the end of his physical poem:

> For if, leaning upon your firm organs of thought [*prapides*]
> With pure efforts you gaze upon them benevolently,
> They [i.e., the roots] will all be present to you throughout
> your lifetime
> And many other good things will come to you from them.
> For these themselves

> Are what makes each thing grow in one's character, according
> to each person's nature.
> But if you yourself covet different things, such as those that
> among men are
> Countless miseries that blunt their thoughts,
> Certainly they will abandon you quickly, as the time revolves,
> In their desire to rejoin the race that is theirs.
> For know that all things feel [*phronêsin ekhein*] and have their
> share of thought [*noêma*].[32]

Roots [*rhizômata*] can thus be referred to by the names of traditional gods—

> Hear first of all the four roots of all things:
> Zeus the gleaming, Hera who gives life, Aidoneus,
> And Nestis, who moistens with her tears the mortal fountain.[33]

This is not, however, the end of the story, which shows an ever-greater level of complexity: for while there are ways of reading Empedocles's poem as being committed to a notion of active matter both at the level of the two demiurgic forces and at the level of the four rhizomatic constituents, it is also the case that Empedocles made room, in the Orphic tradition, for fallen divinities that transmigrated from body to body, animal and human alike:

> For as for me, once I was already both a youth and a girl,
> A bush and a bird, and a sea-leaping, voyaging fish.[34]

The Platonic tradition, in the wake of the *Phaedrus*, whose eschatological myth (Socrates's second discourse in praise of love) is an expanded rewriting of Empedocles's *Purifications*, did not hesitate to assimilate these fallen divinities to souls. In other terms, it took Empedocles to be a radical dualist too. And this is a possibility that one must consider seriously, although the terms in which the dualism in question is couched are certainly different from those we usually operate with.

The situation is simpler with Anaxagoras, the thinker who is probably closest to a matter/mind distinction, although his "mind" (*nous*) is, again, a certain stuff, and Anaxagoras has of course no word for "matter." Here is the most extended prose passage from a Presocratic philosopher that has reached us:

> The other things possess a portion of everything, but mind [*nous*] is unlimited and master of itself, it has not been mixed with any thing, but is the only one to be itself by itself. For if it were not by itself, but had been mixed with some other thing, it would participate in all things, if it had been mixed with any; for in everything is present a part of everything, as I said earlier. And the things that would be mixed with it would prevent it from having control over anything in the same way as it does, being alone by itself. For it is at the same time both the thinnest of all things and the purest, and in particular it retains the full decision [or: understanding] concerning everything and possesses the greatest power; and of the things that have life, whether they are larger or smaller, of these mind is master; and mind has been master of the whole rotation, so that there would be rotation at the beginning. And the rotation began at first from the small, then it rotates more broadly, and it will continue to become even broader. And the things that mix as well as those that are detached and separate out—all these mind decided [or: knew]. And as things were going to be and as all things were that now are not, and as all things are now and as they will be, mind separated and ordered them all, as well as this rotation, which is being performed now by the heavenly bodies, the sun, the moon, the air, and the ether, which are separating out. And the rotation itself caused the detachment. And from the rarefied the dense separates out, from the cold the warm, from the dark the bright, and from the moist the dry. Numerous are the parts of numerous things; yet nothing is completely detached or separates out from one another, except mind. But all mind is similar, the larger and the smaller, and nothing else is similar to anything else, but that of which each thing contains the most, this it is and was most manifestly each thing.[35]

One can understand that Plato's Socrates, at hearing this passage (arguably the one that he alludes in the *Phaedo*), was filled with hope, before experiencing disappointment on further reading.

An Aftermath

I would like to conclude this summary introduction to the question of how to frame the question of active matter in Presocratic thought in a differentiated, nonbinary way by evoking a debate that took place in the seventeenth and eighteenth centuries and that is reflected in the works of Cudworth (*The Intellectual System of the Universe*, 1678) and Diderot (most famously in *The Dream of d'Alembert*, written if not published in 1769). The debate turns around the notion of hylozoism, the view according to which matter/*hulê* possesses life/*zôê*.

As far as I know, the Cambridge Platonist Cudworth was the first to use the term. Hylozoism is for him the great foe, the worst form of atheism, even more dangerous than materialistic atomism. This is because materialistic atomism is easy enough to refute, being incapable of explaining in a satisfactory manner life and life-related phenomena. Hylozoism, on the other hand, claims to explain them by endowing matter with life, instead of recognizing, in the Platonic tradition, that life has an "intellectual" source (Cudworth's word)—meaning the soul, and ultimately God. Hylozoism is what Diderot, without using the name, would embrace one century later in his *Dream of d'Alembert* by endowing matter, and more precisely Epicurean atoms, with sensibility. In so doing, he triggered the nightmare of his interlocutor d'Alembert, who in the real world was a hardcore dualist and passionate defender of Descartes's substantial distinction between body and soul. It should be added for the sake of precision that Diderot distinguished—by analogy with Leibniz's living and dead forces, which d'Alembert referred to in his article "Force" of the *Encyclopédie*—between a living life, as opposed to a dead life, which is nonetheless a life.[36] The only point that interests us here, however, is that Diderot's atomic materialism is not materialistic, but hylozoistic in a sense that would have been a nightmare for Cudworth the Platonist as it was for d'Alembert the Cartesian.

Now, neither Cudworth nor Diderot refers to the Presocratics as representing the hylozoistic position they fought against (in Cudworth's case) or supported (in Diderot's case). It is especially instructive that Cudworth, who had an extended knowledge of the ancient tradition, Presocratics included, does not think of them when referring to the threat raised by hylozoism to true religion.[37] On the contrary, atomists aside, Presocratics are for him respectable heirs of Moses's wisdom. For they posit, like Moses himself (according to an interpretive tradition that one can trace back to Philo of Alexandria at least), a basic distinction between intellectual and material causes, all of them Cartesians (and Cudworthians and Platonicians) *avant la lettre*. By contrast, hylozoism is represented for him by Strato of Lampsacus, a disciple of Aristotle, whose "phantom" he sees revived in the doctrine of Spinoza, which famously identifies God with Nature (*Deus sive Natura*).[38] In Cudworth's words, "his [Strato's] God [is] no other than such a life in matter, as if he was both devoid of sense and consciousness."[39]

So Cudworth, the Cartesian-Platonist, can be counted among those who were able to find in the Presocratics conceptual resources to fight against modern Spinozism. Even if his reading of the evidence of most Presocratics as resolute post-Mosaic dualists belongs to a curiosity cabinet, it can still serve as a warning apologue suggesting that continuities between Presocratics and Plato are stronger than we might be tempted to think and that an important aspect of Presocratic thought has been able to survive, in spite of the joint effects of material and philosophical destruction.

NOTES

1. See, respectively, Cicero, *Tusculan Disputations* 5.3, and, apart from Plato's *Phaedo*, Aristotle, *Metaphysics* 1.987b2–4. By contrast, Socrates is included in the eighth volume of the collection *Early Greek Philosophy*, edited by Glenn W. Most and me (henceforth LM). All translations from the Presocratics are taken from this edition, with occasional minor variations. References are also made to the standard sixth edition of the fragments of Presocratic philosophers by H. Diels and W. Kranz (henceforth DK).
2. Plato, *Phaedo* 96a6–99c8.
3. Ibid., 99d1.
4. Ibid., 99c8–100a8, trans. Emlyn-Jones and Preddy (Loeb Classical Collection).
5. I have explored in Laks, *Concept of Presocratic Philosophy*, the historical and philosophical issues that are at stake in the question of denomination of early Greek thinkers.
6. See Frag. B6 DK, 153 *ad* n. 33.
7. For a dissenting view, Crowley, "On the Use," argues that letters, in relevant texts (in particular *Metaphysics Delta*, 3 1014a26–b15) are only an example among others of ultimate parts constitutive of a given whole.
8. Gadamer, *Gesammelte Werke*, 34. The quotations could be multiplied; see, for example, 59–60.
9. Cassirer's view of the development of Greek philosophy, which is indebted to Hegel, is surely dated, but it offers a principled alternative both to historical positivism and to the criticisms that Dilthey and Yorck had leveled against it, thus opening the way to Heidegger's concept of "historiality." On this constellation, see Laks, *Concept of Presocratic Philosophy*, chap. 6; Laks *Historiographies*, chaps. 1, 8.

10. "And the things out of which birth comes about for beings, into these too their destruction happens, *according to obligation: for they pay the penalty [dikên] and retribution [tisis] to each other for their injustice [adikias]* according to the order of time." 12 B1 DK = D6 LM. In the lines just quoted, the italicized words are the ones that can be attributed to Anaximander himself. The rest of the sentence is tainted with Peripatetic vocabulary. The source for the quote is Simplicius, the Neoplatonic commentator of Aristotle (sixth century CE; i.e., twelve centuries after Anaximander), who has it from Aristotle's disciple Theophrastus (end of the fourth century BCE).
11. 1.2.405a19–21 = D11a LM.
12. Diogenes Laertius, *Life of Eminent Philosophers*, 1.24 (= D11b LM). An original fragment of Hippias's work, the title of which was "Collection" (with the idea of "bringing together," *Sunagôgê*), has been preserved by Clement of Alexandria, a Christian author (second half of the second century CE) who was keen about showing that the Greeks practiced plagiarism. It obviously belongs to the beginning of Hippias's book and summarizes its encyclopedic project (86B6 DK = D22 LM):

    > Of these [ancient opinions] some have doubtless been expressed by Orpheus, others by Musaeus, concisely and by each one in a different place; some by Hesiod, others by Homer, others by the other poets, some in prose treatises, some by Greeks, others by non-Greeks. But I myself have put together from out of all these the ones that are most important and are akin to one another, and on their basis I shall compose the following new and variegated discourse.

    This work by Hippias is a missing link between Aristotle and the Presocratics.
13. Achilles's anger "hurled in their multitudes to the house of Hades strong souls / of heroes, but gave their bodies to be the delicate feasting / of dogs, of all birds," *Iliad* 1.3–5 (trans. Lattimore). In Homer, sôma regularly refers to a corpse, which does not mean that the word *means* exclusively "corpse" in archaic texts. On this interesting semantic issue, see Renehan, "Meaning of ΣΩΜΑ." A straightforward distinction between body and soul features in the Orphic tablets discovered in Olbia, which are the basis of the notion that the body (*sôma*) is a prison (*sêma*) for the soul (of Orphic origin too). For the tablets, see Bernabé, "Etimología platónica," 219. On the destiny of souls after death, see, in addition to the Orphic material edited in Bernabé and Cristóbal, *Instructions for the Netherworld*, Pindar, *Second Olympic Ode*, 56–83, and his Frag. 131b Maehler, as well as Herodotus's description (*The Persian Wars* 2.123) of Egyptian and Pythagorean views on reincarnation. Cf.

more generally Bremmer, "Rise of the Unitary Soul," and (from a broad anthropological perspective) Bremmer, "Body and Soul."

14. Aristotle, *On the Soul* 2.1.412a27f (trans. Smith in the *Revised Oxford Translation,* slightly modified).
15. Explaining the properties of the magnet (also called the "Heraclean stone") is a significant topic of discussion in early (and later) Greek philosophy. Alexander of Aphrodisias, Aristotle's leading commentator (end of the second or third century CE), devotes a chapter to the question in his *Problems and Solutions or Natural Questions* (book 2.23). The first part of the chapter comprises an exposition and refutation of Empedocles's, Democritus's, and Diogenes of Apollonia's accounts of the phenomenon. On Alexander's own explanation (which mobilizes "desire"), see Kupreeva, "Aristotelian Dynamics," 88ff. On magnets in antiquity in general, see Radl, *Der Magnetstein.*
16. I am here using some of the commentary in Laks, "How Preplatonic Worlds," and Laks, "Souls before Plato." T1 = R34b LM; T2 = R38LM; T3 = R35 LM; T4 = R34a LM.
17. Plato alludes to the saying without naming Thales at *Laws* 10, 899b9. Besides T4, Aristotle mentions it in *Generation of Animals* 3.11, 762a21. Cf. also Cicero, *On the Laws* 2.26, Stobaeus, *Anthology* 1.1.29b. It is not included in lists of Thales's sayings that have been transmitted (see MOR. T35 [4] LM and THAL. P16–P17 LM).
18. "[T]he feminine substantive *psukhē,* despite its already diverse uses in Homer, originally signifies the 'breath' that is present in a human while alive and departs temporarily during fainting and permanently in death." Jouanna, "Le souffle, la vie," 203. My translation. Bernabé links T4 with Orphism. "Orfismo," 1142. A bit earlier in the same chapter, Aristotle had mentioned "the so-called Orphic poems" as holding a doctrine that links the soul, cosmic breathing, and winds (410b26–411a2), which may or may not help identify the anonymous authors of T4. In any case, the idea of a cosmic breathing is also attested in a fragment of Aristotle's lost book On the *Philosophy of Pythagoras* (Frag. 201 Rose = ANON. PYTH. D30 LM). The question of the relationships between Orphics and Pythagoreans is obscure and much debated (for an overview of the question, see Bernabé, "Orphics and Pythagoreans").
19. The verb "to pass through" also suggests Stoic influence.
20. Some interpreters, however, take the evidence at face value. See, for example, Naddaf, "Spontaneous Generation," fighting the traditional picture of Thales as a material monist (ultimately derived from Aristotle): "My own contention here is that for Thales the kosmos or the present world order (universe and its contents) originated from a living

conscious primordial substance. Of course, there are those who argue that it originated from a living unconscious primordial substance. But [etc.]." Although I don't belong to the latter category either (for reasons that need not be formulated here), it will be clear that my reading of the evidence is rather different from Naddaf's. It seems fairly clear that Thales was no substantialist in the first place, not even a material substantialist, let alone a psychological one. What needs to be explained is the scholarly drive to make him hold such positions.

21. See Philippson, *Genealogie*, who lists among Hesiodic gods natural entities (sky and Earth, stars, rivers and streams), anthropomorphic powers (Eros, Eris/Quarrel), aspectual characterizations (Thauma/Wonder and Theia/Divine), and normative forces (Themis/Justice, Nemesis/Indignation, Mnemosyne/Memory).
22. Hesiod, *Works and Days* vv. 122–25. "By the plans of great Zeus they are fine spirits upon the earth [. . .] clad in invisibility, walking everywhere upon the earth" (trans. G. W. Most).
23. Plato, *Timaeus* 56c5f.
24. The term *energeia* is attested for the first time in Aristotle's *Protrepticus*; cf. especially Frag. 79 Düring = 14 Ross: "living [*to zên*] seems to be said in two ways, one in the sense of *dunamis* [power, capacity] and the other in the sense of an *energeia*." It originally refers to the activity; that is, the actual exercise of a cognitive power. Menn, "Origins of Aristotle's Concept," 78–81.
25. B8.53–56 DK = D8.58–61LM.
26. B19 DK = D13 LM.
27. B12.3–6 = D14 LM.
28. B17.15–20 = D73.246–51 LM. *Aither* is Empedocles's word for what we call "air." It is translucent and bright. *Aêr*, in Greek, tends to be associated with humidity and obscurity.
29. B23 DK = D60 LM. Speaking of the painters, Empedocles uses a Greek verbal form (the dual) that usually implies that there are two subjects, although it may also serve as an alternate form for the plural. Strife does not "mix" the roots as painters mix pigments, since its function is dissociation. But it definitely has a role to play, even though it is a negative one, in the shaping of the world. The ambiguity of the dual might be intentional.
30. See "equal in length and in breadth" in the first of the two Empedoclean quotes.
31. Theophrastus, *On Sensations* § 3 = PARM. D52 LM.
32. B110 DK = D257 LM.
33. B6 DK = D57 LM. Whereas the four divinities surely represent

Empedocles's four roots, there is a discussion and much confusion, since antiquity, about who is who. Picot, in "Empédocle magique," discusses the intricate evidence and brings strong arguments in favor of the view shared by a majority of interpreters (Zeus = fire, Hera = air, Aidoneus; i.e., Hades = earth, Nestis = water).

34. B117 DK= D13 LM. Whether this happened in a separate poem titled *Purifications* or was part of the poem traditionally called *On Nature* is a disputed question. Most recent commentators incline, on the basis of a papyrological evidence published in 1999, toward the one-poem-one-story theory (most recently Trépanier, "Empedocles on the Origin"). My own view is that there were two distinct but tightly related stories, whether in two separate poems (as other interpreters think and as is edited in LM), or possibly juxtaposed in a single one. For the issues at stake, see Laks, "*Hystéron protéron.*"
35. B12 DK = D27 LM.
36. Cf. d'Alembert, "Force," in *Encyclopédie,* 112. In Diderot's *Dream of d'Alembert,* d'Alembert objects at the beginning of the conversation, "But what is the relationship between movement and feeling? Is it by chance that you would recognize an active feeling and an inert feeling, as there is a live force and a dead force?" (Mais quel rapport y a-t-il entre le mouvement et la sensibilité ? Serait-ce par hasard que vous reconnaîtriez une sensibilité active et une sensibilité inerte, comme il y a une force vive et une force morte ?) My translation. On the reception Leibniz's ideas in eighteenth-century France, see Winter, "Diderot und Leibniz" (specifically on Diderot, 67–69).
37. On Cudworth and the Presocratics, see Osborne, "Ralph Cudworth."
38. Breteau, "Fantôme de Straton."
39. *True Intellectual System,* book 1, chap. 3, 107. Cudworth's main source is Cicero (Cicero, *Academica* 2.121 = Strato Frag. 18 Sharples):

> You [Stoics] say that nothing can [come to be] without god; look, here is Strato of Lampsacus, who gives this god a release from a great task, indeed. [. . .] He says that he makes no use of work by the gods in constructing the world, and teaches that whatever there is is all brought about by nature, and not in the same way as he who says that all these things are composed of rough and smooth and hooked and barbed bodies interspersed with void; he says that there are the dreams of Democritus.

"Strato, the one who is called 'the naturalist,' [. . .] holds the view that all divine power is located in nature, which is responsible for generation, growth and decay, but is without any consciousness and without any shape." Cicero DND I. 35 = Strato Frag. 19 Sharples.

BIBLIOGRAPHY

Aristotle. *On the Soul.* Translated by J. A. Smith. In *The Complete Work of Aristotle: The Revised Oxford Translation*, edited by J. Barnes, 1:641–92. Princeton, NJ: Princeton University Press, 1984.

Bernabé, Alberto. "Una etimología platónica: ΣΩΜΑ ΣΗΜΑ." *Philologus* 139 (1995): 204–35.

———. "El Orfismo y los demás filósofos presocráticos." In *Orfeo y la tradición órfica. Un reencuentro*, 2 vols., edited by Alberto Bernabé and Francesc Casadesus, 1141–60. Madrid: Ediciones Akal, 2008.

———. "Orphics and Pythagoreans: The Greek perspective." In *On Pythagoreanism*, edited by Gabriele Cornelli, Richard McKirahan, and Constantinos Macris, 117–51. Berlin: De Gruyter, 2013.

Bernabé, Alberto, and Ana I. Jiménez San Cristóbal. *Instructions for the Netherworld: The Orphic Gold Tablets.* Leiden: Brill, 2008.

Betegh, Gábor. "On the Physical Aspect of Heraclitus' Psychology (with New Appendices)." In *Doctrine and Doxography: Studies on Heraclitus and Pythagoras*, edited by David Sider and Dirk Obbink, 225–61. Berlin: De Gruyter, 2013.

Bremmer, Jan N. "Body and Soul between Death and Funeral in Archaic Greece." In *Ultimate Ambiguities*, edited by Peter Berger and Justin Kroesen, 227–51. New York: Berghahn, 2015.

———. "The Rise of the Unitary Soul and Its Opposition to the Body: From Homer to Socrates." In *Philosophische Anthropologie in der Antike*, edited by Ludger Jansen and Cristoph Jedan, 11–29. Frankfurt: Ontos, 2010.

Breteau, Jean-Louis. "'Le fantôme de Straton s'est récemment remis à marcher': Les platoniciens de Cambridge et l'hylozoïsme." *Kairos* 28 (2006): 45–72.

Crowley, Timothy J. "On the Use of *Stoicheion* in the Sense of 'Element.'" *Oxford Studies in Ancient Philosophy* 29 (2005): 367–94.

Cudworth, Ralph. *The True Intellectual System of the Universe: The First Part; wherein, All the Reason and Philosophy of Atheism is Confuted; and Its Impossibility Demonstrated.* London, 1678. Reprinted Stuttgart: Fromann Verlag, 1964.

D'Alembert, Jean Le Rond. "Force." In *Encyclopédie ou dictionnaire raisonné des sciences, des arts et des métiers.* Vol. 7. Paris, 1757.

Diderot, Denis. *Éléments de physiologie.* Edited by P. Quintili. Paris: H. Champion, 2004.

Diels, Hermann, and Wilhelm Kranz. *Die Fragmente der Vorsokratiker.* 6th ed. 3 vols. Berlin: Weidmann, 1951–52.

Gadamer, Hans-Georg. *Gesammelte Werke.* Vol. 6, *Griechische Philosophie: II.* Tübingen: J. C. B. Mohr, 1985.

Hesiod. Edited and translated by G. W. Most, 2 vol. Loeb Classical Library. Cambridge, MA: Harvard University Press, 2006.

Homer. The *Iliad* of Homer, translated by Richmond Lattimore; introduction and notes by Richard Martin. Chicago: University of Chicago Press, 2011.

Jouanna, Jacques. "Le souffle, la vie et le froid: Remarques sur la famille de ψύχω d'Homère à Hippocrate." *Revue des études grecques* 100 (1987): 203–24.

Kupreeva, Inna. "Aristotelian Dynamics in the 2nd Century School Debates: Galen and Alexander of Aphrodisias on Organic Powers and Movements." *Bulletin of the Institute of Classical Studies* 47 (2004): 71–95.

Laks, André. "Aristote et Thalès: Un récit allégorique?" In *Histoire, Doxographie, Vérité. Études sur Aristote, Théophraste, et la philosophie présocratique,* 160–66. Louvain-la-Neuve: Peeters, 2007.

———. *The Concept of Presocratic Philosophy: Its Origin, Development, and Significance.* Translated by G. W. Most. Princeton, NJ: Princeton University Press, 2018. Originally published as *Introduction à la "philosophie présocratique,"* 2004.

———. *Historiographies de la philosophie antique. Neuf études.* Paris: Les Belles-Lettres, 2021.

———. "How Preplatonic Worlds Became Ensouled." *Oxford Studies in Ancient Philosophy* 56 (2019): 1–34.

———. "*Hystéron protéron. Des Origines aux Purifications.*" In *La philologie au présent. Pour Jean Bollack,* edited by Christoph König and Denis Thouard, 19–26. Coll. Cahiers de Philologie. Villeneuve-d'Ascq: Presses Universitaires du Septentrion, 2010.

———. "Souls before Plato: Five Short Doxographical Studies." In *Cosmology and Biology in Ancient Philosophy,* edited by Ricardo Salles, 8–28. Cambridge: Cambridge University Press, 2021.

Laks, André, and Glenn W. Most. *Early Greek Philosophy.* 9 vols. Loeb Classical Library. Cambridge, MA: Harvard University Press, 2016.

Menn, Stephen. "The Origins of Aristotle's Concept of *Energeia: Energeia* and *Dunamis.*" *Ancient Philosophy* 14 (1994): 73–114.

Naddaf, Gerard. "Spontaneous Generation and Creationism in Presocratic Monism in Light of Aristotle's Analysis in the *Physics.*" *Anais de filosofia clássica* 5, no. 8 (2010): 23–40.

Osborne, Catherine. "Ralph Cudworth, *The True Intellectual System of the Universe.*" In *The Presocratics from the Latin Middle Ages to Hermann Diels,* edited by Oliver Primavesi and Katarina Luchner, 215–35. Stuttgart: Steiner Verlag, 2011.

Philippson, Paula. *Genealogie als mythische Form. Studien zur Theogonie des Hesiod.* Oslo: A. W. Brøgger, 1936. Reprinted in *Hesiod.* Edited by Ernst Heitsch. Wege der Forschung 44. Darmstadt: Wissenschaftliche Buchgesellschaft, 1966.

Picot, Jean-Claude. "L'Empédocle magique de P. Kingsley." *Revue de philosophie ancienne* 18 (2000): 25–86.

Plato. *Euthyphro; Apology; Crito; Phaedo.* Translated by Chris Emlyn-Jones and William Preddy. Loeb Classical Collection. Cambridge, MA: Harvard University Press, 2017.

Radl, Albert. *Der Magnetstein in der Antike. Quellen und Zusammenhänge,* Stuttgart: F. Steiner Verlag Wiesbaden, 1988.

Renehan, Robert. "The Meaning of ΣΩΜΑ in Homer: A Study in Methodology." *California Studies in Classical Antiquity* 12 (1979): 269–82.

Trépanier, Simon. "Empedocles on the Origin of Plants: PStrasb. gr. inv. 1665–1666, Sections d, b, and f." In *Presocratics and Papyrological Tradition: A Philosophical Reappraisal of the Sources,* edited by Christian Vassallo, 271–98. Berlin: De Gruyter, 2019.

Winter, Ursula. "Diderot und Leibniz. Die Leibniz-Rezeption in der Naturphilosophie der französischen Aufklärung." *Studia Leibnitziana* 36 (2004): 57–69.

EIGHT

Active Matter: A Philosophical Aberration or a Very Old Belief?

Guido Giglioni

Abstract

This chapter provides a short history of Western philosophical views on active matter, from antiquity to the early modern period. The emphasis is on the conserving power of life, associated with matter's ability to either receive energy or even produce it by itself. Three specific domains of philosophical and scientific inquiry are taken into account: the existence of a primeval energy buried in the world of mineral nature; the kind of vegetative activity immersed in matter that is displayed by plants; and the astonishing ability to shape bodies that the animated principle of generation demonstrates through the many complex ways in which animals reproduce themselves. These different occurrences of active and acting nature offer us three paradigmatic scenarios that help shed light on human attitudes toward active matter.

Active Matter: Hylozoism and Panpsychism

To begin this excursion into the history of Western views on living matter, let us imagine a network of primordial forces—of a physical and chemical nature—acting in opposition to each other. Through their unyielding oppositions, they bring about all the material phenomena that fall under our senses. We can call these forces "powers." The most stable and recurring concretions that are being generated by their tensed oppositions become the first "elements," each of them throbbing with inner energy. From the combination of the primordial forces and first elements a whole universe of interacting "fluids" emerges. Reality would thus be the result of a fundamentally liquid and yet self-balancing arrangement of numberless interdependent parts. The explanatory principle that underlies this view of reality is the idea of a relentless amalgamation of different forces and parts: "mixture." In this liquid cosmos, substance—the stuff of which things are made—would be brought to life by the constant action of the original forces. This means that, from an ontological point of view, the category of relation would come before that of being. It is a dynamic account of things in which matter is perceived as a living reality. Historically speaking, this view was in fact elaborated by several observers of natural and cultural phenomena active in fifth-century BCE Ionia (the coastal area in western Asia Minor corresponding to present-day Turkey). They were called φυσικόι/*physikoi*, a term that at the time could mean both "physicians" and "natural philosophers." The Hippocratic school—this is how this particular group of itinerant savants came to be known in antiquity—assumed that the physical forces of matter—(δυνάμεις/*dunameis*)—were the primordial powers shaping the material universe—the world of nature (φύσις/*phusis*)—through the mixing (κρᾶσις/*krasis*) of primary elements (στοιχεῖα/*stoikheia*) and primary fluids (χυμόι/*khumoi*).[1]

To be sure, there is a certain latitude in the Hippocratic literature, but some of the treatises espouse a strongly vitalistic explanation of the universe that influenced the tradition of learned medicine down to the eighteenth century, in cultural settings as varied as the University of Montpellier in France, the Scottish Enlightenment in Britain, and pietist medicine in Germany, to mention only a few examples of the persistence of Hippocratic vitalism.[2] Galen, the extraordinarily influential physician and philosopher who lived in the second century CE, was the author who systematized this way of looking at matter as a flexible system of relations based on the interactions of a limited number of original forces.[3] During the Middle Ages, the universities in Europe codified this approach to the study of matter and nature with the label *via medicorum*, the "way of the physicians," often contrasted with the parallel and complementary "way" of the philosophers, the *via philosophorum*.[4]

Here it is important to stress that the medical dunameis—that is, the physical forces of matter (heat, cold, moisture, and dryness)—frame the cosmos from within, for they are described by the Hippocratic and Galenic doctors in terms of selective propensities capable of distinguishing between what preserves the continuity of their own being and what may jeopardize the regular fulfilment of their vital operations. Galen formulated this point most clearly when he said, "Nature is not posterior to the bodies, but is a long way prior to them and older than they." This nature shapes bodies of plants and animals "by virtue of certain powers which it possesses: they are, on the one hand, attractive and assimilative of what is appropriate, and, on the other, expulsive of what is foreign. Further, nature skilfully moulds everything during the stage of generation."[5]

In this view, the idea of vital conservation is key. If the balance between assimilation and rejection is not consistently preserved, nature ceases to be alive. What is more, no record of previous attempts at self-preservation would be retained and kept in the vast body of nature. Medical hylozoism—the theory that physical forces in nature have the ability to sense all that is different from them and to react accordingly—is only one of the various ways in which Western thinking has conceived of active matter. According to this view, the sense of what one being is (identity) and the sense of what it is not (difference) are inscribed in the very matter of things, and this explains their tendency to self-preservation. The "self" in the locution "self-preservation" indicates that there is room for the idea of a biological selfhood. Life is matter (in that it builds its own bodily reality) and matter is life (in that bodies are lives, although not necessarily fully conscious lives). In a hylozoistic framework full awareness can only be attributed to the whole (prime matter), not to the parts, and as I will suggest at the end of this essay, hylozoistic awareness is a very problematic notion of consciousness.[6]

In order precisely to avoid these consequences, other thinkers and schools have connected life to matter by positing the existence of external influences capable of activating a substratum that is deemed to be originally inert and lifeless. Some historical instances of this approach (panpsychism) include Judeo-Christian-Islamic narratives of the creation; Platonic notions of demiurgy and cosmic animation; Aristotelian varieties of radical hylomorphism, championed (with rather different outcomes) by followers of either Alexander of Aphrodisias (fl. 200 CE) or Ibn Rushd, known in the Latin West as Averroes (1126–98); and a variety of Renaissance philosophers advocating models of universal animation, such as Giordano Bruno (1548–1600), Tommaso Campanella (1568–39), and Francesco Patrizi (1529–97). Atomist philosophies, finally, starting with Leucippus's and Democritus's accounts of nature, developed traditions of vital corpuscularianism in which atoms are described as material elements endowed with motion from all eternity. Ultimately, the key question that differentiates all these various accounts of active matter is whether matter can be said to have life of its own because its nature is inherently active (hylozoism) or whether it is wholly animated by an external source of activity (panpsychism).[7] Obviously, hylozoism and panpsychism postulate different understandings of conservation depending on whether the source of preserving energy comes, respectively, from within or from without. Conservation as the act of retaining information and keeping the memory of things alive is at the core of matter understood as active substance. Conservation is the cultural side of what in early modern medicine and political theory was called self-preservation (*conservatio sui*). Hylozoism, as a philosophical perspective that shuns dualistic solutions, assumes the identity of conservation and preservation. Matter would not be active if it weren't self-sufficient, that is, endowed with the power to retain knowledge, repair injured life, and restore itself. Hylozoism traces the millennial history of life's attempts at expanding knowledge and homeostasis in nature.

In my contribution, to provide a very short history of Western attitudes toward matter and its life, I will focus on three specific domains of philosophical and scientific inquiry in which the issue of active matter has caught the imagination of investigators throughout the centuries: the existence of a primeval energy buried in the world of mineral nature; the kind of vegetative activity immersed in matter that is displayed by plants; and the astonishing ability to shape bodies that the animated principle of generation demonstrates through the many complex ways in which animals reproduce themselves. These different occurrences of active and acting nature provide us with three paradigmatic scenarios that help shed light on human attitudes toward active matter.

Alchemy, or the Matter of Relations

Of the many disciplines through which human beings have tried to unearth the mysteries of matter, alchemy is one of the most ancient and richest in cultural, anthropological, and social meanings. Economic and technological concerns may have enhanced the desire to control and exploit the forces hidden in mineral nature. However, a variety of natural-philosophical, theological, and metaphysical pursuits have also accompanied the history of this time-honored practice.[8] Through artificial means, alchemy aimed at producing imitations of natural things, sometimes transforming these things into each other or even making them anew. For reasons of space, here I will limit my account to a particular moment in the history of chemical enquiries:

Paracelsian alchemy. The reasons are principally two: first, the work of Paracelsus (1493/94–1541) and his followers between the sixteenth and eighteenth centuries testifies to a distinctive confluence of practical, religious, and speculative motives; second, Paracelsus's rethinking of chemical concepts and practices represents a veritable watershed in the history of philosophical and cultural uptakes of alchemical traditions.

The Paracelsian ontology of matter is richly layered and textured. Life pervades its material substratum at multiple levels. Natural bodies are living entities that are activated by invisible seeds and directed by vital principles, which Paracelsus called *archei.* Celestial influences keep the process of animation going without disruptions. Matter is stratified not only from the point of view of its inner life but also in terms of corporeal consistency. Conservation is ensured by a calibrated interplay of external and internal forces established at the very beginning of the universe when God created it. The material continuum is maintained in its condition of vital activity by uninterrupted processes of rarefaction and condensation. Aristotle's four elements (a remnant of the Ionian science mentioned at the beginning of this article) are seen by Paracelsus as matrices, that is, cosmic wombs out of which God extracted living beings by mixing active seeds with the body (the three elements of sulfur, mercury, and salt). Within this picture, matter is alive, but both matter and life have been created by God, who remains the ultimate alchemist. Theologically speaking, prime matter in Paracelsus's cosmology is the *limus terrae* from which God fashioned elemental and sidereal bodies. While the elemental body includes seeds and elements, the sidereal body results from the active force animating everything in nature. One might characterize Paracelsus's cosmos as belonging to a panpsychist view of active matter; historian Dane Daniel, however, has rightly spoken of "spiritual" hylomorphism with respect to this theory.[9] Matter is always accompanied by life in nature, but the combination is not intrinsic and essential in itself. It is rather dependent on an external force—the most powerful force, indeed, being the force of the Creator. Material animation remains therefore contingent on the action of God's infinite will, and conservation rests on the latter's continuous presence.

Paracelsian philosophers, among them the Danish Petrus Severinus (1540–1602), resorted to an oxymoronic expression—*spiritualis materia*—to denote this most hybrid of beings. In keeping with the traditional language of chemical philosophy, in his work titled *Idea Medicinae Philosophicae* (A pattern of philosophical medicine), Severinus calls vital matter by many names: *materia perlata* (perfected matter), *balsamum, manna, elixir.*[10] His aim in this work is to demonstrate that alchemy is a most ancient wisdom, whose founding principles can be found in the writings of the great sages of antiquity such as Democritus and Hippocrates. Indeed, Severinus compares his concept of material power to Hippocrates's natural heat, more specifically to the divine principle that the author of the Hippocratic treatise *Fleshes* said to be capable of "hearing, understanding and perceiving everything" (*omnia audire, intelligere et animadvertere*).[11] Severinus connects this universally sentient *balsam* to the vegetative force that is active in the earth and in the plants that grow from it. To lay stress on the intricacy of chemical and botanical motifs, Severinus appropriates the concept of vital moisture (ἔμβιος ὑγρότης/*embios hugrotes*) discussed by the Aristotelian Theophrastus (ca. 371–ca. 287 BCE) in his *On the Causes of Plants.* Theophrastus, Severinus writes in *Idea Medicinae,* "grasped the nature of this matter and called it the vital principle of nature, by virtue of which all things are alive and vigorous, and all the rest is dead."[12] What is more, through this botanical addition in the imagery, Severinus shifts the emphasis of conservation back to a mechanism that is considered to be endogenous to the system.

Active matter is therefore the vital root (*natura radicalis*) that works as the "governing principle of all natural actions."[13] This *natura,* says Severinus, is "a sort of superior matter, furnished with quite powerful signatures and qualities . . . which through the unwavering conservation of the species preserves the changeable and transient nature of the individuals from dying."[14] Minerals have this principle of life—*validissimum balsamum*—within themselves, but its energy is, as it were, imprisoned in them. It can, however, be released in forceful fashions by expert practitioners, and these are the kind of exploits alchemists were famous—and notorious—for. The great majority of people think that minerals lack life. In fact, Severinus explains, their energy comes from "deeper parents [*altiores parentes*]," as is well known by those "adepts" who are able to properly inspect the "workshops" of mineral bodies. These "grow, have their specific times of maturation, nourish themselves and produce waste that they often reject with powerful outbursts." But why use extra words, Severinus rhetorically asks his readers, when we can rely on our senses? The sense of smell and taste will reveal the inner life of minerals to us: "Who will pronounce dead all those things that are endowed with the robust faculty of flavors and odors? These are the distinguishing features and degrees of their lives."[15] The life of minerals manifested to our senses is further evidence that the energy of matter is conserved from within.

Plant Metamorphosis, or the Matter of Vegetation

Even to the most untrained eye, plants display conspicuous tendencies to grow and to renew parts of their bodies. Their regenerative powers are evident in the ease with which new stems, leaves, and flowers are quickly developed. More than any other body in nature, plants reveal that a vital force is at work in all their parts. Entire specimens can grow anew from cut sections. This unique dynamic between the parts and the whole in plants has often led philosophers and savants to assume that plants are alive (*vegetae*) to the utmost degree. Etymologically speaking, vegetables are from the Latin verbs *vegeto* (to arouse and to invigorate) and *vigeo* (to be vigorous and to flourish); that is, they are especially enlivened and animated. Regardless of contemporary metaphors used to describe people with particularly dull lives, plants are indeed exuberantly active creatures. Between the eleventh and fifteenth centuries, the adjective *vegetativus* spread in medieval science and philosophy to denote that which is capable of growing by itself.[16] The vegetative functions in plants testify to their abilities to process food, grow, and reproduce themselves. They appear to be doing this through an almost direct and mutual involvement of matter and life. For this reason, throughout the history of Western representations of nature, the life of plants has been often used as evidence that matter is alive and capable of conserving itself.[17] Indeed, as Francis Bacon (1561–1626) noted in several of his works, plants offered a true model of how to conserve matter for artificial and technological purposes. In the *Historia Vita et Mortis* (Investigation on life and death, 1623), Bacon concentrated his attention on three particular areas of research in which plants appear to display a greater level of vitality compared to animals: nutrition, grafting, and seeds. Regarding nutrition, plants are defined by two characteristics: they have a direct relationship with the source of their energy—the soil—and do not accumulate waste when they nourish themselves. Their relationship with the earth is so close and intimate that theirs can be called a true cycle of self-nutrition: "In every nutrition process there are two actions: evacuation and assimilation. Vegetable bodies assimilate their food completely without producing any excretion."[18] Second, grafting makes plants stronger and their cultivation more effective, for this is a way for them to re-create a cycle of growth that is based on the similarity of *alimentum* and *alimentatum*, of feeding and fed substances, a similarity that makes the process of nutrition more efficient. Plant seeds, finally, remain active and fertile for a long period of time. Their energy does not expire as quickly as in animal bodies: "It should be noted that the seeds of animals are nutritive only to the extent that they are fresh; by contrast, the seeds of plants remain for a long time a source of nourishment."[19] Regeneration, nutrition, grafting, and seeds are all phenomena that testify to the fact that the active matter of plants favors conservation at various levels: by preventing wasteful uses of resources; by effacing the boundaries between subject (alimentatum) and object (alimentum); and finally, by preserving the activity of nature in seminal records of past life (the seeds).

Galen thought that the root was the primordial organ of vegetative operations in nature, in both plants and animals. Better, he saw the phenomena of plant rooting—ῥίζωσις/*rhizosis*—as a clear indication that a vital transformation took place in all natural beings endowed with life, in the germination of seeds as well as in the rudiments of livers in animal embryos.[20] In the philosophical tradition, the Platonic image of the human being as an inverted plant, whose roots were deemed to be in the heavens and not in the earth, and the Aristotelian description of the roots as the mouth and head of a plant both confirmed the idea that roots were in plants what the head was in human beings.[21] The progress of seventeenth- and eighteenth-century botanical anatomy and taxonomy, from Marcello Malpighi (1628 94) to Carl Linnaeus (1707–78), brought to the fore the importance of several other organs in accounting for the many transformations of vegetable matter, such as leaves, flowers, and fruits.

Here it may be worth briefly recalling Johann Wolfgang von Goethe (1749–1832), for his observations on the nexus of life and vegetal matter are especially relevant to the present study. In his lifelong inquires into natural phenomena, he found the key to understanding the process of material changeability by focusing on the leaves of plants. Goethe was particularly interested in the most recent results of evolutionary and morphological research. By the time he published his *Metamorphosis of Plants* in 1790, morphology had become the "science of organic forms and formative tendencies aimed at discovering underlying unity in the vast diversity of plants and animals," as Gordon Miller sums up in his introduction to a 2009 edition of the work.[22] We can say that from Galen's emphasis on the original root (ῥίζωσις/rhizosis) to Goethe's primordial leaves (the primal *Urblatt* and the experimental *Samenblätter*), investigators of plant matter had pointed to nature's astonishing abilities in shaping bodies out of winding tangles of roots and the ever-complex folding of vegetable forms. Plants, observed Goethe, were capable of wrapping and unwrapping their matter by following orderly and rhythmical paths in their growth. As the primordial organ of the plant, the leaf was for him the "true Proteus," capable of hiding or revealing himself "in all vegetal forms." The original unity of nature was therefore uncovered by the leaf through the countless shapes it could assume in the vegetal kingdom.[23] Leaves contracted into sepals, sepals expanded into petals, petals contracted into stamens, and so on, progressing all the way from seed to fruit. Philosophically speaking, Goethe's metamorphosis of plants, with its emphasis on seamless transitions of forms, brought to the fore the image of a universal material substratum, inherently pliable, always contracting and expanding depending on the laws of intensification and polarity that were at work in every aspect of nature.

As pointed out by Goethe, plants develop through coordinated motions of expansion and contraction. The supple character of matter is laid out by the *Urpflanze,* the archetypal plant that demonstrates the existence of a material life prior to its being articulated into organs. Plants show us the existence of fluid matter caught in imperceptible but steady transitions of reductions and enlargements. Here there are no real organs as in animals, but one single organ that alternatively shapes itself in a great variety of forms to reach fructification "through the modification of a single organ."[24] Goethe's foliar theory of plant metamorphosis corroborates the pliable nature of living matter. "By changing one form into another," plant metamorphosis "ascends—on a spiritual ladder—to the pinnacle of nature," that is, reproduction.[25] The fact is that in reality there is no pinnacle in nature, especially if the view is from matter and matter is alive. In a universal substratum in which every single bodily part is capable of preserving itself while adjusting its growing and shrinking to the growing and shrinking of every other part, generation and reproduction are in fact ceaseless motions of plastic energy. The conservation of the system of matter, as we are going to see in the next section, is closely connected to what one understands by generation.

Reproduction, or the Matter of the Soul

The tenet that nutrition, growth, and reproduction are stages within the active continuum of one plastic substance is a very old assumption that has crossed centuries and traditions of philosophical and medical research. A key text in this transmission of ideas, in which the philosophical and anatomical categories of matter, form, and life seem for a moment to face a true ontological showdown, is Galen's *De foetuum formatione libellus* (Short treatise on fetal formation).

As we have seen so far, the vegetative activities of nutrition, growth, and reproduction are the vital functions of the soul that seem to be the closest to the primal forces of matter. The belief in the growth of minerals and metals as well as phenomena of arborescence in specific chemical amalgams demonstrate that a principle of vegetation lies dormant and seemingly inanimate in ample sections of matter. Alchemical transformations and plant metamorphoses suggest that analogical relationships can be heuristically established in the domain of material changeability. All sorts of ramifications and segmentations in nature refer to the presence of ore veins, tangled roots, and the vascular structures of both plant and animal tissues in the domain of earthly and vegetal life. Since antiquity, these processes of nutrition and growth were deemed to reach their culmination in the natural function of generation. In Galen's treatise, we find enlightening observations regarding the delicate relations between the primal forces of matter—what he calls the natural powers (δυνάμεις/dunameis)—and the ability to shape matter into individual organisms. Just as the formation of mineral "trees" by alchemists could question the boundary lines—species and genera—separating metals from plants, so the phenomenon of animal reproduction in its earliest stages reconfirms the vegetal beginnings of all embryos. Galen compared the vein growing out of the navel in animal fetuses to a tree whose roots are in the liver while "the branches ramify widely towards the heart first and then everywhere in the body." For Galen this was the proof that the "nature that fashions living beings" was a plastic virtue diffused and active in every part of the growing body.[26] He wondered whether the purposive and "artistic" coordination of motions that leads to the shaping of a living body could in fact issue from a type of agency supposed to be devoid of reason (οὐσία ἄλογος/*ousia alogos*), that is, nature.[27] And yet, he insisted, our conscious mind does not know how to move the muscles corresponding to particular motions of the body. Indeed, Galen pointed out that the more than three hundred muscles present in our body were functioning without our necessarily being aware of their activities. No one could seriously think that each muscle was an "animal," that is, some sort of sentient being.[28]

This was Galen's major predicament in his theory of reproduction as an instantiation of plastic activity. Galen referred to anatomists who stated that "the soul builds its own body." In his opinion, if this were the case, then it would be reasonable to think that the soul, having made the parts of the body, knew how to use them.[29] On the contrary, our mind appeared to ignore vast provinces of unconscious and vital life. To solve this apparent contradiction—How can the soul, which has built the body, not know how to use its organs, that is, not be aware of the bodily mechanisms necessary to move the muscles?—some philosophers (among them the Aristotelians) had concluded that there were in fact two types of soul, one that fashions the body while fulfilling involuntary functions and another that is aware of its activity and performs voluntary tasks. Platonist and Platonizing thinkers, Galen went on, had instead assumed that eternal matter in eternal motion had been ordered into a cosmos by a demiurgic power following rational norms. However, if matter had been animated from eternity, then one soul should take care of the whole universe. If that was the case, Galen argued, there was only one soul diffused throughout the universe, in charge of all operations, conscious and unconscious. And yet this lofty and venerable soul of the world should also be depicted as busy making, among many other things, flies and worms. Galen's reasoning did not stop here and continued to explore all the logical possibilities underlying the notion of a formative function of nature. If one preferred to attribute the soul directly to matter, then it was very unlikely that this material entity could have the level of engineering expertise necessary to shape and reproduce all beings of nature.[30]

Through this exegetic and argumentative tour de force, Galen showed how torn he was between Platonic panpsychism (the *animated matter* of the first solution) and hylozoism (the *soul of matter* of the second solution). The first solution was unworthy of the metaphysical status pertaining to soul; the second was too liberal toward matter. And yet Galen's wavering between skepticism and agnosticism (he always admitted he was not able to say what a soul is), together with the cosmological premises of his system (matter is shaped by the original physical δυνάμεις/dunameis of nature), suggests that his mind was tilted toward vital materialism. How can the soul that builds the body be external to the very body that it has turned into one living being? Which of the two agents is the most knowledgeable, the conscious (directive) principle that is capable of voluntary operations or the plastic (constructive) principle that is able to set up all the myriad functions accomplished every second by the body? And if this latter ability is extended to each single organ and particle in the body, wouldn't this very fact be the proof that matter is animate, plastic, and above all, self-conserving throughout?[31]

The Interplay of Medical Dunameis in Early Modern Views of Matter

In the early modern period, the challenge represented by Galen's predicament concerning the formative power of matter was taken on by several physicians and philosophers. Here I will refer to only three of them: the already mentioned Bacon, William Harvey (1578–1657), and Francis Glisson (1598/99–1677). Bacon described nature as a principle of plastic activity. In the work titled *De Sapientia Veterum* (The wisdom of the ancients, 1609) and the unfinished treatise on the principles and origins of things written during the 1610s, the *De Principiis atque Originibus* (On principles and origins), also known as the *Fables of Cupid and Heaven*—perhaps the most metaphysical of Bacon's writings—he called this condition of skillful and yet prereflective and unreflective activity in matter "Cupid." Cupid meant for Bacon "prime matter and its faculties."[32] By criticizing the kind of "imaginary" matter (*materia phantastica*) that he thought was a distinctive characteristic of Platonic and scholastic philosophies, Bacon intended to vindicate the rights of material reality.[33] Like Galen, he regarded the plastic motions shaping the many bodies in nature as inner specifications within the vital continuum of pliable matter: motions of gravity and conflict, of expansion and contraction, of pressure and connection. For Bacon, this variety emerged out of the inexhaustible energy enclosed in each atom of matter: "just as the elements of all bodies are contained in the body of the atom, so the beginnings of all motions and forces are in the motion and force of the atom."[34]

A decade later, in the *Novum Organum* (1620), Bacon departed from this form of vital atomism to embrace a view of matter understood as a pliable continuum (*plica materiae*). By that time, the label of atomism meant for him a plurality of motions and forces, and no longer the particulate state of matter. As a result, the interplay of identity and otherness in nature depended not on the combination and division of material corpuscles but on the unfolding of original tendencies in matter. In the *Sylva Sylvarum* (1626), written at the end of his philosophical career, Bacon characterized these primal tendencies as a special kind of perception:

> when one body is applied to another, there is a kind of election to embrace that which is agreeable, and to exclude or expel that which is ingrate: and whether the body be alterant or altered, evermore a perception precedeth operation; for else all bodies would be alike one to another.[35]

The sentence "else all bodies would be alike one to another" is crucial. It suggests a hypothetical scenario in which the conflict of forces that represented the norm in Bacon's world has come to an end and everything in nature has been assimilated—self-assimilated, I would say. In other words, without a power of plastic specification, matter reverts to a principle of universal and static undifferentiation.

A similar anxiety for undifferentiated matter unable to prevent itself from being wasted away can be found in Bacon's contemporary William Harvey. In this case, too, Galen's notion of plastic power was put to good use. There is a radical idea winding through the text of Harvey's *Exercitationes de Generatione Animalium* (Exercises on the generation of animals), published in 1651: life has the power to organize and shape itself before any organ is available or any form has emerged out of matter.[36] The complex tangle of Platonic, Aristotelian, and Galenic ideas underpinning early modern discussions about the fashioning skills of the vegetative power is at the core of Harvey's discussion of animal reproduction in his *De Generatione Animalium.* In this work, he explored the initial stages in the development of life, what he called the *primordium vegetale* or *principium vitale.* The unique characteristic of this power was deemed to manifest itself above all in the process of reproduction, which, in line with Aristotelian and Galenic tenets, was for him a seamless combination of nutrition, growth, and self-differentiation.[37]

Galen and all his numerous followers in late antiquity, the Middle Ages, and the Renaissance had recognized the "artistic" nature of this formative force. Harvey's original contribution lies in his argument that life's ability to replicate itself by growing and shaping itself betrayed a surplus of knowledge rather than being the mere actualization of a teleologically predetermined potentiality. These reactions, Harvey insisted, belonged to the category of motions and actions that physicians had long called "natural"; they could not be controlled by the will or the mind, and yet, he pointed out, they did not take place without involving a certain degree of sentient engagement, for they were stimulated, irritated, and altered by the senses. Such elusive phenomena as inflammation, infection, nausea, and vomit were for him incontrovertible evidence that there was in the intestines a kind of cognitive capacity (*sensus aliquis*). This capacity did not depend on the brain; in the case of vomit, for instance, a form of obscure perception (*sensus quidam*) located in the stomach was clearly able to discriminate between harmful and healthy substances. Harvey called this ability "a certain kind of sensation" that occurred without the exercise of the senses (sensatio quaedam . . . sine sensu).[38]

Besides revealing that animals grow like plants, anatomical and embryological observations confirmed to Harvey that the vegetative power was not a neatly individuated faculty, but a source of diffuse energy. From this point of view, beyond fecund animals and growing plants, it was the earth as a whole that in the first place appeared to be endowed with the ability to grow and to shape homogeneous matter. Harvey acknowledged that it was necessary to establish a link between the autonomous power of vegetative transformation and a wider background of vital energy.[39] More importantly, Harvey's account of the developing embryo presupposed a type of anatomy in which processes prevailed over structures, where the *contenta* of Galenic anatomy (the fluids "contained" inside the organs) were more important than the *continentia* (the "containing" organs). Indeed, Harvey's is a paradoxical anatomy in which the *contenta* are already in place—the blood, for instance—when the *continentia*—the heart and the vascular system—are not formed yet. In Harvey's anatomical model, the inner force (the Hippocratic and Galenic *dunamis*) shapes the outer form from within. Matter is conserved in the very act of fashioning bodies and materials.

I think it may be fitting to conclude this specific trajectory in the history of reproductive biology from Galen to Harvey by saying a few words about the English physician Francis Glisson, for he was one of the first Galenic anatomists to espouse Harvey's discovery of the circulation of the blood while being an avowed follower of Bacon in philosophy. Above all, though, Glisson is the author of one of the most fascinating and comprehensive books ever written on the topic of active matter, titled *Tractatus de Natura Substantiae Energetica, seu De Vita Naturae* (Treatise on the active nature of substance; that is, The life of nature, 1672).[40] In the dedicatory epistle to Anthony Ashley Cooper, First Earl of Shaftesbury, Glisson presents his endeavor as a metaphysical investigation into the life of matter (*vita materiae*). The phrase "vita materiae" does not appear in the title of the work but recurs inside the book and is the central argument of the whole treatise.[41] What is mentioned in the title is the inner activity of nature (*vita naturae*). Nature as a whole is alive not because of transcendent animating influences in Platonic or Aristotelian terms (i.e., life derived from an immaterial intelligible form), nor because it is animated from the outside through a mechanical organization set up by God at the beginning of the creation (i.e., life as a specific frame of parts). Nature for Glisson is alive because everything derives from active matter. Only by assuming the existence of this matter can reality be deemed to be provided with an ultimate foundation.

We can therefore say that in *De Natura Substantiae Energetica* Glisson established a most intimate and compelling connection between nature, matter, and life. In his hands, these hoary notions become different ways of conceptualizing the same reality, that is, *substantia,* which he calls by the Greek name of βιουσία/*biousia,* "living substance." Nature is βιουσία/biousia in that life pervades every aspect of the cosmos. Matter is alive as it is a self-sufficient and self-preserving being. Life, finally, is the defining characteristic of reality, for every single entity in nature is able to respond to any other entity, whether favorable or inimical to it. Glisson defines this ability to act and react in every circumstance as perception. The "energetic" life of matter is therefore a primordial tendency to sense reality and respond to it. This nexus between nature, matter, and life premised on perception is what allows Glisson to define materiality as the very foundation of conservation.

By making matter the ultimate self-conserving substance, Glisson took the hylomorphism of the Aristotelian framework to its logical limits. If life according to the Aristotelians consists of the form that shapes matter—soul is the act of a body that is differentiated in organs and limbs, following Aristotle's renowned definition (*De anima,* 412a)—and if this organizing principle is meant to coincide with the very essence of a living body (otherwise life would come, in Platonic terms, from outside the realm of nature), then the most intimate principle of life would seem to stem from matter. Aristotle and a long succession of philosophers who over the centuries carried on the peripatetic research program never formulated or accepted any thesis directly arguing that matter is alive in itself. And yet several Aristotelians such as Theophrastus, Alexander of Aphrodisias, and Averroes—to remember here only the Aristotelians I have mentioned in this essay—were fully aware that the hylomorphic model advocated by the peripatetic school had an extraordinary heuristic potential for furthering the study of life in nature. Strato of Lampsacus (ca. 335–ca. 269 BCE), one of the earliest scholarchs of the Lyceum, known in antiquity as the *physicus* for his strong interest in the material aspects of nature, argued that all the life phenomena in the sublunary universe could be explained in simple material terms by resorting to the pneumatic characteristics of matter activated by the energy of heat and cold.[42] In the late seventeenth century, the English philosopher Ralph Cudworth (1617–88) resurrected the name of this celebrated Aristotelian precisely to refer indirectly to Glisson, his contemporary and colleague at the University of Cambridge (both Glisson and Cudworth were *regii professores*). Like his anatomical referent Harvey, Glisson was an avowed Aristotelian (a scholastic one, for that matter), and like Strato, according to Cudworth, he had brought to light the inner tensions of the Aristotelian view of ensouled matter.[43]

A Few Concluding Aporias

With the help of history, we have explored some concrete instances and sets of arguments according to which life is said to be rooted in matter. However, if matter is considered to be naturally and immanently active, we also are presented with a series of striking ontological outcomes. Here I will mention two of them, as they are the most recurrent and notorious in the history of philosophy. First, if we concede that matter is alive, individual living beings will be regarded as local and transient arrangements of energy within an all-encompassing and undifferentiated medium. Second, the conscious states of our minds will be seen as fitful and isolated episodes of self-recollection within an unbroken flow of perceptual experience. In both cases, the notion of individual self seems to become untenable. Historians of ideas, anthropologists, philosophers, and neuroscientists are still working on these prickly issues.

Perhaps our minds need to rely on inert and dead matter as a backdrop against which to demonstrate the existence of individual living and thinking beings. To be sure, where to draw the line in nature to distinguish animate from inanimate matter has been and still is one of the most puzzling questions in the history of human thinking. This problem is all the more compelling now, for it is forcing us to rethink the ways in which we conceptualize the relationship between matter and technology, nature and culture, nonhuman and human agency. In the age that it is becoming customary to name Anthropocene, a sense of philosophical frustration is growing and surrounds the use of worn-out categories and images, such as substratum, prop, receptacle, and black hole, when they are used to approach the idea of matter. The adjectives traditionally associated with these stereotypical views of matter—indeterminate, undifferentiated, passive, inert, and incomplete—are doomed to reconfirm binary divisions that no longer reflect the new reality of life—quantum, fungal, astrobiological, irredeemably murky, and hybrid. An ontology of relational interdependency suits our times better than the habitual array of things and substances, however accessorized they may become with fancy properties. To get out of this metaphysical impasse, we can start by having a look at forgotten conceptions of living matter, from Hippocrates's sentient fire to Glisson's life-sustaining βιουσία/biousia. Life ceaselessly folds matter in and out, as evidenced by the vital arborescence of chemical reactions, the foliar expansions and contractions of vegetal fluids, and the ever-growing tangles of roots, tendrils, fibers, and nerves as they make the stuff of generated bodies. It's a type of matter that is spun and woven out of ever-developing relationships caused by the infinite actions and reactions of conflicting physical and chemical powers. At the beginning was δύναμις/*dunamis*. Only later came οὐσία/*ousia*. The truth of the matter is that substance is conserved and conserves things out of a changeable network of interrelationships, not the other way around.

NOTES

1. On the Hippocratic tradition of natural philosophy, see Temkin, *Hippocrates*; Jouanna, *Hippocrates*; Cantor, *Reinventing Hippocrates*.
2. Huneman, "Montpellier Vitalism"; Schofield, *Mechanism and Materialism*; Brown, "From Mechanism to Vitalism"; Wright, "Materialism and the Life Soul"; Geyer-Kordesch, "Georg Ernst Stahl's"; Helm, "'Quod naturae ipsae.'"
3. On Galen's impact on philosophical discussions concerning matter, see Hirai, *Medical Humanism*; Hankinson, "Galen and the Ontology"; Bigotti, *Physiology of the Soul*; Eijk, "Galen on Soul"; Nutton, *Galen*, 76–97.
4. On the notion of via *philosophorum*, see French, "Gentile da Foligno"; Schmitt, "Aristotle among the Physicians."
5. Galen, *On the Natural Faculties*, 45. With slight changes to the translation.
6. On hylozoism, see Verdenius, *Hylozoism*; Giglioni, "Francis Glisson's Notion"; Skrbina, *Panpsychism in the West*.
7. For a defense of panpsychism in contemporary philosophy, see the chapter titled "Panpsychism," in Thomas Nagel's *Mortal Questions*, 181–95. By the same author see also *Mind and Cosmos*.
8. On views of matter in Western chemical traditions, see Brooke, *Thinking about Matter*; Newman, *Atoms and Alchemy*; Principe, *Secrets of Alchemy*.
9. Daniel, "Invisible Wombs." For a comprehensive view of Paracelsian cosmology, see Webster, *Paracelsus*.
10. Severinus, *Idea Medicinae*, 4, 104. On Severinus, see Shackelford, *Philosophical Path*.
11. Severinus, *Idea Medicinae*, 17 (my translation); Hippocrates, *Fleshes*, 133.
12. Severinus, *Idea Medicinae*, 21: "τὸ ἔμβιον τῆς φύσεως, vitale principium in natura, cuius virtute vivunt vigentque omnia, cetera mortua sunt." See also Theophrastus, *De Causis Plantarum* I.3.

13. Severinus, *Idea Medicinae*, 20–21: "naturam quandam radicalem . . . omniumque naturalium actionum moderatorem."
14. Ibid., 18: "praestantiorem quandam materiam, potentioribus signaturis et qualitatibus ornatam . . . quae mobilem et momentaneam individuorum naturam, stabili specierum conservatione, ab interitu tueretur."
15. Ibid., 25: "Crescunt [mineralia] enim, maturitatis tempora habent, nutriuntur, excrementa pariunt quae potentibus paroxysmis saepe reiciunt. . . . Et quid verbis opus est? Quaecunque saporum et odorum valida facultate sunt, ea quis mortua pronuntiabit? Vitarum discrimina sunt et gradus."
16. Latham, *Revised Medieval Latin Word-List*, s.v. "vegetativus." On the relationship between ideas of organic growth, primordial matter, and representation of vegetal nature in medieval art, see Weinryb, "Living Matter."
17. On this question, see Coccia, *Life of Plants.*
18. Bacon, *Historia Vitae et Mortis*, 194.
19. Ibid.
20. Galen, *De placitis Hippocratis et Platoni*, 418; Hankinson, "Galen's Anatomy," 229.
21. Plato, *Timaeus* 90A; Aristotle, *De partibus animalis* IV.10.686b. See also Chambers, "'I Was But.'"
22. Miller, introduction, xvi. See also Tantillo, *Will to Create*, 58–77; Pfau, "'All Is Leaf.'"
23. Goethe, *Italian Journey*, 366.
24. Goethe, *Metamorphosis of Plants*, 6.
25. Ibid.
26. Galen, *De foetuum formatione*, 86.
27. Ibid., 94.
28. Ibid., 96.
29. Ibid.
30. Ibid.
31. Ibid, 104–6. On Galen's theory of the soul and living matter, see Moraux, "Galien comme philosophe"; Nickel, "Kommentar," 160–61.
32. Bacon, *De Principiis*, 204.
33. Ibid., 206.
34. Ibid., 202.
35. Bacon, *Sylva Sylvarum*, 603.
36. Harvey, *De Generatione Animalium*, 266.
37. Ibid., 476–77, 479, 482–83.
38. Ibid., 452.

39. Ibid., 304.
40. Glisson, *De Natura Substantiae Energetica.* On Glisson's place in the history of early modern living matter, see Giglioni, "What Ever Happened."
41. Glisson, *De Natura Substantiae Energetica,* A6, c3v, c4r.
42. Sharples, "Peripatetic School," 148–49.
43. Cudworth, *True Intellectual System,* 1:203, 1:272, 2:620–21, 2:405–6.

BIBLIOGRAPHY

Bacon, Francis. *De Principiis atque Originibus Secundum Fabulas Cupidinis et Coeli.* In *Philosophical Studies c.1611–c.1619,* edited and translated by Graham Rees, 195–267, 416–36. Oxford: Clarendon Press, 1996.

———. *Historia Vitae et Mortis.* In *The "Instauratio Magna" Part III: "Historia Naturalis et Experimentalis"; "Historia Ventorum" and "Historia Vitae et Mortis,"* edited and translated by Graham Rees with Maria Wakely, 140–377, 418–60. Oxford: Clarendon Press, 2007.

———. *Sylva Sylvarum.* In *Works,* edited and translated by James Spedding, Robert Leslie Ellis, and Douglas Denon Heath, 14 vols., 2:323–672. London: Longman, 1857–74.

Bigotti, Fabrizio. *Physiology of the Soul: Mind, Body and Matter in the Galenic Tradition of the Late Renaissance (1550–1630).* Turnhout: Brepols, 2019.

Brooke, John Hedley. *Thinking about Matter: Studies in the History of Chemical Philosophy.* London: Variorum Reprint, 1995.

Brown, Theodore M. "From Mechanism to Vitalism in Eighteenth-Century English Physiology." *Journal of the History of Biology* 7, no. 2 (Autumn 1974): 179–216.

Cantor, David, ed. *Reinventing Hippocrates.* Aldershot: Ashgate, 2002.

Chambers, A. B. "'I Was But an Inverted Tree': Notes toward the History of an Idea." *Studies in the Renaissance* 8 (1961): 291–99.

Coccia, Emanuele. *The Life of Plants: A Metaphysics of Mixture.* London: Polity, 2018.

Cudworth, Ralph. *The True Intellectual System of the Universe.* 3 vols. London: Thomas Tegg, 1845.

Daniel, Dane T. "Invisible Wombs: Rethinking Paracelsus's Concept of Body and Matter." *Ambix* 53, no. 2 (July 2006): 129–42.

Eijk, Philip van der. "Galen on Soul, Mixture and Pneuma." In *Body and Soul in Hellenistic Philosophy,* edited by Brad Inwood and James Warren, 62–88. Cambridge: Cambridge University Press, 2020.

French, Roger. "Gentile da Foligno and the *Via Medicorum.*" In *The Light of Nature: Essays in the History and Philosophy of Science Presented to A.*

C. Crombie, edited by J. D. North and J. J. Roche, 21–34. Dordrecht: Martinus Nijhoff, 1985.

Galen. *De foetuum formatione.* Edited by Diethard Nickel. Berlin: Akademie, 2001.

———. *De placitis Hippocratis et Platonis (On the Doctrines of Hippocrates and Plato).* Edited and translated by Phillip De Lacy. Berlin: Akademie, 1978.

———. *On the Natural Faculties.* Edited and translated by Arthur John Brock. Cambridge, MA: Harvard University Press, 1916.

Geyer-Kordesch, Johanna. "Georg Ernst Stahl's Radical Pietist Medicine and Its Influence on the German Enlightenment." In *The Medical Enlightenment of the Eighteenth Century*, edited by Andrew Cunningham and Roger French, 67–87. Cambridge: Cambridge University Press, 1990.

Giglioni, Guido. "Francis Glisson's Notion of *Confoederatio Naturae* in the Context of Hylozoistic Corpuscularianism." *Revue d'histoire des sciences* 55, no. 2 (April–June 2002): 239–62.

———. "What Ever Happened to Francis Glisson? Albrecht Haller and the Fate of Eighteenth-Century Irritability." *Science in Context* 21, no. 4 (December 2008): 465–93.

Glisson, Francis. *Tractatus de Natura Substantiae Energetica, seu De Vita Naturae, eiusque Tribus Primis Facultatibus Naturalibus, I. Perceptiva, II. Appetitiva, et III. Motiva.* London: Elizabeth Flesher, 1672.

Goethe, Johann Wolfgang. *Italian Journey, 1786–1788.* Translated by W. H. Auden and Elizabeth Mayer. London: Collins, 1962.

———. *The Metamorphosis of Plants.* Edited by G. L. Miller. Translated by Douglas Miller. Cambridge, MA: MIT Press, 2009.

Hankinson, Robert J. "Galen's Anatomy of the Soul." *Phronesis* 36, no. 2 (1991): 197–233.

———. "Galen and the Ontology of Powers." *British Journal for the History of Philosophy* 22 (2014): 951–73.

Harvey, William. *Exercitationes de Generatione Animalium.* In *Opera Omnia*, edited by the Royal College of Physicians of London, 159–603. London: William Bowyer, 1766.

Helm, Jürgen. "'Quod Naturae Ipsae Sint Morborum Medicatrices': Der Hippokratismus Georg Ernst Stahls." *Medizinhistorisches Journal* 35, no. 3/4 (2000): 251–62.

Hippocrates. *Fleshes.* In *Works*, edited and translated by Paul Potter, 127–65. Cambridge, MA: Harvard University Press, 1995.

Hirai, Hiro. *Medical Humanism and Natural Philosophy: Renaissance Debates on Matter, Life and the Soul.* Leiden: Brill, 2011.

Huneman, Philippe. "Montpellier Vitalism and the Emergence of Alienism in France (1750–1800): The Case of the Passions." *Science in Context* 21, no. 4 (December 2008): 615–47.

Jouanna, Jacques. *Hippocrates.* Translated by Malcolm B. DeBevoise. Baltimore: Johns Hopkins University Press, 1999.

Latham, R. E. *Revised Medieval Latin Word-List from British and Irish Sources.* Oxford: Oxford University Press, 1965.

Miller, Gordon L. Introduction to Goethe, *Metamorphosis of Plants,* i-xxxi.

Moraux, Paul. "Galien comme philosophe: La philosophie de la nature." In *Galen: Problems and Prospects,* edited by Vivian Nutton, 87–116. London: Wellcome Institute for the History of Medicine, 1981.

Nagel, Thomas. *Mind and Cosmos: Why the Materialist Neo-Darwinian Conception of Nature Is Almost Certainly False.* Oxford: Oxford University Press, 2012.

———. *Mortal Questions.* Cambridge: Cambridge University Press, 1979.

Newman, William R. *Atoms and Alchemy: Chymistry and the Experimental Origins of the Scientific Revolution.* Chicago: University of Chicago Press, 2006.

Nickel, Diethard. "Kommentar." In Galen, *De foetuum formatione,* 108–73.

Nutton, Vivian. *Galen: A Thinking Doctor in Imperial Rome.* London: Routledge, 2020.

Pfau, Thomas. "'All Is Leaf': Difference, Metamorphosis, and Goethe's Phenomenology of Knowledge." *Studies in Romanticism* 49, no. 1 (Spring 2010): 3–41.

Principe, Lawrence M. *The Secrets of Alchemy.* Chicago: University of Chicago Press, 2015.

Schmitt, Charles B. "Aristotle among the Physicians." In *The Medical Renaissance of the Sixteenth Century,* edited by Andrew Wear, Roger French, and Iain M. Lonie, 1–16. Cambridge: Cambridge University Press, 1985.

Schofield, Robert. *Mechanism and Materialism: British Natural Philosophy in an Age of Reason.* Princeton, NJ: Princeton University Press, 1970.

Severinus, Petrus. *Idea Medicinae Philosophicae Fundamenta Continens Totius Doctrinae Paracelsicae, Hippocraticae et Galenicae.* Basel: Sixtus Henricpetri, 1571.

Shackelford, Jole. *A Philosophical Path for Paracelsian Medicine: The Ideas, Intellectual Context, and Influence of Petrus Severinus (1540–1602).* Copenhagen: Museum Tusculanum Press, 2004.

Sharples, Robert W. "The Peripatetic School." In *Routledge History of Philosophy.* Vol. 2, *From Aristotle to Augustine,* edited by David Furley, 147–87. London: Routledge, 1999.

Skrbina, David. *Panpsychism in the West.* Cambridge, MA: MIT Press, 2005.

Tantillo, Astrida Orle. *The Will to Create: Goethe's Philosophy of Nature.* Pittsburgh: University of Pittsburgh Press, 2002.

Temkin, Owsei. *Hippocrates in a World of Pagans and Christians.* Baltimore: Johns Hopkins University Press, 1991.

Verdenius, Willem Jacob. *Hylozoism in Early Greek Thought.* Amsterdam: Blikman, Laporte & Dosse, 1977.

Webster, Charles. *Paracelsus: Medicine, Magic and Mission at the End of Time.* New Haven, CT: Yale University Press, 2008.

Weinryb, Ittai. "Living Matter: Materiality, Maker, and Ornament in the Middle Ages." *Gesta* 52, no. 2 (2013): 113–132.

Wright, John P. "Materialism and the Life Soul in Eighteenth-Century Scottish Physiology." In *The Scottish Enlightenment: Essays in Reinterpretation*, edited by Paul Wood, 177–97. Rochester, NY: University of Rochester Press, 2000.

NINE

Oak and Oil, Chalk and Flint—Rood Screens and Churches

Spike Bucklow

Abstract

This paper suggests that most of the conservation profession is currently informed by a mid-twentieth-century understanding of matter and that the chemical classification of artworks' constituent materials has affected conservation practice. It further suggests that the current approach to matter is very different from the understanding of those people who made and originally cared for premodern artworks. Taking the example of the constituent materials of medieval East Anglian churches and their rood screens—oak, oil, chalk, and flint—the paper considers the hybrid and active nature of matter, as understood by modern science and also by the traditional sciences that were contemporary with the artworks' manufacture. It suggests that we can find parallels between Aristotelian and Neoplatonic cosmologies and the new ecological approaches to matter. The paper concludes that ecological approaches to the matter of artworks would be more appropriate to conservation than the current chemistry-based approach. A caveat briefly acknowledges differences between ecologically active matter and divinely active matter, which would be significant for an exploration of social attitudes to artworks and their conservation but is not pursued here.

Currently, conservation is predicated on the idea that the material nature of an artwork is essentially passive and is acted upon by the environment or by people. The fundamentally passive nature of matter is suggested by its description with fixed categories such as organic or inorganic and natural or unnatural (synthetic). These material categories are reflected by divisions in the conservation profession, with its specializations by media.

A while ago I initiated a project on rood screens in medieval churches, in part because they were hybrid objects that straddled those divisions. Having considered rood screens in terms of commonalities of practice between medieval craftspeople and modern heritage professionals, I would like here to consider profound differences in their approaches to matter.[1] The project is currently addressing the relationship between painted wooden rood screens and the stone buildings in which they are housed.[2] In other words, it is explicitly bridging the gap created by the professional separation of painting and building conservation. This paper is intended to suggest that such divisions in conservation practice are very recent and would probably have perplexed those who made, and initially cared for, the churches and rood screens.

Weight by weight, East Anglian rood screens are predominantly oak. Many of the churches housing them are predominantly chalk and flint. Modern science considers one of these materials to be organic, two to be inorganic, and all three, natural. This paper briefly considers the cultural significance of these materials to suggest that such modern distinctions would have been neither clear-cut nor useful for the people who built, used, and cared for churches and rood screens.

Oak and Oil

Analysis of paintings (and building timbers) by dendrochronology has enabled insights into the origins of panel supports (and structural beams). As the name suggests, dendrochronology can reveal details about the age of the tree from which it was derived. A comprehensive dendrochronological survey of rood screen timbers has yet to be undertaken; however, a wide visual survey of the timber's figures suggests that the usual construction material for East Anglian rood screens is oak (fig. 1). In addition to providing information about dating, dendrochronology can also provide information about the location in which a tree grew. With a practiced eye, guided by dendrochronology, the geographic origin of a piece of oak can often also be broadly recognized from the timber's figure. Of course, the eye's ability to judge oak's origins is far less specific than dendrochronological analysis but, for English medieval timbers, two main origins suggest themselves—homegrown and imported. Homegrown oak tends to have a wavy grain, while imported oak's grain tends to be much straighter. While this simple characteristic is not completely reliable, it is sufficient to support a working hypothesis that East Anglian rood screens were made from

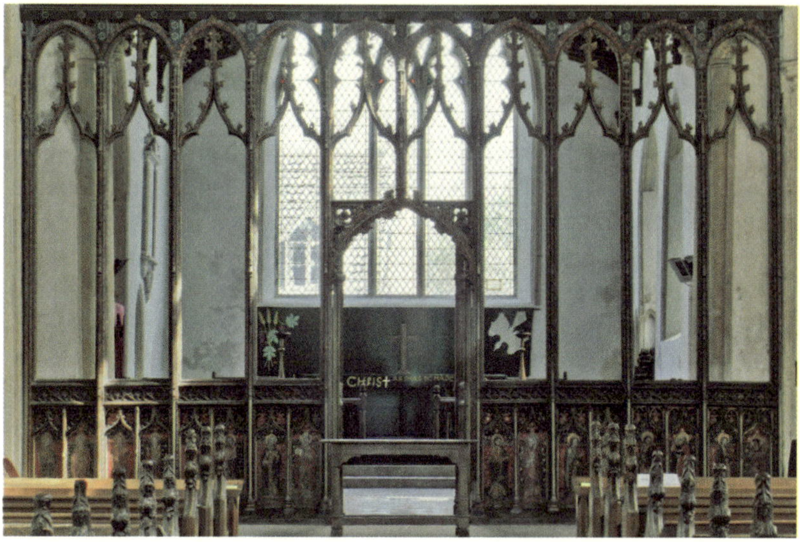

Fig. 1. English, Cawston rood screen, ca. 1485–1505. Oil on oak, approximately 7.85 × 5.24 m. Photo: Lucy Wrapson, Hamilton Kerr Institute, University of Cambridge.

both local and imported oak. Wide dendrochronological studies indicate that from the thirteenth to seventeenth century, English carpenters used local oaks together with significant quantities of oak from northern Europe and the Baltic. For example, in the late fourteenth century, about 100,000 boards of up to two and a half meters were imported through Hull and Lynn annually, and 250,000 came into the whole of England annually.[3]

Timber for rood screens was sourced from two species, *Quercus petraea* (sessile oak) and *Quercus robur* (pedunculate oak) as well as hybrids of the two. Sessile and pedunculate oak trees can be distinguished by small differences in their acorns and leaves, but their timbers cannot be distinguished, so it is not possible to say whether individual timbers came from sessile or pedunculate oaks. The geographic ranges of these oaks overlapped in both England and much of northern Europe, so the use of local or imported wood was not determined by preference for one species or the other. The visual differences between oak timbers in rood screens (either wavy or straight grain) were due not to the species of tree but to the way in which the tree grew.

The correlation between local and wavy, on the one hand, and imported and straight, on the other, implies differences in the way oaks grew in England and northern Europe between the twelfth and sixteenth centuries. In northern Europe harvesting timber involved going into a mature forest and felling a single tree, then moving to a different part of the forest and felling another. In this way, the forest canopy remained intact and young trees on the forest floor grew straight toward the gaps in the canopy left by felled trees. The timber derived from such trees had straight grain and few knots, since the trees did not waste energy by growing many side branches. By contrast, in England, trees were harvested by chopping down whole forests and replanting as required. In this style of forest management, the young trees all grew at the same time, competing with each other for space and light and exposed to the winds. The timber derived from such trees had wavy grain and knots since the trees opportunistically produced side branches.

The straight grain of oak from trees grown under intact forest canopies—corresponding to northern European imports—provided timber that was easy to split and appropriate for making thin, flat panels such as in rood screens' dados. The wavy grain of oak from trees grown in open countryside—corresponding to local English wood—had to be sawn, which was more labor-intensive but would not split catastrophically under duress so was good for rood screens' structural members. Different styles of forest management of the same species of trees created timbers with different working properties.

Oak trees growing in fifteenth- or sixteenth-century English or northern European forests were therefore natural organisms that responded to culturally influenced environments. Before the trees were even felled and sawn or split, they were already hybrids, part natural and part cultural. The rood screen maker's raw materials were therefore not particularly "raw," bearing as they did the different surface (saw or adze) marks of those who rendered trees into planks as well as the different anatomical (straight or wavy grain) marks of foresters' husbandry processes. (It is by these marks that conservators can visually hypothesize both local and imported oak in rood screens.)

Yet the oak carcass was not the only rood screen component that manifest the influence of cultural activity on vegetable matter, thereby blurring the distinction between the natural and unnatural. The rood screen carpenters' discrimination between different oaks for different components—local for structural, imported for panels—indicates their awareness of cultural variables associated with ostensibly natural resources. There may have been similar geographic variations in the sourcing of linseed oil for their paint (fig. 2). However, since oil was traded in much smaller quantities, the historical record does not allow any conclusions to be drawn, and scientific methods are currently unable to discriminate between different sources of historic linseed oil. Whether or not there were regional variations in the cultivation of linseed or in the processing of its oil that may have been relevant to painters, all medieval flax or linseed was derived from *Linum usitatissimum.* This plant was domesticated from the wild *Linum bienne,* which was a source of fibers that were spun and dyed from thirty thousand years ago.[4] But at some point, the plant was domesticated to increase its yield of oil so that, by six thousand years ago, it had been genetically changed into the one we know today.[5] However, flax was still also being used for its fibers, and while rood screens were being painted, linen was employed alongside wool in the textile industry that made East Anglia one of the wealthiest regions in England.[6] Hence the numerous rich churches.

Like oak, linseed oil was a hybrid that blurred the distinction between the natural and unnatural. Rood screens were constructed with a mix of contemporary nature-culture hybrids and painted with a prehis-

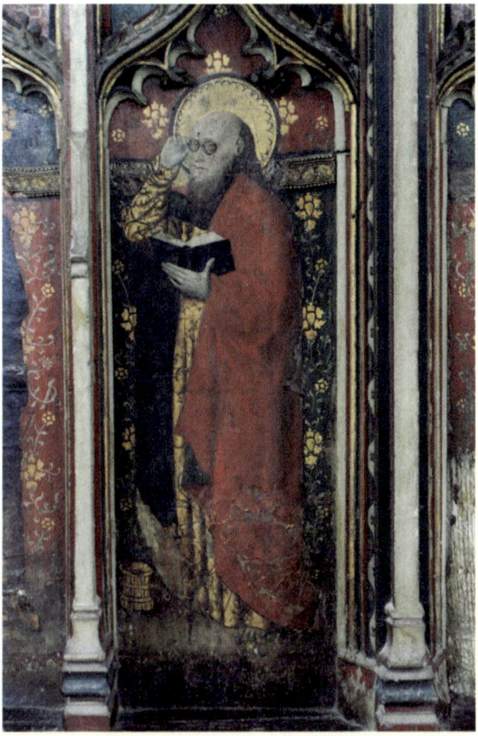

Fig. 2. English, Cawston rood screen (detail of St. Matthew), ca. 1485–1505. Oil on oak, panel sight size 0.31 × 0.91 m. Photo: Lucy Wrapson, Hamilton Kerr Institute, University of Cambridge.

toric nature-culture hybrid. While the cultural factors that determined the properties of oak were known, the cultural factors that had created the oil were not known. So, to the medieval painter, flax plants were ostensibly natural. On the other hand, their chalk and flint—major components of church walls—were unambiguously natural (fig. 3). Yet the following section will briefly suggest that, while natural, neither chalk nor flint was considered exclusively inorganic (or inanimate) as opposed to organic (or animate).

Chalk and Flint

When the rood screen–containing East Anglian churches were under construction, the mineral realm was not perceived as exclusively inanimate. It was obvious that crystals "grew"—indeed, tree-shaped mineral structures are still called "dendritic"—but neither chalk nor flint exhibits obvious crystal growth. However, they both participated in other phenomena that were perceived as evidence of mineral matter's

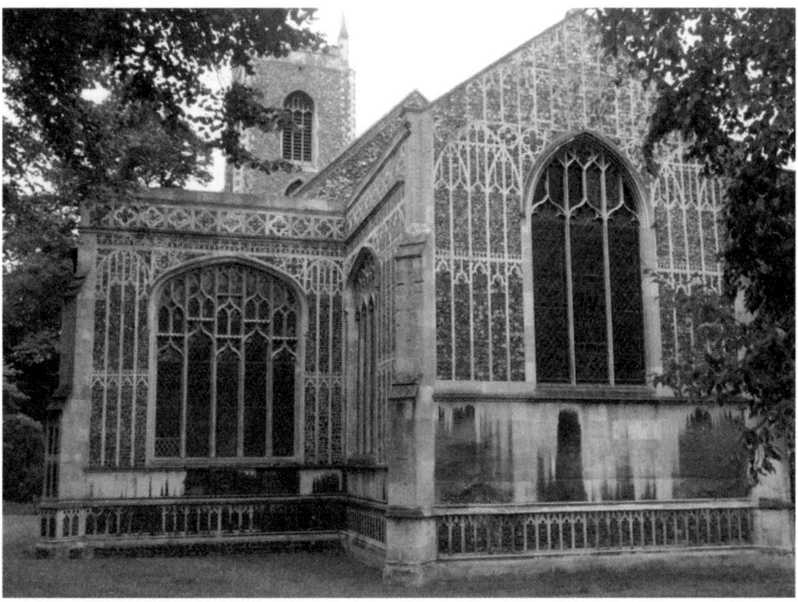

Fig. 3. St Michael Coslany, Norwich, from the southeast, early sixteenth century. Flint, lime mortar, and flushwork. The east wall of the chancel was rebuilt in the 1880s by William Hubbard of Dereham. Photo: Spike Bucklow, Hamilton Kerr Institute, University of Cambridge.

engagement with organic life. Fossils, for example, are stones in the shapes of plants or animals. According to Albertus Magnus, some were formed by terrestrial "petrifying forces," a process likened to the generation of coral from seaweed exposed to the Gorgon's head.[7] However, Albertus thought most fossils were the result of heavenly influences. Some conjunctions of planets could impede the germination of plants or conception of animals, yet other heavenly conjunctions could facilitate generative processes to the extent that "even in stones hardened by vapours, there is impressed upon the material the shape of a man or that of some other species that nature produces, either by painting, or by making it partly or wholly in relief."[8] Such fossils could be found in both chalk and flint.

Modern science may take issue with the heavenly mechanisms that Albertus thought responsible for fossils, but the fact remains that mineral matter can bear the visible trace of life forces. And one of the stones used in church building, chalk, had a continuing cultural relationship with life forces. Just as oak and flax each served more than one purpose, so, too, chalk served many purposes. By far the most significant use of chalk (and its close relative, limestone) was as an agricultural additive to boost soil fertility. Depending on location, initial soil deacidification could require up to three to five tons of chalk per acre, and an additional one to two tons per acre were added every four to six years to replenish the soil after leaching. In agrarian communities such as East Anglia, chalk (which, unlike limestone, was available locally) increased crop yields and was therefore a prerequisite for life.[9] To those who made churches and rood screens, fossils and the influence of chalk on plant life could have suggested an intimate relationship between the inorganic and the organic, categories that modern science and conservation practice have separated.

Much smaller quantities of chalk were used as a building material—hidden in wall cavities and, a bit more visibly, as a component of mortars. Weight by weight, the historic use of chalk as a building material was almost insignificant, but it looms disproportionately large in the modern imagination because this use created durable structures whereas in the other, much more significant use, the chalk was modified by various organisms then consumed by people (its calcium resides temporarily in our teeth and bones). The chalk in mortars was also modified, in processes that were both natural and unnatural. First, chalk—or again, limestone, if more convenient; hence "lime mortar"—was "calcined" or treated with *fire*; then it was "slaked" or treated with *water*. Finally, through its natural interaction with *air*, the resultant material turned back into a solid *earth*, capable of binding walls together. Together, these combined natural and unnatural processes would have been understood in terms of cycling around—or, more accurately, crisscrossing—the four Aristotelian elements.[10]

Especially in Norfolk churches, chalk-based mortars were most visibly evident in conjunction with another local stone, flint. As was obvious to those who collected the flints for the construction site, East Anglian flints were naturally distributed through a matrix of chalk. This was also evident for all to see in river- or sea-cut cliffs and ledges. In building, after the mortars reverted to chalk, the flints became embedded in a matrix of chalk, so the two stones' relationship in the church wall came to echo their original relationship in nature. The combination of flint and chalk in buildings had great utility, but the complex patterns decorating church walls—juxtaposing hard, glossy, black flint with soft, matte, white chalk—suggest that the combination also possessed significant aesthetic value (fig. 4).

This correspondence between nature and artifice—flint removed from a natural chalk matrix to become embedded in an artificially transformed chalk matrix—was a characteristic of medieval technologies in general. And the co-occurrence of culturally significant materials was not restricted to Norfolk's chalk and flint. For example, ancient and classical Mediterranean mines could provide both the marble required to make sculptures and the emery to finish them. In the vegetable kingdom, analogous co-occurrence is evident in the folklore of stinging nettles and dock leaves, the proximity of toxin and antidote presumably being providential.

The correspondence between original natural occurrence and final cultural usage can be seen in other East Anglian building materials. For example, where sedimentary rocks were used in structural roles or as flushwork—as in those East Anglian churches that used materials imported from the neighboring Midlands—their orientation in the building generally echoed their original orientation in the earth.[11] Even in good-quality freestones, where the orientation of strata is difficult to see, the blocks are still generally laid with their strata horizontal.[12] Such architectural mimicking of geological orientations and combinations is in keeping with Nicholas of Cusa's understanding of art or culture as an "elaboration" of nature, as opposed to a "transcendence" of nature.[13] Just as craftspeople's understanding of their materials could blur the categories of organic and inorganic, so their practices could blur the categories of natural and unnatural.

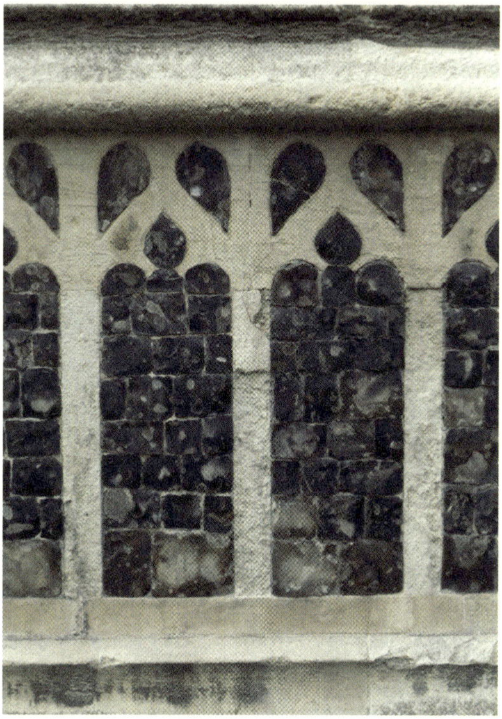

Fig. 4. St Michael Coslany, Norwich (detail of dado panel motif), early sixteenth century. Flint, lime mortar, and flushwork, single panel size 0.23 × 0.63 m. Photo: Spike Bucklow, Hamilton Kerr Institute, University of Cambridge.

Conservation and Ecologies, Old and New

Medieval ideas that fossils were signs of stone's (admittedly limited) potential to support life and that one stone could be produced in a matrix (literally, "womb") of another suggest that inanimate matter was viewed as possessing some degree of animation or activity. Blurred boundaries between animate and inanimate matter were reinforced by theoretical blurring between mineral, vegetable, and animal matter. Lapidaries, herbals, and bestiaries all contained examples of entities that were part mineral and part vegetable, or part vegetable and part animal. Such transitional entities were certainly not due to vagueness about classifying the world—classical and medieval natural taxonomies were very rigorous—but may instead have been expressions of the wholeness of creation. Abrupt distinctions between mineral, vegetable, and animal matter could have signaled gaps in the fabric of the universe, and it was said that "nature abhors a vacuum." The stairway to heaven—ascending through mineral, vegetable, animal, human, and angelic realms—had a continuous, ramplike quality.

A sympathetic modern mind might find such ideas poetic, but they are not accepted by a modern science in which categories—such as organic or inorganic—provide little room for poetic boundary crossing. And until very recently, scientifically informed conservation rested upon the assumption that the materials of artworks and buildings were inert, albeit subject to slow chemical reactions that modified their properties. This understanding of matter as inherently divided and devoid of agency has had consequences for professionalized conservation. It encouraged a fragmented and interventive approach with the assumption that conservators alone have agency to manage change in diverse materials. But already, such a view has become rather old-fashioned, hence the concept of (from the conservator's perspective) "passive" control, with internal environments buffered by appropriately chosen (active) building materials. The shift of focus within conservation toward an understanding of the natural interactions between relatively fluid materials—as opposed to the identification of relatively defined materials—reflects a shift in emphasis, away from chemistry and toward ecology.

Perhaps surprisingly, this new ecological approach lends some support to the poetic medieval visions of matter. The interactions that ecology is able to accommodate—and that are beyond the scope of chemistry—throw light on the nature of chalk and flint, for example. Geologically, East Anglia's chalk is the accumulated residue of countless life-forms, including foraminifera, that once thrived in the ocean that covered what is currently western Europe. Similarly, its flint is the residue of other life-forms, including diatoms, that once shared the same ocean. Some life-forms were even the source of both chalk and flint—calcite from the skeletons and silicate from the spicules of various sponges, including *Porifera demospongiae.* Both the stone matrix (chalk) and the stone within the matrix (flint) were of organic origin. And the shape of some flints was even determined by organic activity—maybe less obvious than a dinosaur's petrified footprint, but nonetheless, a tangible trace of life. Some of East Anglia's strangely shaped silicate nodules were formed within cavities that had been dug out by creatures, including *Thalassinoides,* burrowing through chalk.[14] So the matter that now resides in church walls was once animate, just as the matter that is now in rood screens was once animate. Ecologically, matter is not fixed in animal, vegetable, or mineral form but is in circulation. Boundaries are not static. They are dynamic.

Matter is active over many timescales. From the time for oaks to develop structurally (centuries), through the time for flax to develop genetically (millennia), to the time for marine organisms to thrive, die, and be deposited or dissolved and precipitated (eons). And then, for churches and rood screens to be constructed and decay (years and centuries). Chalk and flint's change from organic to inorganic was geological, governed by natural processes. But the dawn of the Anthropocene shows that geological eras can also be nature-culture hybrids. In the twentieth century, conservation attached itself to chemistry—origin of the profession's organic-inorganic divisions—and the industrial application of chemical concepts constitutes the Anthropocene's cultural component. For example, the nature-culture hybrid climate change results from the failure to recognize or respect cycling between organic and inorganic matter.[15]

Modern conservators are currently guided by chemical metrics—including relative humidity, temperature, light levels, and rates of air change. Over the past few decades it has been demonstrated that such positivist frameworks for addressing the physical world can effectively address the welfare of buildings and the objects they contain. Yet it should be acknowledged that buildings and contents were originally made—and, for many centuries, conserved—by people who had different, more active and dynamic, frameworks through which to approach the material world. And, provided regular maintenance ensured that church gutters and drains were kept free and gargoyles threw stormwater far from the foundations, those more flexible and poetic ways of seeing the world were nonetheless also demonstrably practical.

I would suggest that historic practices that created enduring buildings with self-regulating environments appropriate for their contents owed much to worldviews that did not see the inorganic as entirely separate from the organic, or culture as completely divorced from nature. The late-medieval approach to the materiality of East Anglian churches and their rood screens might therefore be characterized as ecological. And the originally joined-up nature of rood screens and church buildings is potentially instructive for a conservation profession that has been inherently divided by its inheritance of nineteenth- and twentieth-century concepts of passive and static matter.

Caveat and Coda

Premodern approaches to materiality may have had more in common with modern ecological relationships than with modern chemical identities. However, we should not overlook very significant differences between the ontological and epistemological foundations of the premodern and modern worlds. Constructing a church and rood screen with geologically altered marine sediments and photosynthetic eukaryotes of the subgenus *Quercus* covered in extracts of genetically altered *Linum* is one thing. But raising and equipping the house of God with the materials of His creation is quite another. And people's relationships with materials also differ when their knowledge of them comes from a distant scientific laboratory or from one's own, or one's neighbor's, labor.

The transformation of matter by manual labor was a tangible expression of power, and active relationships between matter and people underlie the status of artworks as "objects of power."[16] It follows that the medieval aesthetic, "the workmanship surpasses the materials," was not the exclusive preserve of those who had read Ovid.[17] It was also available to—and possibly more profoundly appreciated by—those who labored. It would have been understood by those who cleared the land of flints and replenished it with chalk, by those who harvested flax and processed its fiber and oil, and by those who made and used a wide variety of wooden items from yokes to coffins (of local and imported oak, respectively). In other words, the people who commissioned and sat in front of oil-painted oak rood screens in chalk-and-flint churches across East Anglia had intimate relationships with their workmanship and materials. These humble rural parishioners were therefore in the best position to appreciate, in Cusa's terms, the devotional "elaboration" of active matter, as achieved by their local builders, carpenters, and painters.[18]

NOTES

1. Bucklow, "Science and the Screen."
2. Curteis, Wrapson, and Berry, "East Anglia's Medieval Rood Screens."
3. Tyers, "Aspects of the European Trade," 45.
4. Kvavadze et al., "30,000-Year-Old."
5. Allaby et al., "Evidence of the Domestication."
6. Oldland, "Economic Impact of Clothmaking."
7. Albertus Magnus, *Book of Minerals*, 52–53; Ovid, *Metamorphoses*, 114.
8. Albertus Magnus, *Book of Minerals*, 131.
9. Gardner and Garner, *Use of Lime*, passim.
10. In modern terms, calcining involved $CaCO_3 \rightarrow CaO + CO_2\uparrow$; slaking, $CaO + H_2O \rightarrow Ca(OH)_2$; and aerial reversion, $Ca(OH)_2 + CO_2 \rightarrow CaCO_3 + H_2O$.
11. This need not be the case for metamorphic rocks (like marble, naturally heat-treated $CaCO_3$) where strata can be used for decorative effects.

12. This generally only becomes obvious to the untrained eye after significant weathering, as in the ruins of Whitby Abbey, or where blocks are mistakenly laid with vertical strata, whereupon exfoliation occurs.
13. Lyons, *Signs in the Dust*, 75.
14. Bromley and Ekdale, "Trace Fossil Preservation"; Bromley and Ekdale, "Comparative Ichnology."
15. Bucklow, *Red*, 200–205.
16. Gearhart, *Theophilus*, 15–41; DeMarrais and Robb, "Art Makes Society."
17. Ovid, *Metamorphoses*, 50; Panofsky, *Abbot Suger*, 26–27.
18. "And in the handiwork of their craft is their prayer." Ecclesiasticus, 38:34 (Revised Version with Apocrypha, 1885, 1895). If the Eucharist was holy matter, then all matter could be holy since it is in the nature of matter to change. And if the process whereby bread became body was a mystery, then so, too, were the processes whereby the stone that replenished the fields became the stone that held together church walls.

BIBLIOGRAPHY

Albertus Magnus. *Book of Minerals*. Translated by Dorothy Wykoff. Oxford: Clarendon, 1967.

Allaby, Robin G., Gregory W. Peterson, David A. Merriwether, and Young-Bi Fu. "Evidence of the Domestication History of Flax from Genetic Diversity of the *sad2* Locus." *Theoretical and Applied Genetics* 112, no. 1 (2005): 58–65. https://doi.org/10.1007/s00122-005-0103-3.

Bromley, R. G., and A. A. Ekdale. "Comparative Ichnology of Shelf-Sea and Deep-Sea Chalk." *Journal of Paleontology* 58, no. 2 (1984): 322–32.

———. "Trace Fossil Preservation in Flint in European Chalk." *Journal of Paleontology* 58, no. 2 (1984): 298–311.

Bucklow, Spike. *Red, the Art and Science of a Colour*. London: Reaktion, 2016.

———. "Science and the Screen." In *The Art and Science of the Church Screen in Medieval Europe*, edited by Spike Bucklow, Richard Marks, and Lucy Wrapson, 30–44. Woodbridge: Boydell and Brewer, 2017.

Curteis, Tobit, Lucy Wrapson, and Janet Berry. "East Anglia's Medieval Rood Screens: Conserving Sensitive Painted Artworks in Uncontrolled Church Environments." *Studies in Conservation* (2020): https://doi.org/10.1080/00393630.2020.1752970.

DeMarrais, Elizabeth, and John Robb. "Art Makes Society." *World Art* 3, no. 1 (2013): 3–22.

Gardner, Harold, and Harold Garner. *Use of Lime in British Agriculture*. London: Farmer and Stockbreeder, 1957.

Gearhart, Heidi. *Theophilus and the Theory and Practice of Medieval Art.* University Park: Pennsylvania State University Press, 2017.

Kvavadze, Eliso, Ofer Bar-Yosef, Anna Belfer-Cohen, Elisabetta Boaretto, Nino Jakeli, Zinovi Matevich, and Tengiz Meshveliani. "30,000-Year-Old Wild Flax Fibers." *Science* 325, no. 5946 (2009): 1359. https://doi.org/10.1126/science.1175404.

Lyons, Nathan. *Signs in the Dust.* Oxford: Oxford University Press, 2019.

Oldland, John. "The Economic Impact of Clothmaking on Rural Society, 1300–1550." In *Medieval Merchants and Money*, edited by Martin Allen and Matthew Davies, 229–52. London: Institute of Historical Research, 2016.

Ovid. *Metamorphoses.* Translated by Mary Innes. Harmondsworth: Penguin, 1975.

Panofsky, Erwin, ed. and trans. *Abbot Suger and the Treasures of Saint Denis.* Princeton, NJ: Princeton University Press, 1979.

Tyers, Ian. "Aspects of the European Trade in Oak Boards to England 1200–1700." In *Trade in Artists' Materials*, edited by Jo Kirby, Susan Nash, and Joanna Cannon, 42–49. London: Archetype, 2010.

TEN

Bread and Wine, Body and Blood

Lee Palmer Wandel

Abstract

At the center of the Christian tradition are the words Jesus was held to have said to his disciples at the meal the night before he was crucified, "Take, eat; this is my body," and "Drink ye, all of it; for this is my blood of the new testament" (Mt 26:26–28 KJV). These words have evaded all efforts at stabilizing the relationships at their core: between bread and body, wine and blood, a moment in the past and the present. In 1215, the Fourth Lateran Council sought to set the relationship, as "transubstantiation," the essential change of the matter from bread and wine into body and blood. This, in turn, engendered anxieties about the matter of the Eucharist: When exactly was Christ's body, his blood, present? The sixteenth century divided over what the matter of the bread and wine were after the words of consecration were spoken, from John Calvin's "sign," through the Council of Trent's affirmation of the transformation of the substance of bread into the living body of the present Christ and the wine into life-giving blood. For all churches, what was received in the mouth differed depending on the recipient, a moment of singularly anti-inert materiality.

How one reads this image (fig. 1) depends very much on how one understands the doctrine of the Incarnation, captured succinctly in the words of the Nicene Creed:

> We believe in one God the Father all-powerful, Maker of all things both seen and unseen. And in one Lord Jesus Christ, the Son of God, the Only-begotten begotten from the Father,

> that is from the substance of the Father, God from God, light from light, true God from true God, begotten not made, consubstantial with the Father, through whom all things came to be, both those in heaven and those in earth; for us humans and for our salvation he came down and became incarnate, became human, suffered and rose up on the third day, went up into the heavens, is coming to judge the living and the dead. And in the Holy Spirit.[1]

It is not a question of belief. One need not believe the Creed for the doctrine's overthrow of ancient binaries—God and humankind, life and death, spirit and matter—to inform how we approach medieval conceptions of matter and its potentialities.[2]

Nowhere were the implications of the Incarnation for matter more urgent or intimate than in the Eucharist, the central act of worship for medieval Christians. For more than a millennium, for all Christians in the West, the Mass culminated in the Eucharist, in that moment when the priest, as celebrant,[3] spoke the words Jesus was held to have spoken

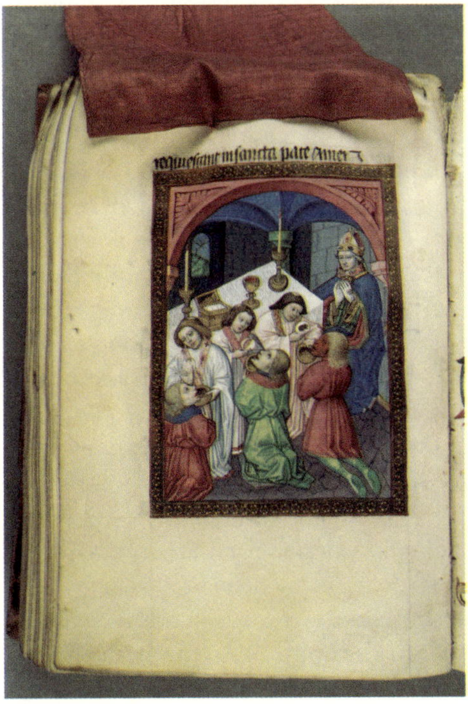

Fig. 1. Lehrbuchmeister, *Siebenhirter Hours*, ca. 1470. Illumination, 9¼ × 6 in. (23.4 × 15.0 cm). National Library of Sweden, MS A 225.

at the Last Supper, the night before he was crucified and died: "Hoc est enim corpus meum" (For this is my body) and "Hic est enim calix sanguinis mei novi et aeterni testamenti" (For this is the chalice of my blood of the new and eternal covenant).[4] In that moment, the Fourth Lateran Council (1215) decreed:

> [Jesus Christ's] body and blood are truly contained in the sacrament of the altar [the Eucharist] under the forms of bread and wine, the bread and wine having been changed in substance, by God's power, into his body and blood.[5]

In the moment when the priest spoke the words, first the small white wafer, the host, was changed into the body of Christ, and then the wine mixed with water in the chalice was changed into the blood of Christ. The word set in Fourth Lateran's decree, *transsubstantis*, "transubstantiated," formally named the change in matter that was to take place in every Mass.[6] Fourth Lateran's promulgation of the word "transubstantiate" placed "substance" at center. Medieval theologians drew on Aristotle, who provided distinctions useful in making sense of the central mystery: Why, if the host and the wine changed into the body and blood of Christ, did the matter of the Eucharist continue to look like a small, thin wafer and have the viscosity and taste of wine? Aristotle had distinguished between "substance" and "accidents."[7] "Accidents" included what the senses perceived: the shape, the weight, the texture, the taste. In the sixteenth century, Martin Luther, John Calvin, and dozens of other theologians and pastors rejected the entire edifice of transubstantiation, as specious to the point of immorally deceptive.[8] But between the early thirteenth century and the early sixteenth century, Fourth Lateran's cryptic formulation—itself a carefully formulated and formal response to an earlier challenge—engendered rich discussion of the relationship of matter to divinity understood as both eternity and presence. The doctrine of transubstantiation implicated the mundane in the divine and posited that the divine could be made present and experienced somatically through the matter of daily foodstuffs, through the human voice speaking the words Christ was held to have spoken.

The concept of transubstantiation destabilized any ready definition, any clear boundaries for all four forms of matter at its center: bread, wine, body, blood. The substance of bread, that is, wheat and water, became the substance of Christ's body, flesh, the flesh of John 1:14, "the Word became flesh and lived among us."[9] The substance of the wine mixed with water, grapes and water, became the substance of the blood that was depicted in Crucifixions and images of the Passion. Both body and blood had not ended with the death on the cross, but were living and present in every Mass. The substance changed, but the appearances did not. Body and blood were present. They were visible to some, but not all, perceptible to some, but not all—the matter did not appear the same to all Christians, nor was their experience of the matter homogenous or uniform. Transubstantiation was no simple transformation; it was not a chemical change.

The transubstantiated body and blood were to be consumed, at every communion by the celebrant; at least once a year, Fourth Lateran also decreed, every Christian was to receive the body of Christ at Easter. The body was consumed but was present with every consecrated host.[10] Christ's blood was to be drunk after each time the wine was consecrated, but it, too, was present each time. Christ's body was single, Christ's blood was unique to him, but they were physically neither absent nor eternally present. Present in the moment of consecration at every altar throughout Christendom each time the words of consecration were spoken correctly, they were neither infinite nor multitudinous. They confounded any conception of "matter" as finite, bounded, sited, of location, dimension, weight, or measurability of any kind. And theologians also explored what it meant for each "body," Christ's and the recipient's, in the moment of communion: Mingling? Incorporation? Incarnation?

For more than three hundred years, Western Christians explored the ever-increasing number of questions transubstantiation engendered. Some did so in formal theological treatises or within universities. Others did so, reaching into the very spaces where all Christians received the Eucharist, as altarpieces resting on the altar and panels; or in the books of hours and psalters devout Christians read daily, such as the illumination that opens this article. The most studied of genres of Eucharistic images—rendered in altarpieces, panels, devotional images, and illuminations—is known as the Mass of Saint Gregory, which depicted Christ rising from the altar before Pope Gregory I (fig. 2).[11] Scholars remain divided as to what these images were intended to do devotionally and theologically, but the entire genre focuses the viewers' attention on the doctrine of the Incarnation as instantiated in the moment of consecration of the Eucharistic elements. Christ, his crucifixion signaled by his wounds, is there present on an altar; he is depicted not as a corpse but in the act of rising out of the altar, living not dead. In depicting him thus, these images suggest that their viewers see

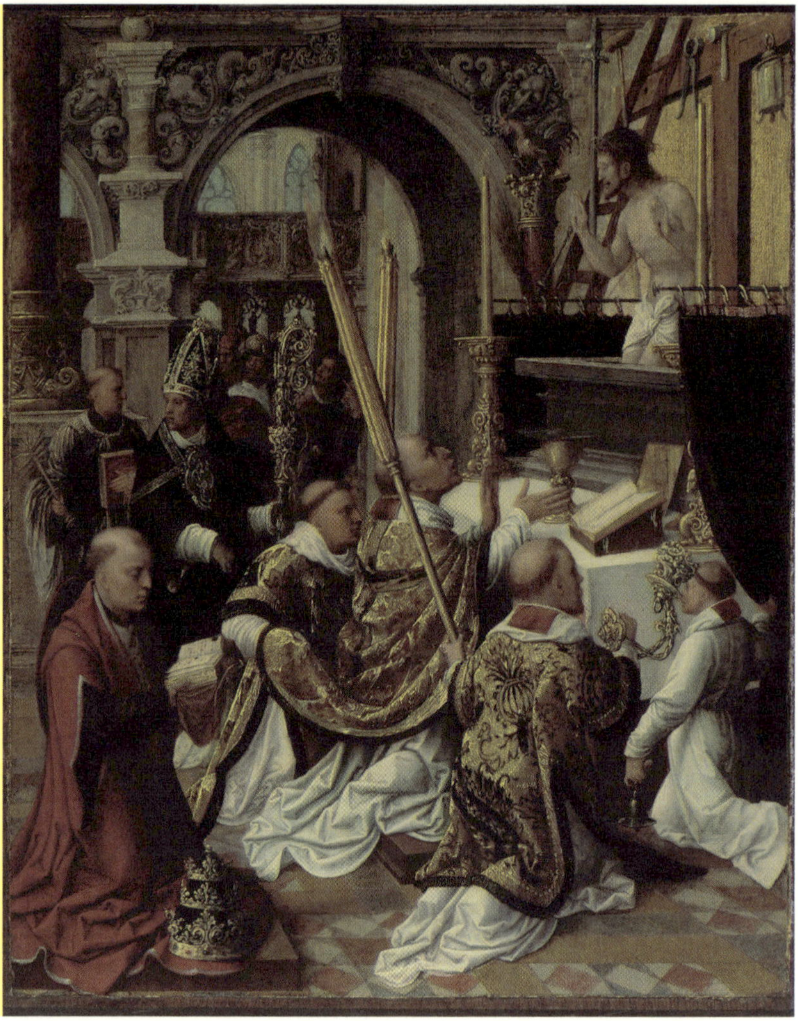

Fig. 2. Adriaen Ysenbrandt, *The Mass of Saint Gregory the Great*, 1510–50. Oil on panel, 14¼ × 11½ in. (36.2 × 29.2 cm). J. Paul Getty Museum, 69.PB.11.

what the celebrant, Gregory, saw, Christ as living and rising. In some, Gregory holds the host before him, between him and the rising Christ; in many, the host is absent; in some, Christ bleeds into the chalice on the altar. Gregory wears the vestments of the celebrant; in many of the images, his papal tiara has been removed for the moment of greatest divine presence at the altar. In this genre of images, Christians saw the Christ who had died but lives forever; the altar on which that transubstantiation took place, again and again, within the rhythms of the liturgy and liturgical time; brought forth by a celebrant whose clothes designated him as belonging to a line of priests reaching back through time to Christ; whose words and actions were directed to the matter of the host and the wine.

> But suppose that when a priest elevates the host, there should appear the species of a child or of flesh. What should he do? I say that either it appears this way to everyone or just to the priest or just to the people. If it appears thus to everyone, namely, to the people and to the priest, prayers ought to be poured out to God that it be turned back into the species of bread. And if that happens, it ought to be consumed. But if not, other matter ought to be taken and consecrated. But if it does not appear to the priest, but just to the people, it ought to be consumed by the priest. (Guido of Monte Rochen, *Handbook for Curates*)[12]

As images of the Mass of Saint Gregory make so eloquently clear, for the priest celebrating the Eucharist, the handling of the matter of the host and the wine presented terrifying potentialities of desecration. Because the medieval Church held that God Incarnate's body was contained in the form of the bread, and God Incarnate's blood, in the wine, the person handling the accidents of bread and wine was at risk of errors or accidents, contingencies, that concerned the very person of God Incarnate. These came to be gathered in their own category, "defects." Guido of Monte Rochen devoted a chapter in his *Handbook for Curates* to some of them. In his *Rationale Divinorum Officiorum*, William Durand offered a rich sense of the many mutabilities the doctrine of transubstantiation set in motion:[13]

> What happens, if between the pronunciation of words over the bread and wine, there is a space of time: is there still transubstantiation? . . . [W]hen these words are pronounced: "This is my Body," the bread is changed into the flesh, and at these words: "This is my Blood," the wine is changed into blood, it seems that the bread can be consecrated without the wine, or that the wine can be consecrated without the bread, as was previously stated. So, if after saying the first words, and before saying the second words, the priest has some accident that prevents him from proceeding, it seems that the bread

> has been changed into the flesh, but the wine has not been changed into the blood. In this case, should another priest repeat all the words from the beginning, and thus redo the consecration over the bread, or should he begin in the place where the first priest left off, and thereby divide this mystery of unity? (*Rationale Divinorum Officiorum* IV, 41, 47)[14]

Here Durand touches on the problems engendered by the intersection of the material separation of flesh from blood with the temporal separation of the words of consecration. What was "body" if "blood" was not in it? Could there be, on the altar, body but not blood? Did the words effect only one change in one element at a time, or was the change complete with the first words and only acknowledged with the second? In the sections immediately following, he pursued the question, Could there be bread without wine (*Rationale* IV, 41, 48–49)?[15]

Implicit in Durand's discussion is another variable: the celebrant's body. Durand's focus is the celebrant's physical ability to complete the consecration of both bread and wine, but lurking is an anxiety shared among many commentators, that the celebrant might vomit the consecrated elements. Commentators were clear that this was to be avoided at all costs, that the celebrant risked mortal sin. The consecrated host remained the body of Christ, the wine, the blood of Christ, and it was never to land on the ground, to be expelled, to be put at physical risk of desecration.

> What, therefore, must be done by the priest if, after the consecration, he discovers that he forgot the wine? I respond: if this was done consciously, in order to introduce heresy, there is no sacrament of the Body; but if he did this through forgetfulness or negligence, the consecration of the Body is nonetheless completed. (*Rationale* IV, 41, 49)[16]

A separate category of defects arose from the celebrant's mind, extending from a faulty memory or inattention to heretical intent. Some of these also illumine the nature of the matter at the center of the doctrine of transubstantiation.[17] Transubstantiation did not occur if the celebrant intended heresy; intent affected the efficacy of the words of institution. A priest who forgot a mortal sin and therefore did not confess it did not affect the elements or transubstantiation but did put his own soul in danger. Water might be mistaken for wine, and Guido offered advice on how to correct in the midst of the Mass such a serious error, as well as considering yet another problem this mistake engendered, that the priest, in taking water, was no longer fasting in preparation for receiving the blood of Christ. Another error of oversight: if the priest forgot to mix water with wine before the consecration, when exactly in that temporal sequence he made that discovery determined what he was to do, whether he was still handling wine or had spoken words of consecration that left the substance in the chalice theologically disputed. The matter of the Eucharistic element, in other words, determined the celebrant's course of correction: if it was still wine, then it was to be handled in one way; if blood, then it was divine and to be handled in another way entirely.

Transubstantiation, doctrine and physics, put the body of God at risk of accidents, contingencies. Guido addressed a danger that troubled a number of commentators, an insect falling into the wine:

> But what if the priest notices that there is poison in the chalice or that a spider or a fly has fallen in? Should he receive it? Certainly, as far as the poison of the spider goes, I say no, lest the cup of life become the cup of death. But this blood ought to be put among the relics in some clean and suitable vessel, with some written notice [saying] that here is the blood of Christ and that it may not be consumed by anyone, because poison is mixed with the species. . . . As for the fly, I say that on account of a fly, he ought not to throw it out without consuming it. . . . And if he fears vomiting, let him eat something that prevents vomiting. And let the fly be burned and the ashes kept in a shrine.[18]

Determinative consistently across all these accidents was the status of the matter: Had it been consecrated and thereby changed in substance into the blood of God Incarnate? That blood could, in the case of such an accident, be treated as a holy relic, but it could never be handled as wine that had gone bad. Consecrated, it was no longer wine. But it remained vulnerable to the same accidents as wine: insects, spilling, vomiting.

When he turned to decay, however, Guido suggested that the change in the matter of the host and the wine was not necessarily permanent.[19] Decay posed its own problems, beginning with clerical neglect. But decay also posed specific problems with regard to the substance of the matter:

> But suppose that the priest reserved the species so long that they are now rotten and decayed. What ought to be done with these species? I say this matter is to be considered according to the general rule that the body and blood of Christ remain under these species for as long as the substance of the bread and wine would remain under them before consecration, and the body and blood of Christ stop being under the species as soon as the species of bread and wine stop being there. And therefore if those species were to come to such decay that the substance or appearance of bread and wine does not or could not remain under them, then the body and blood of Christ are no longer present.
>
> . . . But what if worms come from these decayed species? What ought to be done with the worms? I say they ought to be burned and the ashes concealed within the altar, and the priest, by whose negligence this has happened, ought to be severely punished.[20]

The scope and the specificity of anxieties about defects suggests that the doctrine of transubstantiation was neither abstract nor theoretical. These are discrete problems that arise from its praxis. They also help us to grasp just how viscerally sixteenth-century Western Christians reacted against the nexus of God Incarnate and the bread and the wine of the Eucharist.[21] For them, too, that relationship was neither abstract nor theoretical, but intimate and somatic.

Those who broke with the medieval church in the sixteenth century, whom scholars tend to call "Protestants," offered a range of ways of thinking about the relationship between the bread and the body of Christ and the wine and the blood of Christ. They all rejected the doctrine of transubstantiation. They rejected that separation of the body from the blood that had troubled Durand and other medieval commentators. They all rejected as well as any notion that the celebrant effected any change himself, even in speaking the words Christ spoke at the Seder the night before he was crucified.

Most Evangelicals offered bread—rather than a host—and wine. Martin Luther was untroubled by the Eucharistic vessels—patens and chalices that were often made of precious metals and adorned with precious stones.[22] Other Evangelicals and the churches they founded rejected the artisanal attention and the material magnificence of medieval patens, pyxes, chalices, and other liturgical objects. Ulrich Zwingli and John Calvin called for simple plates and cups.[23] For Zwingli especially, the Seder at which Christ had spoken the words of institution was a meal; for him, worship therefore was the sharing of those simple foodstuffs among the faithful.[24]

Nothing was more divisive in the sixteenth century than the question of the relationship of the bread and the wine to the body and blood of Christ. Luther rejected any effort to make human sense of the words Christ spoke—from the doctrine of transubstantiation with its concepts of "accidents" and "substance" to any notion that Christ's words were "a figure of speech" or a "trope."[25] For him, Christ's words, "This is my body," were sovereign. So, too, Christ's body was unique, not defined by the materiality of the human body. As he stated in 1528, in his fullest expression of his position (which led to the Colloquy of Marburg in 1529):

> Our faith maintains that Christ is God and man, and the two natures are one person, so that this person may not be divided in two; therefore, he can surely show himself in a corporeal, circumscribed manner at whatever place he will. . . . Here you must take your stand and say that wherever Christ is according to his divinity, he is there as a natural, divine person and he is also naturally and personally there.[26]

For Luther, Christ's body was like no other body, so the medieval formulation subsumed Christ's words to human imagination, in an effort to ground a mystery. So, too, Luther held that the ways other Evangelicals understood the words were also not authorized by scripture and failed to acknowledge divine omnipotence and Christ's unique union of human and divine.

If Luther rejected any effort to articulate the relationship between bread and wine and Christ's body and blood, Zwingli and Calvin each posited a notion of "sign"—distinctive from one another and engendering new constellations of relationship among the human faithful, the matter of the bread and the wine, and God Incarnate. For Zwingli, Christ's spatial location next to God the Father meant that the matter of the bread and the matter of the wine remained mundane food substances; Christ was not physically linked to either Eucharistic element.[27] Calvin's understanding of "sign" cannot be so quickly summarized.[28] While he did publish a pamphlet on what he called the Supper, he laid out his understanding most fully in the major work of his entire life, the *Institutes of the Christian Religion.*

> To summarize: our souls are fed by the flesh and blood of Christ in the same way that bread and wine keep and sustain physical life. For the analogy of the sign applies only if souls find their nourishment in Christ—which cannot happen unless Christ truly grows into one with us, and refreshes us by the eating of his flesh and the drinking of his blood.[29]

For Calvin, the bread and the wine were "signs," but signs grounded in the Incarnation.[30] Luther, as well as Catholics, held Calvin's notion of "sign" to be empty. Calvin's conceptualization, viewed not through their lenses, but in his own terms, pointed toward a sense of matter neither inert nor "objective," whose significance God had determined and which humankind might perceive or not, and which worked physically as well as cognitively to "nourish" through the growing "coalescence" of human and Christ.[31]

In closing, let me turn to the image with which this essay begins (fig. 1), which renders the most ancient vulnerability of the matter of the Eucharist: the recipient. On one point, all churches, whether Protestant or Catholic, were agreed: the unworthy did not receive Christ, whether corporeally, as Catholics and Lutherans held, or spiritually, as the Reformed Tradition held. Paul had raised the question of the worthiness of the recipient of bread and wine, body and blood, in I Corinthians 11:

> Whoever, therefore, eats the bread or drinks the cup of the Lord in an unworthy manner will be answerable for the body and blood of the lord. Examine yourselves, and only then eat of the bread and drink of the cup. For all who eat and drink without discerning the body, eat and drink judgment against themselves.[32]

Reformation churches differed as to what each held worthiness to encompass. For the Jesuit Peter Canisius, following Aquinas, "worthiness" encompassed the sacrament of penance, that is, the preparation of the recipient through self-examination, confession, penance, and absolution.[33] Luther's *Enchiridion* taught: "a person who has faith in these words, 'given for you' and 'shed for you for the forgiveness of sins,' is really worthy and well prepared."[34] Calvin's *Genevan Catechism,* as well as the *Heidelberg Catechism,* taught self-examination as preparation for reception and that unworthiness was discernible.[35]

The volatility of the matter of the Eucharist is captured in figure 1. The bishop to the right of the altar in the image visually signals consecration has taken place; the deacons are offering communion to laymen, marked by their clothing. In the left hand of each deacon is a paten, to capture any crumbs that should fall. The lips of each recipient have been rendered in red, clearly marking the mouths that are about to receive what each deacon holds in his right hand. In its accidents, what each deacon holds is identical: a white disk. But what is on each disk dramatically differs: an infant Christ, a piece of bread, and a toad. What differentiates the hosts, the image suggests through visual clues, is not the words of consecration: the liturgical vessels, the albs of the deacons are identical, and the bishop presides over all. As the host on the left visualizes, Christ is present—but not for everyone. The recipients differ. The one on the left is rendered in the same pale white and pink as the deacons. The one in the middle is slightly darker. The one on the right is singularly red.[36] His face signals something other than devotion—perhaps anger, perhaps heresy; he receives a toad.[37]

All churches agreed that Christ had changed how the faithful were to think about bread and wine, body and blood; for them, these four would never be discrete from one another. There was, for each church, an ontological, "real," relationship between the matter of the bread and the wine and the body and blood of Christ. It is perhaps more pertinent to a volume on active matter that, after the sixteenth century, Christians did not agree on what, exactly, the matter of the bread and the wine were after the words of consecration. Equally pertinent, however, is that, for all churches, what was received in the mouth differed, depending on the recipient. This was no mechanical or chemical transformation, but a moment of singularly anti-inert materiality.

NOTES

1. Pelikan and Hotchkiss, *Creeds & Confessions*, 159.
2. For a recent exploration, see Melion and Wandel, *Image and Incarnation.*
3. This term will refer to the priest who reads the canon in the Mass, speaks the words of institution, consecrates and elevates the host and chalice.
4. These are known as the words of institution. On their history, a good place to begin is Jungmann, *Missarum Sollemnia,* 2:243–51. Jungmann also remains the most complete and detailed history of the Mass itself.
5. "[C]uius corpus et sanguis in sacramento altaris sub speciebus panis et vini veraciter continentur, transsubstantiatis pane in corpus et vino in sanguininem potestate divina." Tanner, *Decrees,* 1:230.

6. On the history of the term, see Goering, "Invention of Transubstantiation."
7. An excellent introduction to medieval discussions of transubstantiation is Levy, Macy, and Ausdall, *Companion to the Eucharist.*
8. For an introduction to the sixteenth-century debates, see Wandel, *Companion to the Eucharist.*
9. The *New Oxford Annotated Bible* is referred to here and below.
10. For a recent and important exploration of "presence," see Orsi, *History and Presence.*
11. This genre of images has a rich literature. For a sense of the range of recent approaches, see Meier, Die Gregorsmesse; Bynum, "Seeing and Seeing Beyond"; Gormans and Lentes, *Das Bild der Erscheinung.* The image is available through the painting's page in the J. Paul Getty Museum online collection, accessed August 13, 2021, http://www.getty.edu/art/collection/objects/533/adriaen-ysenbrandt-the-mass-of-saint-gregory-the-great-flemish-about-1510-1550/.
12. Guido of Monte Rochen, *Handbook for Curates,* 104.
13. "William Durandus of Mende (c. 1230–1296), undoubtedly the best known and most widely read liturgical expositor of the late Middle Ages, composed his *Rationale for the Divine Office* (c. 1292–1296) at the very close of the century that saw the widespread dissemination of the liturgy of the papal court." Thibodeau, "Western Christendom," 230.
14. The numbers given in the body refer to the numbering in the modern editions, so, book IV, chapter 41, paragraph 47. *Durand, Rationale IV,* 353.
15. Durand, *Rationale IV,* 41, 48–49.
16. Ibid., 355.
17. Guido of Monte Rochen, *Handbook for Curates,* 102–8.
18. Ibid., 103–4.
19. On anxiety about decay, see, for example, Bynum, *Christian Materiality,* esp. 177–94.
20. Guido de Monte Rochen, *Handbook for Curates,* 107–8.
21. For the fuller development of this analysis of Reformation debates, as well as the relevant literature, see Wandel, *Eucharist in the Reformation.* See also Davis, *This Is My Body*; Wandel, *Companion to the Eucharist.*
22. Heal, *Magnificent Faith.*
23. For recent studies of Zwingli's eucharistic thinking, see Wandel, "Zwingli and Reformed Practice"; Wandel, "Body of Christ"; Gordon, "'It Is the Lord's Passover.'"
24. Zwingli, *Action oder Bruch des Nachtmals.*
25. Luther, "Confession Concerning Christ's Supper," 313. "Although the

oft-cited formula, 'within, with and under' ['in, mit und unter'], is not actually found in Luther's own writings, but in the Formula of Concord of 1577, it echoes his sense," Leppin, "Martin Luther," 55.

26. Luther, "Confession Concerning Christ's Supper," 218.
27. Köhler, *Zwingli und Luther.*
28. For a most promising new reading of Calvin's understanding of the Eucharist, see Bridges, "'In the Flesh a Mirror.'"
29. *Institutes* IV, 17, 10, *Calvin: Institutes,* 2:1370. The Latin reads as follows:

    > Summa sit, non aliter animas nostras carne et sanguine Christi pasci, quam panis et vinum corporalem vitam tuentur et sustinent. Neque enim aliter quadraret analogia signi, nisi alimentum suum animae in Christo reperirent. Quod fieri non potest nisi nobiscum Christus vere in unum coalescat, nosque reficiat carnis suae esu et sanguinis potu.

    Calvin, *Institutio Christianae Religionis,* col. 1009.
30. On Calvin's understanding of "sign," see Soulié, "Signes et sacrements"; Wandel, "Incarnation, Image, and Sign."
31. Calvin, *Institutio Christianae Religionis,* book IV, chap. 17.
32. I Corinthians 11.27–29.
33. Wandel, *Reading Catechisms,* 232, 263.
34. Luther, "Small Catechism," 363, quoted in Wandel, *Reading Catechisms,* 265.
35. Wandel, Reading Catechisms, 270–73. See also Calvin, *Institutio Christianae Religionis,* IV, 17, 41.
36. On red hair and ruddy skin as a sign of otherness, see Mellinkoff, *Outcasts,* vol. 1, chap. 7.
37. On the possible meaning of the toad, see Hamburger, "Bosch's 'Conjurer.'"

BIBLIOGRAPHY

Bridges, AnnMarie M. "'In the Flesh a Mirror of Spiritual Blessings': Calvin's Defense of the Lord's Supper as a Visual Accommodation." In *Quid Est Sacramentum? Visual Representations of Sacred Mysteries in Early Modern Europe, 1400–1700,* edited by Walter S. Melion, Elizabeth Carson Pastan, and Lee Palmer Wandel, 105–24. Leiden: Brill, 2020.

Bynum, Caroline Walker. *Christian Materiality: An Essay on Religion in Late Medieval Europe.* Brooklyn: Zone Books, 2011.

———. "Seeing and Seeing Beyond: The Mass of St. Gregory in the Fifteenth Century." In *The Mind's Eye: Art and Theological Argument in the Middle Ages,* edited by Jeffrey Hamburger and Anne-Marie Bouché, 208–40. Princeton, NJ: Princeton University Press, 2006.

Calvin, John. *Calvin: Institutes of the Christian Religion.* Edited by John T. McNeil. Translated by Ford Lewis Battles. Philadelphia: Westminster Press, 1960.

———. *Institutio Christianae Religionis in Libros Quatuor . . . 1559.* Vol. 2 of *Ioannis Calvini Opera Quae Supersunt Omnia.* Edited by Wilhlem Baum, Edward Cunitz, and Edward Reuss. Braunschweig: Schwetschke & Filium, 1864.

Coogan, Michael D., ed. *The New Oxford Annotated Bible.* 3rd ed. Oxford: Oxford University Press, 2001.

Davis, Thomas J. *This Is My Body: The Presence of Christ in Reformation Thought.* Grand Rapids: Baker Academic, 2008.

Durand, William. *Rationale IV: On the Mass and Each Action Pertaining to It.* Translated by Timothy M. Thibodeau. Turnhout: Brepols, 2013.

Goering, Joseph. "The Invention of Transubstantiation." *Traditio* 46 (1991): 147–70.

Gordon, Bruce. "'It Is the Lord's Passover': History, Theology, and Memory in the Liturgy of the Lord's Supper in Reformation Zurich." In *Liturgy's Imagined Past/s: Methodologies and Materials in the Writing of Liturgical History Today,* edited by Teresa Berger and Bryan D. Spinks, 172–200. Collegeville, MN: Liturgical Press, 2016.

Gormans, Andreas, and Thomas Lentes, eds. *Das Bild der Erscheinung: Die Gregorsmesse im Mittelalter.* Berlin: Reimer, 2007.

Guido of Monte Rochen. *Handbook for Curates: A Late Medieval Manual on Pastoral Ministry.* Translated by Anne T. Thayer. Washington, DC: Catholic University of America, 2011.

Hamburger, Jeffrey. "Bosch's 'Conjurer': An Attack on Magic and Sacramental Heresy." *Simiolus: Netherlands Quarterly for the History of Art* 14 (1984): 4–23.

Heal, Bridget. *A Magnificent Faith: Art and Identity in Lutheran Germany.* Oxford: Oxford University Press, 2017.

Jungmann, Josef Andreas. *Missarum Sollemnia: Eine genetische Erklärung der Römischen Messe.* Freiburg: Herder, 1962. Reprinted in 2003 by Nova & Vetera (Berlin).

Köhler, Walther. *Zwingli und Luther, ihr Streit über das Abendmahl nach seinen politischen und religiösen Beziehungen.* 2 vols. Leipzig: M. Heinsius Nachfolger, 1924 and 1953.

Leppin, Volker. "Martin Luther." In Wandel, *Companion to the Eucharist,* 39–56.

Levy, Ian Christopher, Gary Macy, and Kristen van Ausdall, eds. *A Companion to the Eucharist in the Middle Ages.* Leiden: Brill, 2012.

Luther, Martin. "Confession Concerning Christ's Supper, 1528." In *Luther's Works,* vol. 37, translated by Robert H. Fischer, 151–372. Minneapolis: Fortress Press, 1961.

———. "The Small Catechism." In *The Book of Concord: The Confessions of the Evangelical Lutheran Church*, edited by R. Kolb and T. Wengert, 345–75. Minneapolis: Fortress, 2000.

Meier, Esther. *Die Gregorsmesse: Funktionen eines spätmittelalterlichen Bildtypus.* Cologne: Böhlau, 2006.

Melion, Walter S., and Lee Palmer Wandel, eds. *Image and Incarnation: The Early Modern Doctrine of the Pictorial Image.* Leiden: Brill, 2015.

Mellinkoff, Ruth. *Outcasts: Signs of Otherness in Northern European Art of the Late Middle Ages.* 2 vols. Berkeley: University of California, 1993.

Orsi, Robert. *History and Presence.* Cambridge, MA: Harvard University Press, 2016.

Pelikan, Jaroslav, and Valerie Hotchkiss, eds. *Creeds & Confessions of Faith in the Christian Tradition.* Vol. 1. New Haven, CT: Yale University Press, 2003.

Soulié, Marguerite. "Signes et sacrements dans la théologie de Jean Calvin." In *Les Signes de Dieu aux XVIe et XVIIe siècles*, 27–35. Clermont-Ferrand: Faculté des Lettres et Sciences humaines de l'Université Blaise-Pascal, 1993.

Tanner, Norman P., ed. *Decrees of the Ecumenical Councils.* 2 vols. London: Sheed & Ward, 1990.

Thibodeau, Timothy. "Western Christendom." In *The Oxford History of Christian Worship*, edited by Geoffrey Wainwright and Karen B. Westerfield Tucker, 216–53. Oxford: Oxford University Press, 2006.

Wandel, Lee Palmer. "The Body of Christ at Marburg, 1529." In *Image and Imagination of the Religious Self in Late Medieval and Early Modern Europe*, edited by Walter Melion and Reindert Falkenberg, 195–213. Turnhout: Brepols, 2007.

———, ed. *A Companion to the Eucharist in the Reformation.* Leiden: Brill, 2014.

———. *The Eucharist in the Reformation: Incarnation and Liturgy.* Cambridge: Cambridge University Press, 2006.

———. "Incarnation, Image, and Sign: John Calvin's *Institutes of the Christian Religion* and Late Medieval Visual Culture." In Melion and Wandel, *Image and Incarnation*, 187–202.

———. *Reading Catechisms, Teaching Religion.* Leiden: Brill, 2016.

———. "Zwingli and Reformed Practice." In *Educating People of Faith: Exploring the History of Jewish and Christian Communities*, edited by Dorothy Bass and John Van Engen, 270–93. Grand Rapids, MI: Eerdmans, 2007.

Zwingli, Huldrych. *Action oder Bruch des Nachtmals, Gedechtnus oder Dancksagung Christi.* Zurich: Christoph Froschauer, 6.IV.1525.

PART 3

Indigenous Ontologies

ELEVEN

Introduction: For the Lives of Things—Indigenous Ontologies of Active Matter

Aaron Glass

Abstract

This short essay introduces the Bard Graduate Center working group on Indigenous ontologies of active matter, defines its operative concepts, and frames the four contributing essays that follow. The so-called ontological turn in the humanities and social sciences has encouraged recognition of culturally specific theories of being that depart from those in the dominant Western tradition. These essays consider Indigenous conceptions of "objects" that privilege dynamic activity, multiple agencies, and relational networks and explore the implications of such notions for responsive museum conservation, care, and custodianship.

In the spring of 2008, I encountered a remarkable being at the University of British Columbia's Museum of Anthropology (MOA). Arriving late to an event titled *T'xwelátse Visits MOA*, which I knew little about, I took my place at the back of a sea of people listening to area First Nations welcoming a visiting dignitary to the university, to traditional unceded Musqueam territory, and to Vancouver. The figure these speakers were addressing initially appeared to me as a four-foot-tall, six-hundred-pound granite sculpture wreathed in a cedar-bark cape, seated in a small dugout canoe, and surrounded by photographs of people embracing him (fig. 1). He was Stone T'xwelátse, a person who had been repatriated, two years earlier, from the Burke Museum in Seattle to the Nooksack people, then returned to his Stó:lō relatives in the Fraser Valley, just east of Vancouver. Stó:lō families have an ancestor named T'xwelátse—a name that has been passed down through various lineages—who was turned to stone in a duel with X̱a:ls, the Transformer (there are numerous such Transformed Beings and Places among Coast Salish peoples). Stone T'xwelátse lived for centuries, cared for by his female descendants as X̱a:ls mandated. He relocated within the Fraser Valley over subsequent generations and was "found" in 1892 by a settler farmer in Nooksack territory, over the border in Washington State, and sold to a small museum that in turn transferred him to the Washington State Museum (now the Burke Museum). After Herb Joe, the current name carrier, reconnected with him at the Burke in the early 1990s, collections staff at the museum asked how they should properly care for the ancestor. In response to learning that Stone T'xwelátse was alive, they began a protocol, which continued until the repatriation, of putting him to bed every night and waking him up every morning by speaking to him and covering and uncovering him with a muslin blanket.[1] In 2008, he was on a diplomatic visit (not "on exhibit") at MOA, where staff adopted the same slumber routine.

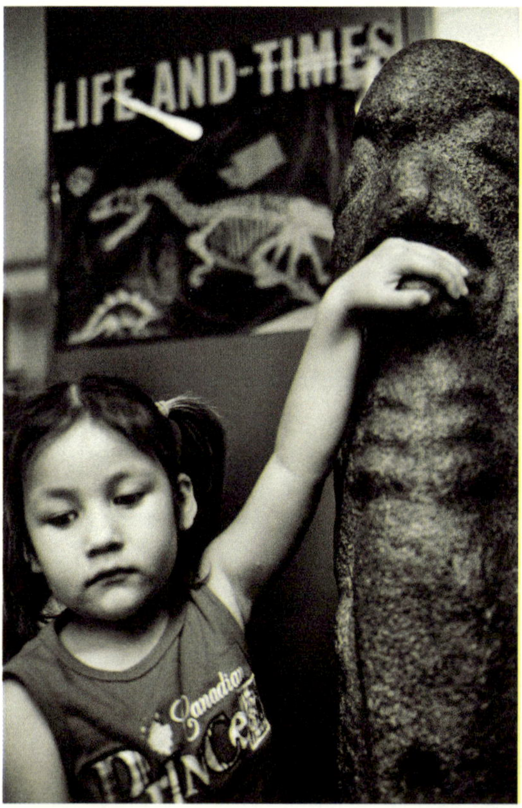

Fig. 1. Georgia Joe, age three, visiting Stone T'xwelátse at the Burke Museum of Natural History and Culture, Seattle, WA, 2006. Georgia's grandfather Herb Joe carries the name T'xwelátse. Courtesy of David Campion.

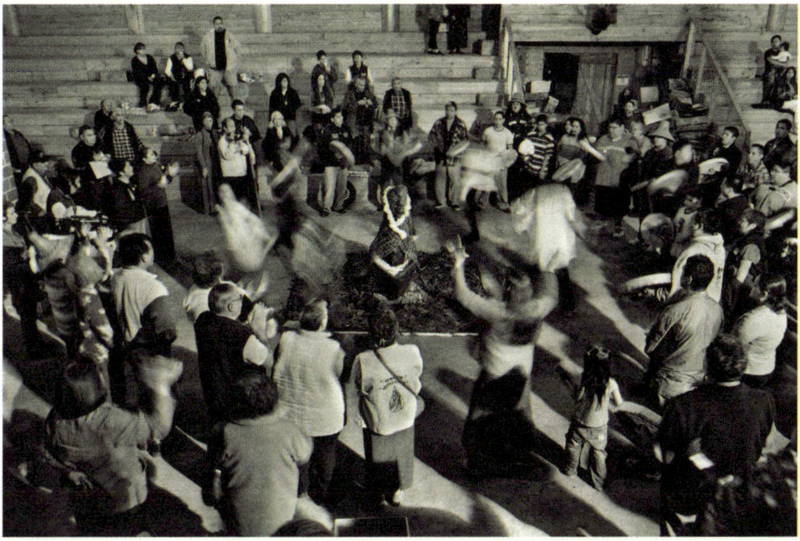

Fig. 2. Stó:lō and Nooksack community members welcome Stone T'xwelátse home in the Semá:th Longhouse, Chilliwack, BC, 2006. Courtesy of David Campion.

Stone T'xwelátse's status as an ancestral being, and not simply a *sculptural representation* of a relative, was formally signaled when the repatriation claim was submitted under two separate sections of the Native American Graves Protection and Repatriation Act, one pertaining to "objects of cultural patrimony" and the other to "Native American human remains." According to the repatriation report submitted by the Stó:lō and Nooksack, Stone T'xwelátse is, among other things, "the physical remains of a human in granite form, retaining his life force (*shxweli*) and soul (*smestíyexw*)"; "centrally important in directly connecting to and attesting to the creation of the world"; "associated with a specific place in the landscape, forming part of a broader physical-spiritual landscape transformed by the Transformers"; "a physical object containing a living, ancestral human soul, and carrying a specific name of a transformed individual"; and "linked to a form of shared oral history that informs the broadest level of cultural identity attached to the shared understanding of the origin of the world and behavioral rules, binding people together across tribal groupings."[2] Thus, he is not only a physical ancestor ("human remains") and a materialization of ancestral knowledge and practice ("cultural patrimony"); he is also a still-living being enmeshed in ongoing social relationships with both relatives and others across community, institutional, and international borders. Stone T'xwelátse is now back at home among the Stó:lō, surrounded once again by kin, re-embedded in his ancestral landscape and network of relations, and cared for by stewards at the Stó:lō Research and Resource Management Centre on behalf of the T'xwelátse family (fig. 2). His story—his very being—challenges conventional Western categories of what "objects" really are and prompts revisions to long-standard museological approaches to the care and conservation of collections.

The Working Group on Indigenous Ontologies

My encounter with Stone T'xwelátse, and his capacity to exist at once on multiple material, spiritual, and temporal planes, returned to me a few years later when Dean Peter Miller first proposed the Cultures of Conservation initiative at Bard Graduate Center. Until then, I had worked in various capacities (docent, volunteer, researcher, curator) at museums of different sorts (art and anthropology; university and metropolitan; gallery and cultural center), and I had thought about a range of museum practices, especially collecting, exhibition, and database design.[3] I had done just enough reading about contemporary conservation to know that Indigenous efforts to reform it play a key role in the larger struggle to decolonize museums by redistributing curatorial power, repatriating objects, and shifting from regimes of property ownership to those of care and custodianship of collections.[4] From our earliest conversations at Bard, we felt it important that the pluralization of "cultures" in the project title should signal the topical breadth of our approach (see "Preface: Report on a Research Project," in this volume). This has entailed bridging the "culture of academia" and the "culture of the museum" through close attention to the material properties of objects as a method of both research and conservation practice. But we also wanted to stress that diverse peoples of the world bring their own cultural values and orientations to the material realm, and thus to conservation concerns: their own fundamental notions of what "the object" is and isn't; what its proper relationship to "persons" might be; and what factors contribute to protocols surrounding its care and repair, its treatment and preservation—if indeed it ought to be preserved at all.

Within the larger initiative, my role in the Indigenous Ontologies working group dates back to a fascinating 2016 presentation. Wolfgang Schäffner, from the Helmholtz Center for Cultural Techniques of the

Humboldt University in Berlin, shared with us his materials science research unit's investigations into the chemical and biological properties of certain materials that are particularly "active" (unstable, mutable, transformational) at the molecular or structural level. His focus was on the materials' potential to be harnessed in forward-thinking design, although he mentioned the philosophical challenge they pose to traditional Western views of the material world. After his talk, both I and a visiting research fellow, Beth Piatote (who teaches comparative literature and Native American studies at University of California, Berkeley, and who is Nez Perce), responded that for many Indigenous people, objects—or at least certain objects and materials—have never been thought of as static in the first place, but rather as both *active subjects* and as *subject to* interconnecting webs of animating force and relationality. Citing the Lakota precept *Taku Skanskan*, "What Moves-moves" (i.e., that which gives motion to everything that moves), Piatote suggested the advantage of starting from Indigenous ontologies of human and nonhuman beings within an animate and animating universe. Schäffner expressed interest in the potential overlap in our perspectives but said that such considerations were beyond his lab's immediate purview.

This conjuncture provided one of many prompts for the establishment of the interdisciplinary, collaborative project Conserving Active Matter and for the inclusion of Indigenous ontologies as one of its research vectors (see Appendix: Events of the Research Project Conserving Active Matter for a list of related events). Our collective starting point was the fact that there are many disparate discourses of something like "active matter" swirling around at the moment. We agreed that we should attempt not to reconcile them but rather to explore the possibility of productive conversation between people who rarely speak to one another at all, much less from positions of shared disciplinary training, cultural orientation, philosophical proclivity, or vocabulary use.

Throughout the project, we continually posed a general question to ourselves and to our colleagues, and we had this question repeatedly posed back to us: When we talk about "active matter," what are we really talking about? This provocation came to entail a series of related queries: Are we talking about molecules in motion, visible under a microscope or some other tool, or the capacity of objects to act as effective agents in the world, whether or not they are capable of temporarily or permanently manifesting animate presences (such as Stone T'xwelátse clearly is)? Is "object" even a universally applicable term? Many Indigenous people reject this word as enacting a limited Western dualism, instead referring to treasured items as Subjects, Beings, Belongings, Medicine, and

Ancestors, or forgoing European languages altogether and simply calling them *Dlugwai* (Kwak̓wala), *At:oow* (Łingit), *Taonga* (Māori; see Rose Evans's chapter in this volume). What do scholars, curators, conservators, and custodians do with museum collections that need to be fed, talked to, or cleansed with purifying smoke? To what, and to whom, do we imagine the "matter" of things being intrinsically and actively related, especially over time: Atomic components? Raw materials? Completed objects? Intentional subjects? Indigenous (and other "non-Western") value systems often insist that these levels or scales must be kept in direct and active *relationship*—to one another; to human makers and users and caregivers of the past, present, and future; and to other-than-human beings, both terrestrial and celestial. When we talk about "conserving" active matter, what is it we are concerned to preserve, why, and for whom: The historical object in something like its "original" state? The raw matter or materials from which it was made? Knowledge of that origin, identity, and embedded place in an ever-expanding nexus of genealogical relations? Or the very preconditions for making, using, and caring for similar things in the future—a future currently threatened by continued colonial disruption and impending ecological disaster?

The Indigenous Ontologies working group took as our mandate the exploration of these questions and the importance of bringing varied Indigenous ways of being and knowing into dialogue with other realms of academic and museum practice—both mainstream and avant-garde—that are invoking similar terms in often quite divergent and siloed contexts. Midway through the project's 2017 launch symposium, my working group colleague Jolene Rickard (a Tuscarora artist and art historian at Cornell University) turned to me and said that rather than imagining Indigenous ontologies as a peripheral—or, worse, token "diversity"—contribution, our perspectives have the potential to unite the concerns of the other working groups as they encompass materials science, philosophy, history, and art. This insight is especially important for the development of mainstream conservation practice, as training in "ethnographic" materials and Indigenous cultural concerns is often restricted to specialist courses or programs but deserves to have much wider influence on the field given its development of progressive and responsive protocols for the care and repair of particularly active cultural materials (see the chapters by Kelly McHugh and Rose Evans, in this volume).[5] It was thus in the dual spirit of intellectual convergence and institutional critique that we initiated our three-year investigation into diverse Indigenous ontologies of active matter and their implications for the future of conservation in and beyond museums.

Working Terms and Section Contributors

Over the past two decades, scholarly developments in a number of disparate fields within the humanities and natural and social sciences have coalesced into what is sometimes called an "ontological turn." This phrase is used to characterize and promote a turn *away from* epistemological critique of knowledge production and *toward* a purportedly radical rethinking of the ontological basis of reality itself and the multiple forms of life (beings, relations, and materialities) that inhabit and constitute reality.[6] While many proponents of an ontological perspective are engaged in a process of mining marginal strains in Western thought and theory in order to posit alternate ontologies, a number of scholars within anthropology and Indigenous studies have brought philosophical interest in ontology to ethnographic theory and methods and to contemporary political concerns over decolonization and sovereignty.[7] The subset of this work that was of primary interest to us examines the implications of Indigenous ontologies for an understanding of material culture and for museum practice.[8]

For the purpose of our working group, we use the term "ontologies" to refer to culture-specific theories of being, especially the beings and belongings that have material presence in museums (a more in-depth treatment of the concept of Indigenous ontologies can be found on the exhibition's companion website).[9] We ask ourselves, According to originating communities, what *are* the items in museum collections, ontologically speaking? What are their *fundamental and intrinsic* identities as certain kinds of beings in relation to other kinds of beings (be these other "objects," "persons," or other kinds of agents), as well as in relation to specific cultural practices and knowledges (something we might gloss as cultural epistemologies)? In the realm of conservation science, how can we move from standard concerns about inherent vice and the prevention of decay to recognition for, if not restoration of, an intrinsic capacity in things for growth and kinship and transformation—for *life* itself? The ultimate goal of our working group has been to examine the practical ramifications for conservation work (as well as exhibition practice and repatriation) that follow from taking seriously Indigenous ontologies of "the object" as a dynamic assemblage of active and emergent materialities, of multiple subjectivities and temporalities, of diverse forms of cultural knowledge and practice (which might be hard to reconcile fully), and of persistent human and other-than-human relations.

A theoretical perspective on the ontological status of objects-as-beings (or of the specific *being* of objects) lends itself particularly well to growing interest in forms of matter or materials that can be understood as especially "active." But how are we to relate the insights of materials scientists like Wolfgang Schäffner to those of Stó:lō descendants of Stone T'xwelátse? Are new "smart textiles" *active* in the same, or at least in comparable, ways as animate beings in sculptural form? Or are we simply pushing the semantic limits of terms like "active" (not to mention "matter") to the breaking point, collating radically different phenomena under the flimsy banner of similar terminology? Such questions have animated our entire project. If prompted, I think we would all admit quite freely that we *are* talking about different sorts of things, and that respecting, acknowledging, and limning those distinctions is important, both theoretically and methodologically. While notions of activity in/of/by matter are philosophically, culturally, and materially divergent, they share more than mere susceptibility to common descriptive language: they all present a range of practical and conceptual challenges to conventional Western conservation principles and protocols. It is this aspect of their converging application that our larger initiative has tried to focus on without erasing or collapsing important distinctions.

Rather than attempt to produce a unified field theory of active matter, we have tried to characterize a range of ways in which the *particularly* active qualities of *certain* objects and materials force an expansion in traditional conservation thinking and practice (see the other section introductions to this volume for parallel approaches to the "activity" of matter). Returning to Rickard's provocation that Indigenous ontologies of the material world encompass materials science, philosophy, history, and art, our working group formulated a provisional set of four variable and overlapping "ways of being active"; these are not discrete types or classifications to which specific objects belong but different qualities of activity we might recognize in them. None is unique to Indigenous peoples, but they help us to organize the various ways in which many Indigenous things can be active, activated, or activating. In particular, these four categories *locate the activity* in question variably in materials; in and between materially diverse persons; across space or cultural divide; and over time:[10]

1. Items made from **materials that are particularly mutable or unstable** at the molecular, chemical, or structural level. This use of the term "activity" easily accords with conventional conservation knowledge and practice but also encompasses the ways in which Indigenous science and traditional ecological knowledge might mobilize the active qualities of certain materials.

2. Items constructed from relatively stable materials but manifest in **forms that demand intentional change over time**. This is the case for many kinds of things, both vernacular and ceremonial, that are kept in active circulation within communities of origin and are subjected to protocols of care, repair, or veneration. One extreme variety comprises objects meant to be "deactivated" or destroyed completely after their intended function has been served.
3. **"Power" objects or animate beings** that may be constructed of relatively stable materials and may maintain relatively consistent form, but that are imbued with or capable of channeling active spiritual presence and agency—the ability to act in and on the world, whether to heal or protect, to cause harm, or to enact genealogical connection (like Stone T'xwelátse). Many such items require regular feeding, human touch, or ritual treatment by specialists and are especially challenging for standard conservation practice.
4. **Items whose activation depends on remaining embedded in persistent networks and relations of knowledge, genealogy, and practice**. This has less to do with intrinsic material conditions than with immaterial properties, temporal continuities, and social contexts of access, awareness, and use. Without connection to specialist knowledge and performance, many such items remain sleeping or dormant in museums.

Taken together, these four approaches to "active matter" draw on certain Indigenous ontologies of the object in order to challenge professional Western museum conservation to adapt and expand its conventional values and protocols, which are by and large grounded in the European Enlightenment.[11] Conservators have long been charged with prioritizing the preservation or stabilization of the material thing, arresting its change and decay, but what if the fundamental nature/culture of the "object" is change and transformation, liveliness, or even outright disappearance? Over the last half century, the reversibility (or at least visibility) of the conservation treatment has been paramount to the professional code of ethics, but what if practices of care and repair should be made intrinsic to the object, altering it essentially and appropriately? Conservators often consider the physical and conceptual "integrity" or "authenticity" of the "original" item when deciding between treatment options, but are such qualities inherent in form and material or in historic contexts of production, or do they rather reside properly in unfolding contexts of use, changing functions or values, or efforts to actively maintain human knowledge, embodied skill, and social practice into the future? Curators and conservators alike frequently speak of the need to respect the intention of the original maker or artist, but what if intentionality and agency are extended beyond the individual human to the social group, to communities of consumption, to the material "thing" itself, and to other-than-human beings (embedding the thing in expansive relational networks)? And what if such agency is extended temporally from the moment of creation both backward and forward in time (embedding the thing in expansive genealogies of knowledge, practice, value, and affiliation)?

Many museums with "ethnographic" collections have been considering such questions for decades now, and our project is intended in part to hasten such considerations, to extend them beyond the realm of Indigenous concerns alone, and to promote an expansion in the conceptualization of what all conservation can be in and for the future.

The four contributors to this section have extensive experience wrestling with such complex questions in the context of transforming the relationship between mainstream conservation practice and the Indigenous peoples whose cultural patrimony has long been collected, preserved, and exhibited under largely Western scientific, aesthetic, and philosophical paradigms. All four are museum professionals (two conservators, Kelly McHugh and Rose Evans; and two curators, Jamie Jacobs and Sven Haakanson) who participated in our Cultures of Conservation programming and adapted their presentations to this exhibition companion volume. McHugh and Evans provide broad overviews of the history and current state of conservation in their respective regions and national contexts (North America and Aotearoa–New Zealand, respectively), focusing on changes to the field that are responsive to Indigenous ontologies. Jacobs and Haakanson offer focused case studies that explore particular kinds of things and modes of their community activation (wampum among the Haudenosaunee of upstate New York and the revitalization of open boats among the Alutiiq in coastal Alaska). Some authors refer explicitly to specific objects or object types that were featured in the exhibition, and all speak more broadly about the varieties of active matter conceptualized in Indigenous ontologies.

NOTES

In addition to the section authors and my Bard Graduate Center colleagues on the Conserving Active Matter project, I'd like to thank the following people for valuable feedback on earlier drafts of this essay: Lucas Bessire,

Fred Myers, Maia Nuku, Beth Piatote, Helen Polson, Jolene Rickard, Susan Rowley, and David Schaepe.

1. Personal correspondence with Laura Phillips, archaeology collections manager, Burke Museum, September 18, 2020.
2. Schaepe, "*Stone T'ixwelatsa* Repatriation Report," 1, 23–24. My thanks to David M. Schaepe and Susan Rowley for additional information. For the complete story of Stone T'xwelátse's life and journeys, visit the Stó:lō Research and Resource Management Centre's comprehensive website Man Turned to Stone: T'xwelátse, 2012, http://www.srrmcentre.com/StoneTxwelatse/1Home.html.
3. See, for instance, Glass, *Objects of Exchange*; Glass, "Indigenous Ontologies, Digital Futures."
4. This is not the place to review the extensive literature on the decolonization of conservation and museums, which the essays in this section discuss and contribute to, but for key texts in the field, see Simpson, *Making Representations*; Clavir, *Preserving What Is Valued*; Peers and Brown, *Museums and Source Communities*; Sullivan and Edwards, *Stewards of the Sacred*; Sully, *Decolonizing Conservation*; Sleeper-Smith, *Contesting Knowledge*; Phillips, *Museum Pieces*; Lonetree, *Decolonizing Museums*; Golding and Modest, *Museums and Communities.*
5. A number of conservators have been promoting just such cross-fertilization within the profession; see Hornbeck, "Intersecting Conservation Approaches"; McHugh and Gunnison, "Finding Common Ground"; Pearlstein, "Conserving Ourselves"; Balachandran and McHugh, "Respectful and Responsible Stewardship."
6. The turn toward such ontological concerns can be seen as one component of a larger and even more diffuse "material turn" that loosely assembles numerous cross-disciplinary movements under an open meshwork banner often called the new materialisms: "object-oriented ontologies" and "speculative realism" (philosophy); "actor-network-theory" (science and technology studies); "agential realism" (physics); "vibrant matter" or "new vitalism" (political ecology); "thing theory" (literature).
7. For useful overviews, see Scott, "Anthropology of Ontology"; Salmond, "Transforming Translations (Part 2)"; Bessire and Bond, "Ontology"; Kohn, "Anthropology of Ontology."
8. Scholars who have addressed ontologies of art and material culture include Myers, "Ontologies of the Image"; Henare, Holbraad, and Wastell, *Thinking through Things*; Salmond, "Transforming Translations (Part I)"; Cipolla, "Taming the Ontological Wolves."

9. Glass, "What Are Indigenous Ontologies?"
10. For the *Conserving Active Matter* website, these four categories are explicated in greater detail and related directly to the featured loan items, both Indigenous and otherwise, 2022, https://exhibitions.bgc.bard.edu/cam/.
11. Clavir, *Preserving What Is Valued.*

BIBLIOGRAPHY

Balachandran, Sanchita, and Kelly McHugh. "Respectful and Responsible Stewardship: Maintaining and Renewing the Cultural Relevance of Museum Collections." In *Preventive Conservation: Collection Storage,* edited by Lisa Elkin and Christopher A. Norris, 3–24. New York: Society for the Preservation of Natural History Collections, 2019.

Bessire, Lucas, and David Bond. "Ontology: A Difficult Keyword." In "Ontology in *American Ethnologist,* 1980–2014." Virtual issue, *American Ethnologist* (2015): https://anthrosource.onlinelibrary.wiley.com/hub/journal/15481425/ontology.

Cipolla, Craig. "Taming the Ontological Wolves: Learning from Iroquoian Effigy Objects." *American Anthropologist* 121, no. 3 (2019): 613–27.

Clavir, Miriam. *Preserving What Is Valued: Museums, Conservation, and First Nations.* Vancouver: UBC Press, 2002.

Glass, Aaron. "Indigenous Ontologies, Digital Futures: Plural Provenances and the Kwakwaka'wakw Collection in Berlin and Beyond." In *Museum as Process: Translating Local and Global Knowledges,* edited by Raymond Silverman, 19–44. London: Routledge, 2015.

———, ed. *Objects of Exchange: Social and Material Transformation on the Late Nineteenth-Century Northwest Coast.* New York: Bard Graduate Center, 2011.

———. "What are Indigenous Ontologies?" Conserving Active Matter (website), 2022, https://exhibitions.bgc.bard.edu/cam/.

Golding, Viv, and Wayne Modest, eds. *Museums and Communities.* London: Bloomsbury Academic, 2013.

Henare, Amiria, Martin Holbraad, and Sari Wastell, eds. *Thinking through Things: Theorising Artefacts Ethnographically.* London: Routledge, 2007.

Hornbeck, Stephanie. "Intersecting Conservation Approaches to Ethnographic and Contemporary Art: Ephemeral Art at the National Museum of African Art." *Objects Specialty Group Postprints* 20 (2013): 207–26.

Kohn, Eduardo. "Anthropology of Ontology." *Annual Review of Anthropology* 44 (2015): 311–27.

Lonetree, Amy. *Decolonizing Museums: Representing Native America in National and Tribal Museums.* Chapel Hill: University of North Carolina Press, 2012.

McHugh, Kelly, and Anne Gunnison. "Finding Common Ground and Inherent Differences: Artist and Community Engagement in Cultural Material and Contemporary Art Conservation." *Studies in Conservation* 61, sup. 2 (2016): 126–29.

Myers, Fred. "Ontologies of the Image and Economies of Exchange." *American Ethnologist* 31, no. 1 (2004): 5–20.

Pearlstein, Ellen. "Conserving Ourselves: Embedding Significance into Conservation Decision-Making in Graduate Education." *Studies in Conservation* (2016): 1–10. https://doi.org/10.1080/00393630.2016.1210843.

Peers, Laura, and Alison Brown, eds. *Museums and Source Communities: A Routledge Reader.* London: Routledge, 2003.

Phillips, Ruth. *Museum Pieces: Toward the Indigenization of Canadian Museums.* Montreal: McGill-Queen's University Press, 2011.

Salmond, Amiria J. M. "Transforming Translations (Part I) 'The Owner of These Bones.'" *HAU: Journal of Ethnographic Theory* 3, no. 3 (2013): 1–32.

———. "Transforming Translations (Part 2): Addressing Ontological Alterity." *HAU: Journal of Ethnographic Theory* 4, no. 1 (2014): 155–87.

Schaepe, David M. "*Stone T'ixwelatsa* Repatriation Report." Prepared for the Nooksack Indian Tribe, Stó:lō Nation Archives, Chilliwack, BC, 2005.

Scott, Michael W. "The Anthropology of Ontology (Religious Science?)." *Journal of the Royal Anthropological Institute* 19, no. 4 (2013): 859–72.

Simpson, Moira. *Making Representations: Museums in the Post-Colonial Era.* London: Routledge, 1996.

Sleeper-Smith, Susan, ed. *Contesting Knowledge: Museums and Indigenous Perspectives.* Lincoln: University of Nebraska Press, 2009.

Sullivan, Lawrence, and Alison Edwards, eds. *Stewards of the Sacred.* Cambridge, MA: American Association of Museums and Harvard University Press, 2004.

Sully, Dean, ed. *Decolonizing Conservation: Caring for Maori Meeting Houses outside New Zealand.* Walnut Creek, CA: Left Coast Press, 2007.

TWELVE

Living Knowledge in Cultural Collections

Sven Haakanson

Abstract

What happened to the original boats from the Kodiak region of Alaska? What can we learn from historical, ethnographic, and living knowledge to revive boat construction and use in our community? Unlike the kayak, the angyaaq (open boat) from the Kodiak region is virtually unknown. In 1991 I saw my first model angyaaq at the Kunstkamera (MAE) in Leningrad, Russia. Since then, I have found only fifteen other models scattered in museums across the world. In 2014, working with the Burke Museum and Community of Akhiok, we made thirteen model angyaat—the first to be made on Kodiak in over 150 years. By 2016 we constructed two full-size angyaat, reverse engineering the available information to bring back knowledge that was forgotten because of colonial practices. People who work for museums not only are caretakers of collections, but have an important role in preserving cultural knowledge and ensuring it is accessible to the communities where it originated. Indigenous communities are taking ethnographic accounts and the knowledge embodied within collections to heal, reawaken, and grow their cultural knowledge once again. The example of the angyaaq is just one of many ways we can start to heal colonialism's impacts on Indigenous communities.

1. Apaka qasqangqertellria iwarpagaqa	**Apaqaa angiaqlia (angyaaq) Amiaq iwarpagaqa**	*I am seeking my Abba, Amiak, who had many* **angyaat**.
2. Apaka angiaklia Amiak iwarpagaqa	**Apaqaa qasqangnqertellria iwarpagagaqa arnaq arniaqlia.**	*I am seeking my Abba who had lots of sea otter skins.*
3. Apaka qasqangnqertellria iwarpgagaqa arnaq akhniaklia.	**Apaqaa Kigiliak narscarlia.**	*My Abba, Kigiliak, who was a great provider.*
4. Apaka Kigiliak nakhschchaglia	**Apaqaa urtayaq kashliaklia**	*My Abba sang of his wealth.*
5. Apaka ukhtaiak kashliaklia	**Apaqaa qasqngqertellria iwarpagaqa.**	*I am seeking my wealthy Abba.*

Above is a mourning song sung by Aleksei Ikuik, Sugpiaq from Alikhtalik on the southern end of Kodiak Island, in 1804.[1] Ikuik shared this song with Hieromonk Gideon, who was a Russian Orthodox priest traveling around to survey the people of Kodiak and how they were being treated by the Russians. Ikuik was lamenting about prosperous times before the Russians arrived, twenty years prior. Gideon documented this song as part of his report back to the tsar in Russia. This lamentation song shows by this time what the Sugpiat had already lost at the hands of the Russian American Company, which was managed by Alexander Baranoff.[2] It is one of the few references that mention an "angiaklia" (*angyaaq*), a traditional open boat from this region, in the written record. Without living knowledge, that word's meaning would have been lost.[3]

In 1833, Vasilii Ivanovich Kashevarov surveyed the Alutiiq people, their jobs, names of the villages, and their equipment to document what the Russian American Company was claiming to control. Kashevarov's mention of angyaat is the last one that I have seen, and this is only found in an unpublished manuscript, translated in 1990 by Dr. Lydia Black.[4]

The angyaaq (open boat) existed on Kodiak, but quickly disappeared from the living memory of the Sugpiat people from Alaska's

Kodiak region in the mid-1800s. This knowledge was lost to our living memory and only survived in obscure archival documents and in models housed in museums scattered across the world. It would take me over thirty years to learn about angyaat. It was not until 2014, nearly two centuries later, that many of our community members would realize their ancestors constructed and used angyaat, open boats, in the past. The story of the angyaat embodies, to me, the relevance of museums today because they preserve both objects and their knowledge, but more importantly how our collections are meaningless without living cultures. It is important to understand how to synthesize these knowledges that have been dispersed across the world in museum collections, conservation reports, archives, publications, oral histories, languages, personal journals, and images. The angyaaq project would never have happened without the help of community members, museum staff, and collections. With only fifteen original models in existence in our world, I feel lucky to have been part of this reawakening.

As a scholar and Sugpiaq I am interested in the history and role that boats played in our past and today. It is important for our community to understand and for us to teach the next generation about our environment and how to care for it. Knowing about and teaching how to make and use these boats is one part of this ongoing process. We are using this knowledge in an ongoing project at the Akhiok Kids camp to teach our participants about kayak making, from the gathering of the materials to the curing, carving, and construction of the boats.

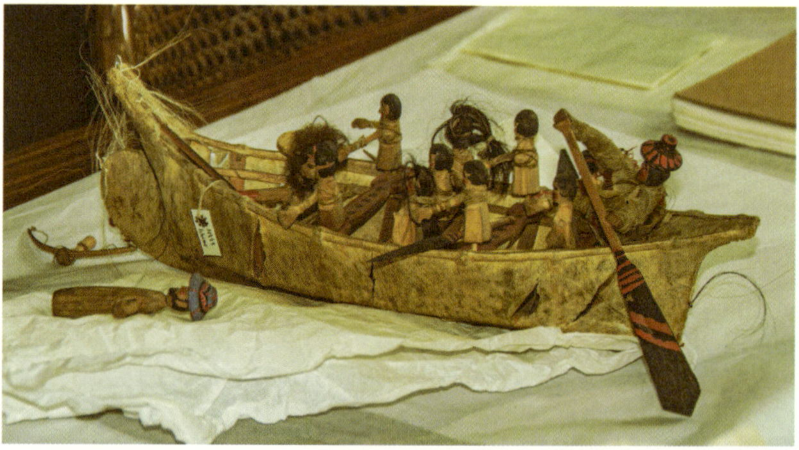

Fig. 1. Model angyaaq collected from Alaska in the 1800s. Now residing at the Lübeck Museum, Germany. Photo by Sven Haakanson.

I grew up in the village of Old Harbor, on the east side of Kodiak Island, and we did not learn about our people having open boats, only single, double, and three-person kayaks. In fact, my first encounter learning about angyaat from my region was in 1991, when I was living in Magadan, Russia. I was on an exchange program, and we had traveled to Leningrad (now Saint Petersburg), where I first saw on display two model open boats in the Museum of Anthropology and Ethnography. They were labeled as Baidara, 1840 Kodiak Island, Alaska. I did not learn the real name of this vessel from our region until 2006, nor did I realize how important these models would become until I started to do research into open boats from Kodiak in 2006. I was drawn to these vessels because I grew up on the water and wanted to learn more about how my ancestors used and made them. They represent a powerful part in our cultural reawakening on the island as seagoing people.

What I learned from a Siberian Yupik elder was that open boats north of Nome to Greenland are called "umiaks," and south of Nome to Kodiak they are called "angyaat" in Yupik and Alutiiq. I realized that these boat models were usually labeled as "umiaks" in museum collections, and no distinction was made either in scholarly work or by collectors about where they were from. I have not seen any scholars talk or write about this distinction, and this made research into this vessel type held in museums even more challenging.

Our elders rarely spoke about traditional boats, yet every museum that I have visited and has an Arctic collection has model boats from Alaska. Through these models I have been slowly piecing together bits of information as I encountered them over the last thirty years. For example, when I was staying at the Captain Cook Hotel in Anchorage, Alaska, I saw prints of John Webber's drawing *A View in Prince William Sound May 1778.* They were of Sugpiat people in these bulbous-bowed open boats. This reminded me of the two model open boats I had seen at the Museum of Anthropology and Ethnography in 1991. This then sparked a quest for me to find out what happened to these open boats and why we did not know about them or have any examples on or from Kodiak.

When I start researching a museum's collection, I share with the staff what I am doing and why. I tell them I will share what I learn about each cultural piece with my community. I never hide my intentions about taking this knowledge and giving it back to our community to have and for us to use, share, and celebrate. I am one part of a process in the repatriation of knowledge that was nearly forgotten. Thanks to the preservation of museum collections, we can use this knowledge to understand and celebrate once again.

For me personally, every person who works in a museum is important because they are the caretakers of Indigenous cultural histories and knowledge. It is not just the curators who make the decisions; it is collection managers and conservators who are involved in the care of collections. I originally never saw myself as a conservator when I started working with collections; however, as an Indigenous person, I realized how I and my community were working together to help conserve and preserve the "intangible knowledges" that are embodied within "tangible pieces." As our community engaged in this process, the roles within conservation change from passive to active in this ongoing process of caring for our cultural heritages at multiple levels. This involves understanding an agreed-upon ethics of care for the use and sharing of collections. The example of the angyaaq demonstrates, for me, how conservation is care for not just the tangible preservation, but the intangible preservation and understanding of all the pieces in our institutions, for our communities. Conservation is not just repairing and stabilizing a piece; it is actively engaging with communities, their cultural protocols, and the living knowledge of a community's history. Conservators should always ask, even if it is uncomfortable or seen as being insubordinate, whose directions they are following and whether this is what the communities want done to the piece. If yes, how do they ensure that the pieces are safe and that what they are sharing will help bring this knowledge into a living context and one that helps the community, in other words, that is relevant to them, not just to outsiders? It is shifting our awareness of what conservation signifies and means to the communities they are engaged with through the pieces. Conservators, even if they are behind the scenes, are still responsible for understanding the ethics of care of and for the cultural pieces from communities.

Sharing what I do, and letting museum staff know how important the collections are to us as Natives, opens up the hearts and minds of the caretakers of our histories. As museum staff experienced and saw this work, they realized I was there to learn and not to threaten anyone with repatriation or otherwise. I was always provided access to the collections storage areas where items were placed, and I could identify them based on what they were, at times assisting with updating identifications in museum records.

What opened these doors was a federal law enacted in 1990, the Native American Graves Protection and Repatriation Act (NAGPRA).[5] At every museum I go to, we talk about NAGPRA at some point during our visits. I have started major projects with museums from first getting access to collections through NAGPRA consultations. However, I wanted to learn more about and from the collections that are not considered under this law. Many of the cultural pieces like our clothing, tools, and transportation (such as the angyaaq) are not seen as sacred, funerary, or objects of cultural patrimony and thus have been largely ignored under NAGPRA as potential candidates for physical repatriation. But each and every cultural piece has knowledge embodied within it, and we are culturally responsible for bringing this out and sharing it with our communities. This open and honest approach allowed me to conduct further research in collections because museum staff learned and saw that I was not doing this for myself but for my community. These preserved pieces from the past, when studied and "reawakened" in the present by and for communities, become active resources for cultural futures.

In smaller museums I was able to communicate directly with the director, curator, and/or collections manager, and they were often eager to facilitate this process of reawakening the objects. It was harder in the larger museums because of limited staff availability and the difficulty of setting up the time and the bureaucratic procedures involved in getting access to multiple storage areas. I had to specify exactly what I wanted to see in advance, and this was only what I was given access to. When things are mislabeled, which was the case for the angyaat, they are not easily identified from the database or catalogue until seen in person, especially if no photograph accompanies the collection records. It has taken me over twenty years to locate fifteen model angyaat from our region thus far, and I have hopes of maybe one day finding a full-sized angyaaq in a collection to document and learn from.

One related project that ties to our reawakening of traditional vessels was when I was invited to collaborate on a kayak restoration project with the Peabody Museum at Harvard University in 2010.[6] I researched and documented two of their Alutiiq kayaks with Alfred Naumoff (traditional kayak maker) and Susie Malutin (skin sewer), which culminated in the loaning of a single-person warrior's kayak back to the Alutiiq Museum and our community.[7] This was a collaborative effort with conservators. In 2003, prior to this project being funded, I was invited, along with elder Ronnie Lind, to do NAGPRA consultation on the Peabody Museum collections. After we completed this part, we asked whether we could see a three-person kayak in their collections. While looking at this kayak, Ronnie noticed another kayak, which I assumed was an Unangan kayak from the Aleutians, because of the bow share, which at that time I did not know was broken. Ronnie quickly corrected me and pointed out that it was an Alutiiq kayak, and more importantly, it was a warrior's/whaler's kayak because of the human hair sewn into the seams of the skin on the bow.[8] If he had not pointed out this kayak and corrected the label, we would not have known about its significance today. This was an important discovery for us, and the staff heard this. Dr. Patricia Capone, over the next several years, maintained our conversation about this kayak and pursued grants to conserve and collaborate with us.

This started another very important project that has identified the cultural importance and value of this kayak to the Alutiiq community. This was a several-years-long conversation between me, Dr. Patricia Capone, and the conservation staff at the Peabody Museum about what we could do to bring back and share this knowledge. By collaborating with the conservators, we were able to examine in detail the sewing techniques, the skins, and the wood the makers used to construct the kayak.[9] For example, we did not know they were using harbor seal skins for covering the kayaks. Because of the size of the skins, we thought they were sea lion, but we were wrong. What has happened is the size of these animals has changed drastically in the past two centuries because of the reduction in their foods from overharvesting by humans. Dr. Capone and the Peabody staff took the initiative to get funding and helped carry this project to its culmination of sending the kayak back to our community for a ten-year renewable loan. This is one of only four full-sized original single-man kayaks in the United States.

In 2007 I was visiting the Burke Museum in Seattle and, with the help of Dr. Nadia Sethi, found within its collection one of only thirteen model angyaat I had seen until then. All the angyaat I knew of were either in Europe or on the East Coast and inaccessible to us. In 2013 I was hired as the curator for North American anthropology at the Burke Museum, and it provided me with ability and time to research, photograph in detail, reverse engineer, and create a working model angyaaq (like the one on loan for this exhibit), based off the original Burke angyaaq model. That fall, I called Mitch and Judy Simeonoff and their son Speridon Simeonoff, the managers of the Akhiok Kids camp, to talk about making model angyaat at the camp in the summer of 2014. We all agreed this would be an important project to do because we could tie in all the other cultural practices we have been teaching at the camp over the past fifteen years and together try to bring the knowledge back to our communities.

The camp was created by Mitch and Judy in the 1990s to give the kids a safe place to learn about Sugpiat cultural practices and ways of being and knowing, and for them to learn on their own terms. The camp allows the participants—the kids—to decide and choose what they want to do each day and learn as they want to, not as we counsellors want them to, thereby following a traditional way of teaching and learning

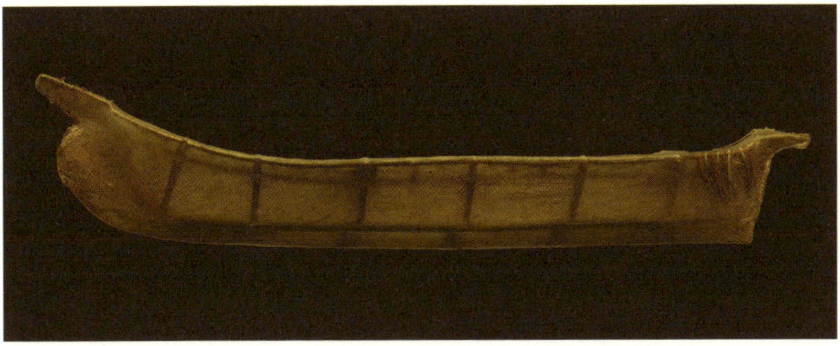

Fig. 2. Model angyaaq collected from Alaska in the 1900s. Now residing at the Burke Museum, Seattle, Washington. Photo by Sven Haakanson.

where the participants are responsible for themselves. The adults have set things such as carving, sewing, beading, hunting, and gathering during the day, and participants choose their activities and what skills they want to learn.

I spent the fall and spring of 2013–14 making a model angyaaq frame, based off the Burke's model angyaaq. That spring I made several angyaaq model kits for the camp participants to construct themselves. During this time, I did more research into the boat's distinctive bulbous bow, learning how this design helped the boat go faster and reduced the energy input needed to move it. It is a brilliant bow design created by Indigenous peoples who understood more than what Europeans give them credit for. By the end of summer camp we made thirteen models, using no glue or nails and following only traditional ways of construction. It was at that time that Mitch Simeonoff asked, "Why don't we build a full-sized angyaaq?" I thought about this for a second, and we laughed about why we didn't do this sooner. I spent the next year trying to figure out how to do this, with the help of museum collections, books, and talking with elders from Alaska.

In the fall of 2014, I started gathering materials to construct a full-size angyaaq. It took seven months to gather the wood and fabric covering (donated by George Dyson) to start the construction of a full-size frame. With the help of Dr. Peter Lape and Burke Museum staff, we constructed a twenty-seven-foot angyaaq frame in the back of the old Burke Museum parking lot. Having an idea of what it took to make the frame, I then took it apart and stored it to finish later. In the summer of 2015, I went to Akhiok with an understanding for how we would construct a full-sized angyaaq frame.

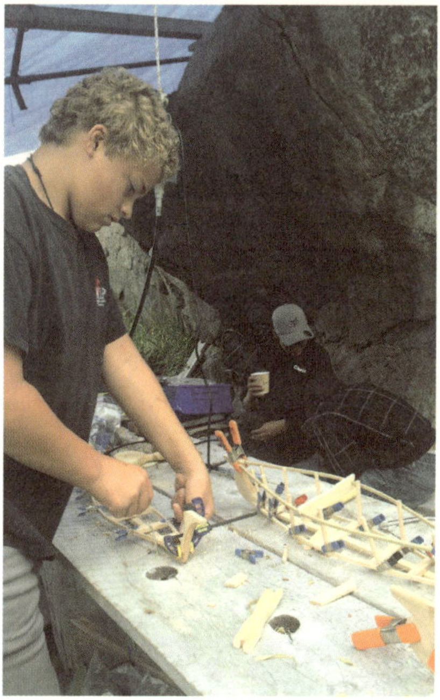

Fig. 3. Elijah Lake constructing a model angyaaq in 2014. Photo by Sven Haakanson.

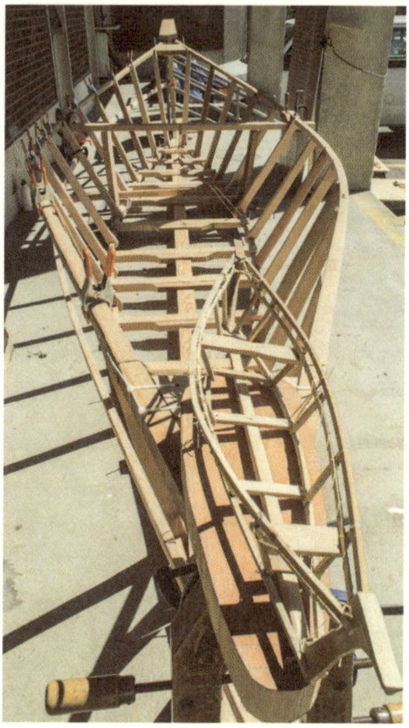

Fig. 4. Angyaaq frame being constructed at the Burke Museum parking lot in the spring of 2015. Photo by Sven Haakanson.

Working with the community of Akhiok at Cape Alitak, we were able to construct a frame within the week we were there. It was very challenging because we had to gather the wood off of the beach and then split and carve the wood into parts to construct the frame of the angyaaq. Once we were done with the frame, I asked Speridon about his impressions of constructing the angyaaq, and he said he "was very happy to help [his] father Mitch's dream come through in making an Angyaaq on Kodiak again, as well as involving all the other community members in building it." He then went onto share how "[i]t was a very enjoyable process for sure. It is always fun to do things with your hands and have a result that changes all our views and understanding of our heritage. We were learning to preserve our heritage through lost knowledge. I never imagined that it would only take twine to hold together such a solid seagoing vessel. You don't need nails or pegs to do this." We both agreed that this experience changed how we fundamentally understand our ancestors' ability to thrive in our region.

At every museum I visit, I ask if they have any “umiaks” to see whether they might have an angyaaq hidden in their collections. None of this would have happened if it had not been for the museum staff giving me access to the collections to rediscover and learn about this vessel from our region. Given the fact that these models are so rare, it is a miracle we were able to do this. I am grateful to the caretakers and museums for keeping this knowledge alive for us to use today and to inspire our next generation.

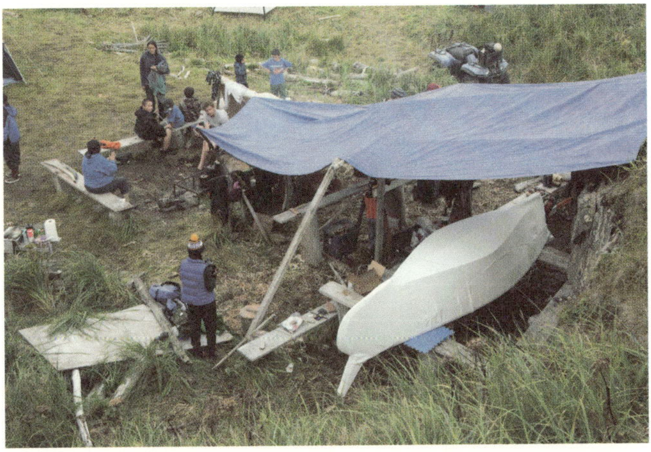

Fig. 5. The construction of a full-size angyaaq at the Akhiok Kids camp in the summer of 2015. Photo by Sven Haakanson.

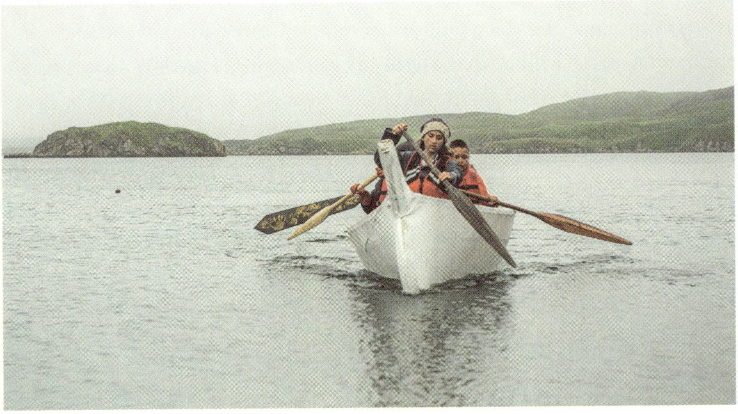

Fig. 6. Youth at the Akhiok Kids camp paddling the angyaaq around in the summer of 2016. Photo by Sven Haakanson.

I am grateful to individuals like Dr. Capone and the conservation staff at the Peabody Museum who actively collaborated with us to reactivate cultural knowledge within our community through the single-person warrior's kayak once again. This collective work happens at all levels of an institution that cares for Indigenous collections. Conservators who maintain the integrity of the pieces, collections managers who care for the collections, and curators who share this knowledge all have an important role in the ongoing care of our histories.

I started with a lamentation song to show what we thought was lost. But I end with a focus on how we are all caring for collections, how we can share this knowledge and support communities to use it once again in a living context. We are creating new stories that celebrate the return of angyaaq and more. The angyaaq shows us how brilliant our ancestors were and how we can build off that knowledge once again. We have a long paddle ahead to change what has been done to us culturally. We can't change our past, but we can change our future by returning knowledge to the communities, re-embedding it into a living context and into the hearts and minds of our children, our next generation of cultural bearers.

NOTES

1. This song was written down in 1804, and in order for us to understand the Alutiiq today, we had to edit the spelling and retranslate it into modern-day Alutiiq and English. The first column is the original text from the book, the second column is our retranscription into the current orthography, and the third column is our new translation. The original text had "apaka," which we thought meant "grandfather"; however, as Sperry Ash and I retranslated this text, we soon realized it was Abba and not "appa" (grandfather). Apaka/Apaqaa/Abba is a term of respect in our culture for a person whom you are meant to emulate and follow throughout your life as a mentor and role model.
2. Over the past two centuries the Natives of Kodiak have been called by many names, from Aleut to Pacific Eskimo, Koniag, and Alutiiq. Now we are asking to be called by our ethnonym Sugpiat, which means "the human beings." In the Alutiiq language when the last letter of a word is a *q*, it is singular, and when it is a *t*, it is plural. "Sugpiat" means "human beings." "Sugpiaq" means "one human." Same goes for *angyaaq* ("one boat") and *angyaat* ("many boats").
3. Gideon, *Round the World Voyage.*

4. Black, "Descriptions of the Kad'iak District."
5. The Native American Graves Protection and Repatriation Act of 1990 became law on November 16, 1990. Pub.L. 101-601; 25 U.S.C. 3001-3013; 104 Stat. 3048-3058.
6. "Conserving Alutiiq Cultural Heritage."
7. "Warrior Kayak Is Welcomed in Alaska."
8. "Museum to Conserve."
9. Buckley et al., "Species Identification"; Holdcraft et al., "Collaborative Study."

BIBLIOGRAPHY

Black, Lydia, trans. "Descriptions of the Kad'iak District and Subdivisions." Unpublished manuscript by Vasilii Ivanovich Kashevarov. Originally written 1833. Translated by Dr. Black in 1990.

Buckley, M., S. Fraser, J. Herman, N. D. Melton, J. Mulville, and A. H. Pálsdóttir. "Species Identification of Archaeological Marine Mammals Using Collagen Fingerprinting." *Journal of Archaeological Science* 41 (2014): 631–41.

"Conserving Alutiiq Cultural Heritage: Conservators at Work; Alaska's Historic Kayaks Renewed." Peabody Museum of Archaeology & Ethnology. Harvard University. October 26, 2011–August 28, 2013. https://www.peabody.harvard.edu/node/747.

Gideon, Hieromonk. *The Round the World Voyage of Hieromonk Gideon 1803–1809.* Edited by Richard A. Pierce. Translated by Lydia Black. Ontario: Limestone Press, 1989.

Holdcraft, T. Rose, Sven Haakanson Jr., Ellen Promise, Judy Jungels, Fran Ritchie, and Patricia Capone. "Collaborative Study and Preservation of Coastal Alaskan Native Watercraft and Material Culture." *Journal of the American Institute for Conservation* 60, no. 1 (2021): 2–17. It was also published online April 6, 2020.

"Museum to Conserve Rare Alutiiq Warrior Kayak in Gallery." Press release, October 17, 2011. Peabody Museum of Archaeology & Ethnology. Harvard University. https://www.peabody.harvard.edu/node/742.

"Warrior Kayak Is Welcomed in Alaska." Inside the Peabody Museum. Peabody Museum of Archaeology & Ethnology. Harvard University. June 2016. https://www.peabody.harvard.edu/node/2658#story_3.

THIRTEEN

The Orator's Dilemma: Wampum as Material, Media, Medicine, and Memory

Jamie Jacobs

On November 16, 1990, the United States passed the Native American Graves Protection and Repatriation Act, otherwise known as NAGPRA. The passing of this law made it possible for Indigenous tribes in the United States to file claims of repatriation for sacred items of cultural patrimony and any human remains that were excavated for scientific study. The tribes explained to the holding institutions not only that these items of cultural patrimony were viable for the continuation of their sacred practices but also that the objects being repatriated were considered "living" beings whose spirits assisted in their perpetual continuation of their culture. They believe strongly that their objects made from wood, shell, and bone, and even those made of synthetic materials, are living matter. The dilemma that tribal officials and medicine men faced to convince modern science and scholars that objects that didn't audibly speak, move, or produce sound of their own accord became exhaustive. It is only now that we can come together and bring understanding to each other and know that objects do in fact carry with them "living matter."

My name is Jamie Jacobs, and in this essay I will do my best to relate oral traditional and cultural information from my community with hope that it will be helpful in broadening the understanding of how the Haudenosaunee people utilize aquatic and marine shells as "active matter" in the form of a small bead that we call wampum. Drilled and shaped from what some would call inanimate layers of calcium carbonate and proteins, the beads that are formed are a kind of media carrying voices of the ancestors and vital messages of history, prophecy, healing, and truth. Traditional orators use wampum to convey these messages, and the potential absence of shells or objects made from them presents a challenge to the transmission of cultural memory in our communities.

Oceanic and freshwater shells begin forming in water by absorbing elements from their surrounding environment, mainly calcium. The clam grows the mantle, which is the holding area for the clam itself. As it grows, it releases a gluelike substance and adds material to form layers of shell. Once the shell is produced, there is no life-flow within the layers of calcium, which only act as a protective vessel for the clam, thereby making this material inanimate. Humans have somewhat similar behavior. We build standing structures to protect ourselves and our offspring from the outdoor elements. Even though 20 percent of this structure is produced from living trees, many people tend to believe that there is no life force within the structure: it does not breathe, it doesn't have blood, it doesn't have a mind of its own, and its sole purpose is to stand and provide shelter from the elements.

This belief, however, doesn't exist in the minds and hearts of the Iroquois people. We believe that the structure that provides us warmth—whether from the chemical process of burning wood or from the love that families share with one another—does in fact have a soul or a spirit. The house speaks. It tells its owner when it needs tending to, and it breathes the air that the inhabitants release into it. It stands tall but becomes dilapidated when it's abandoned. It's no wonder, then, that the Haudenosaunee's whole philosophy is based around the symbol of a house. A longhouse—a structure that would seem as inanimate as the clam's shell. But in fact, this house would become the unifying symbol of fifty families in what is now known as the Iroquois Confederacy. In fact, a cultural practice among the Iroquois, or Haudenosaunee, is that when a longhouse or sanctuary has become too old or weak to provide shelter, we burn sacred tobacco and send a prayer, address the "old house" and explain, and ask it not to be offended that no more ceremonial gatherings will be performed from that day on inside the house. When the replacement structure is built and finished, we burn sacred tobacco inside the new sanctuary and address the new house and ask that the structure stand tall and steadfast for as long as it can to provide a roof for the new ceremonies that will take place inside for generations to come. Wampum beads, and the strings and belts made from them, are alive for us in parallel ways, including the need to be activated by the voice.

The Beginning of Voices

Haudenosaunee males are trained at early stages in their lives in the art of public oratory, speaking to relay historical information, religious speeches, and messages sent from one nation to another. I was born and raised in a Native American community in upstate New York, located just outside Buffalo. Since the day I was born, I have been immersed in Iroquoian ways, culture, oratory, song, and dance. Today I am a ritual custodian of the Tonawanda Seneca Community longhouse or sanctuary. My mother's side of the family come from a matrilineal line of women from the Turtle Clan, thereby making me of the same clan and lineage. Within our family there are individuals who are given the sole responsibility of ensuring the perpetuation of the sacred rituals that take place throughout the calendrical year. It is their responsibility to learn the repertoire of songs and dances and to speak the words to the crowd of people as to how the religious ceremonies will go and release encouragement in the minds of the people. My father's family comes from the lineage that we call the Snipe Clan. His grandmother, who lived to be ninety-six years of age, was also a member of the Keepers of the Faith. She was a first-language Seneca speaker, and she didn't speak English until she attended the third grade in the nearby public school in Akron, New York. She attended Akron Public School from 1920 to 1923, until her father insisted that her culture was more important than Western education and refused to let her attend anytime after that. Sadly, she passed away in the year 2010, taking with her a vast collection of knowledge that was orally handed down for many generations.

To the Iroquois, or Haudenosaunee as we traditionally prefer to be called, along with other groups in the northeast located around the Great Lakes, wampum has always played an important role in diplomacy, ceremony, and aesthetics. In our language the traditional word for wampum is *Otgoä*, which translates to "shell bead." This word is not used for any other kind of bead, whether it be made from glass, ceramic, or a synthetic material. It's only used for wampum beads formed from the freshwater or oceanic shells harvested from the waters of the earth. The shape and tools used by the ancestors have evolved over time. Prehistoric beads, some of which can be found in the collections of the Rochester Museum and Science Center, started off very large and barrel shaped. Then, as the quality of tools improved and experience was gained, the beads became smaller and smaller until the introduction of iron brought from the Old World by early Europeans, when the beads took on a much narrower length and a more consistent shape. The beads in figure 1 were excavated from known historical Seneca sites located along the Genesee River Valley in western New York. They date from the 1550s well into the 1660s.

Wampum beads at that time were made and used in different ways. They were used for ceremonies; they were used for aesthetics, worn around the neck, and oftentimes buried with individuals of high standing. Larger beads and beads with a darker rich purple were manufactured from the quahog shells harvested from the Atlantic Ocean because the shells grew to a much larger size. Overharvesting for food and to fill the demand for factory-produced beads for the eighteenth-century trade era caused the shells to no longer grow to their original size. The beads made by Atlantic coastal tribes were often made to trade to nations located farther inland. From what we know from the amount of wampum belts, strings, and beads that were being used by the Haudenosaunee in the eighteenth century, we were most likely the best customers. To the Seneca peoples who live in western New York, the many miles that these oceanic beads had to travel made them that much more valuable.

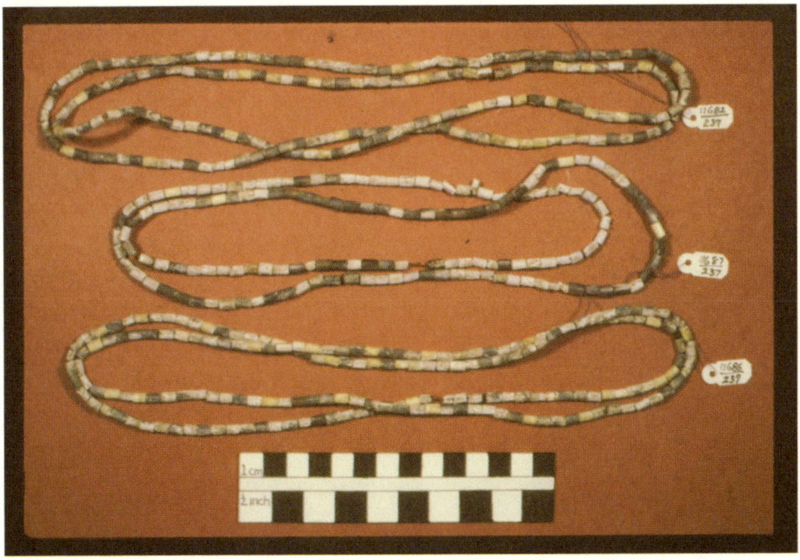

Fig. 1. Seneca maker, six strands of excavated beads of shell wampum, late 17th century. Each strand 32 in. (81.28 cm) length. The Rock Foundation. Photo courtesy Rochester Museum and Science Center.

The term "Haudenosaunee" translates to "those who have extended the house." Linguistically, "those" would refer to the original Five Nations that inhabit what we now call New York State. The Mohawk Nation, Oneida, Onondaga, Cayuga, and Seneca Nation, with the addition of the Tuscarora Nation in 1722, now make up the Six Nations of the Haudenosaunee Confederacy. Oral tradition states that when the Five Nations came together in unity, an individual known only as the Peacemaker visited these Five Nations to bring an end to perpetual warfare. Each nation had vowed to conquer one other to become the masters of the environment and the natural resources that were available to sustain the conquering nation for the future. The Peacemaker brought a message of peace and unity based on three principles: peace, power, and friendship. To convey his message to the people and to affirm the truth in his message, he used wampum as his media. When he spoke words to the people, they would know that what he was telling them was the truth.

The main shell used by the ancient Iroquois was the quahog clamshell, or the *Mercenaria mercenaria.* The quahog naturally produces two colors within its layers, which are purple and white. The dark, rich purple color is due mainly to the shell's genetics with a combination of sediments absorbed during shell formation. In Iroquoian beliefs, the white shell was looked upon to represent peace or friendship and also as a symbol of power. The purple beads that were produced from the purple part of the shell represented darkness caused by death. Families who suffered a loss were said to be in a state of darkness and unable to mentally sustain their livelihoods.

Throughout our history, wampum have served as mnemonic devices to remember the stories and diplomatic agreements between nations and families. We didn't have a written language to write them down and preserve them. We did not develop any kind of orthography nor any kind of alphabet, although our ancestors certainly had the intelligence to do so if they chose. They saw the importance of human interaction. Two human beings speak to one another to exchange not only information but emotion and conviction, thereby holding both the speaker and the listener accountable for the information. A higher respect is gained for one another, and open communication forms a holistic bond between the different parties.

My great-great-grandfather was a chief of the Tonawanda Senecas. His name was Edward Black Squirrel, and the Seneca name that he carried was Gaentëh, which means "Out in the Middle of the Field." (The word for the state of Kentucky may actually be based on an Iroquoian word from this same base, which is pronounced "Gaentëhgeh.") When the photograph of him in figure 2 was taken in 1936, my great-great-grandfather was holding short strands of wampum. He, along with famed Iroquois anthropologist William N. Fenton, was visiting the different Haudenosaunee communities and was relaying a message to them. In the photo, he's inviting the people to a convention to hear the recitation of the Code of Handsome Lake, which is now the belief system of a majority of current Haudenosaunee communities (see below). What he's holding in his hand are short strands of wampum strung on small cotton threads and tied to a stick that has been carved with notches. In the Seneca language we call these small wooden sticks *Yötnëshadiyödahgwa'*, which translates to "it tugs them by the arm." The number of carved notches correlates to the number of nights or sleeps until a convention is to be held. The wampum is used to pull invited parties' arms in the direction of where the meeting is going to take place. It also acts as a protective talisman to safely escort them to the meeting. Once they arrive, they return the wampum to the original owner along with words of thanksgiving and greetings. The same practice is also used to

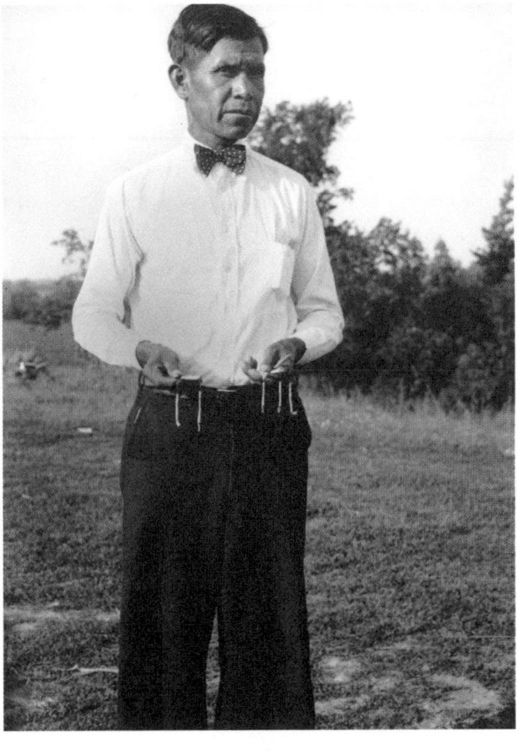

Fig. 2. Tonawanda Seneca chief Edward Black [Squirrel] delivering the "Invitation Sticks," 1936. William N. Fenton Photo Collection. Photo by William N. Fenton. The photographer's label for this image gives the name Edward Black, but thanks to family history we know that his full name was Edward Black Squirrel.

invite the grand sachems and clan mothers to a Grand Council meeting to discuss political issues that affect the confederacy as a whole. The Onondaga Nation sends out messengers or "runners." They are given notched sticks with tied wampum, and they deliver them to the rest of the Five Nations to attend the large council meeting on a certain day. The Yötnëshadiyödahgwa' plays an active role in inviting the chiefs and the clan mothers to the council meetings that are still held today on the Onondaga Nation territory in Syracuse, New York. In fact, in order for nations to speak at the Grand Council and to be considered invited, they must receive and return a Yötnëshadiyödahgwa'.

Beads of Condolence and Confederacy

As I mentioned, the unification of the Iroquois was the result of the work of a man known only as the Peacemaker. His traditional name is only to be mentioned when the laws of the Haudenosaunee are recited to the people. As it is told, the Peacemaker came to the Five Nations around a thousand years ago. Together with his ally, Hiawatha (not to be confused with the character in the poem "Song of Hiawatha" by Henry Longfellow), he formed the Five Nations Confederacy. In the Seneca language, the ally's traditional name is pronounced Haye:watha', which translates to "He Keeps Them Awake." When the Peacemaker first encounters Haye:watha', Haye:watha' is in perpetual grief because his daughters suffered homicide as the result of perpetual warfare among the people. He then becomes a wanderer of the forest, and every time he comes to a place, or every time he looks at a tree or watches the sun rise and set, all he can think about is his daughters who are no longer at his side. He thinks to himself about how to bring an end to the constant bickering and killing that brought about the doom of his family, so that no one else shall suffer the heavy grief that he does. He sends thoughts and prayers on a daily basis until he finally comes to the reasoning where he says to himself, "If any individual that I come upon is suffering the same sorrow and grief, I will console them with the use of these sacred beads that I have formed from a special shell found at the bottom of the lake." The sacred beads that Haye:watha' is referring to are beads that he produced from freshwater quahog shells that he collected while walking across a dry lake bed. It is not known which lake he might have crossed, but we know that this is the beginning of the use of wampum beads for ceremonially condoling an individual who has suffered the loss of a loved one.

Haye:watha' meets the Peacemaker while he is wandering the land in search of an answer for the death of his seven daughters. The Peacemaker was at the same time visiting the Five Nations and bringing a good message of unity and an end to the constant killing and bloodshed. The Peacemaker was believed to possess the ability to read thoughts of the people in the hope of understanding their thinking and spreading his message of peace, power, and brotherly love. As the Peacemaker greets Haye:watha', he says to him, "I know exactly what you're thinking, so I will console you." And he, too, brings these wampum strings, and he says, "Now I wipe away the tears and I use a short strand of wampum so that now you can see all the things around you." He picks up another wampum string, and he says, "The grief has gotten stuck into your ears. You can't hear anything or the people who wish to speak to you. So now I use this wampum and I dislodge the grief from your ears and now you can begin to hear even if it is for the period of one day. Now you can hear the birds singing. You can hear the people talking among you." He picks up yet another wampum string, and he says, "The grief has gotten stuck into your esophagus to the point where you can't breathe, and I use this short strand of wampum beads and I remove this grief from your throat. Now you can breathe the fresh air, and you can swallow the good food the Creator has provided." The Peacemaker soothes Haye:watha's cloudy mind and brings a sense of purpose back into his being. The Peacemaker continues on, and he metaphorically removes the yellow spots in Haye:watha's stomach. He prepares a special tea made from medicinal herbs and roots, and with another wampum string he says, "Now I clear away the grief that has had an effect on your stomach, which has become twisted." He also metaphorically replaces Haye:watha's sun because he can't see or feel its brightness. Once the two perform this reciprocal consolation, they visit the Five Nations as allies and bring them together to form the league, or what we now know in history as the Iroquois Confederacy.

This condolence ritual is still performed today among the Haudenosaunee, and it is an integral part of the installation of the male leaders known as sachem chiefs. In the Iroquois language we refer to them as *hoyane:h*, which translates to "he is of goodness." The female leaders carry the responsibility of choosing and also deposing a chief if he is not working together with the rest of the leaders. They have a lot of power, and their voice in political matters is always heard. When we speak of women's rights, the women have the same rights as men, and it is difficult to imagine that in Western society this has only been achieved in contemporary times, within the last fifty years.

In Haudenosaunee culture, the women have always had a voice. They have always had the symbol of authority, and the idea of matrilineal hereditary leadership was fortified by the Peacemaker with the use of wampum. When a female leader elects her male candidate, he is stood up in the position of leadership and a strand of white wampum is used to symbolize his authority. This string is called the Antlers of Leadership because the Peacemaker compared the sachem chiefs to the deer. When a deer is threatened, he puts down his head to defend himself and his family, so a sachem chief carries a white strand of wampum until his ultimate death, and these "antlers" are passed on to another member of the family to carry on the position of leadership.

Death is personified as a crawling entity that points its staff or mace at the person who is to leave the earth. We call death *Osöndowëhgowa:h*, or the Great Darkness. Wampum acts as a device to console the living members of the family and to wipe away the falling tears of grief. Grief is too heavy for any person to carry for a certain period of time, and if the burden is too heavy, the consequences may be dire and cause more calamity. The tears must be removed and wiped away so that the soul can see what is going on around him or her. Everyone must have this ability to see clearly in life, and it is even more important for a person in a leadership role. A council seat becomes covered in red stains because the chief has not survived. The people metaphorically wipe away those red stains of blood on the council seat, so when the new leader comes in, he's on a clean slate. When people suffer grief, they lose their sky. So in our language, we say, "Wa'agojö:yahdö'," literally meaning "Her sky has become lost." So now we replace her sky, and we use wampum to do that.

In the legend of Peacemaker offering condolence to Haye:watha', the Peacemaker removed the tears from his eyes using a short wampum string. The wampum took the place of a soft white deerskin and was used to gently wipe away the grief so that Haye:watha' could clearly see that there was more to living than being in a perpetual storm of darkness. They called this first string of many that are used in the condolence process *Ogahsä'*, or "the Tears." Today, in the elaborate ceremony of installing a chief of the confederacy when the clan mother loses her chief to death, wampum is used to remove the tears of grief so that she may keep up her noble responsibility as a leader for the family. The second string is named *Aöhda'*, which translates simply to "Ear," to remove the grief from the ear. The third one is called *Deyagonya'do:gë:h*, and the translation of this would be "It is between her esophagus," referring to the grief or "lump" that gets caught in the throat, and once again wampum is used to metaphorically remove the grief in the throat. It continues on, and the Peacemaker and Haye:watha' end up with fifteen strings of condolence. The condolence strings of wampum (fig. 3) are named as follows:

1. The Tears
2. The Ears
3. The Throat
4. Yellow Spots Formed on the Stomach
5. Blood Stains on the Council Seat
6. Removing the Darkness
7. Loss of Sky
8. Loss of Sun
9. Mind on the Grave
10. The Scattered Council Fire
11. The Twenty Matters
12. The Ceremonial Leaders Also Experiencing Grief
13. The Niece and Nephew (assistants to the chief)
14. Preventing Suicide
15. The Torch

The artist's interpretation in figure 4 portrays Haye:watha' and the Peacemaker in what we call the Combing of the Snakes from Tadodaho's Hair. Tadodaho was an adversary of the work of the Peacemaker and Haye:watha' as they were forming the league. In fact, it was Tadodaho who in our mythology was the individual responsible for the death of Haye:watha's daughters. He was not in support of a unification of the Five Nations and did everything in his power to prevent the admission of the Onondaga Nation into the confederacy. The snakes in Tadodaho's hair are symbolic of the way in which he did not agree with the Peacemaker's message of peace, power, and friendship. They are a metaphor for the negative thoughts that were spewing from his mind and any person who came close to him; the snakes would latch on, and that person would become influenced by his negative way of thought. The Peacemaker and Haye:watha' used wampum beads to calm his mind and restraighten his body, and this act of convincing Tadodaho is expressed in the metaphor of combing the snakes from his hair. Wampum had such power that it rid his mind of evil, the white color of the beads casting away the snakes from his hair. Tadodaho became one of the main sachems of the Iroquois Confederacy at that time, and the Onondaga Nation was brought in as the fifth nation of the Iroquois Confederacy.

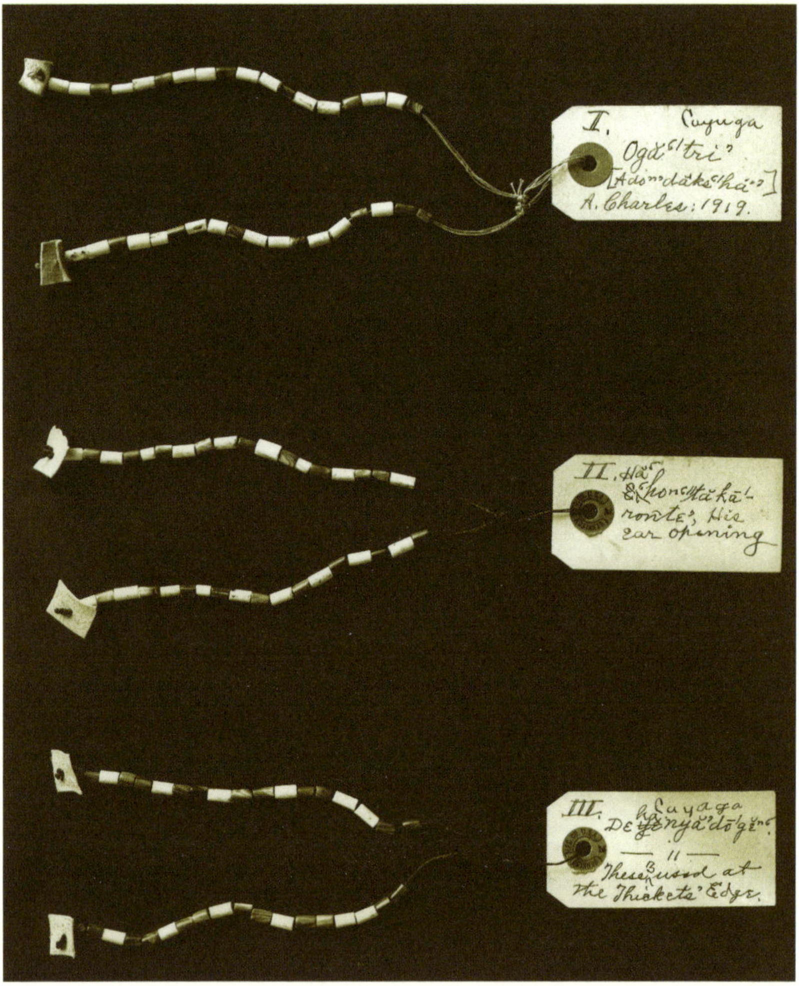

Fig. 3. Maker unknown, wampum strings representing the Three Bare Words, 1919. Black-and-white photograph in the John N. Hewitt Collection. Photo by National Archives Smithsonian Institution.

Now that all Five Nations are unified, the Peacemaker says to the sachem chiefs, "You chiefs are like trees. You stand tall, and your roots will go deep into the earth. Your branches reach toward the heavens. Your arms shall all lock together to form a chain that is unbreakable and will shelter your people and future generations. Your roots will go deep into the earth, meaning that when you make a decision, you must look as far as what we call seven generations. If you make one wrong decision, that root might go deep into the earth, and the root may become rotten." This means that when you make a wrong decision, it will affect people seven generations to come. So he stood the sachem chiefs in a circle and he made them lock arms, and he said, "You are now the fifty sachems of the Iroquois confederacy. Standing in the middle is the tree of peace, and at the top of the tree is a symbol for freedom and liberty, which is the eagle." They metaphorically uprooted the tree, and underneath there in the dark chasm they threw their weapons of war, no longer to be used against one another.

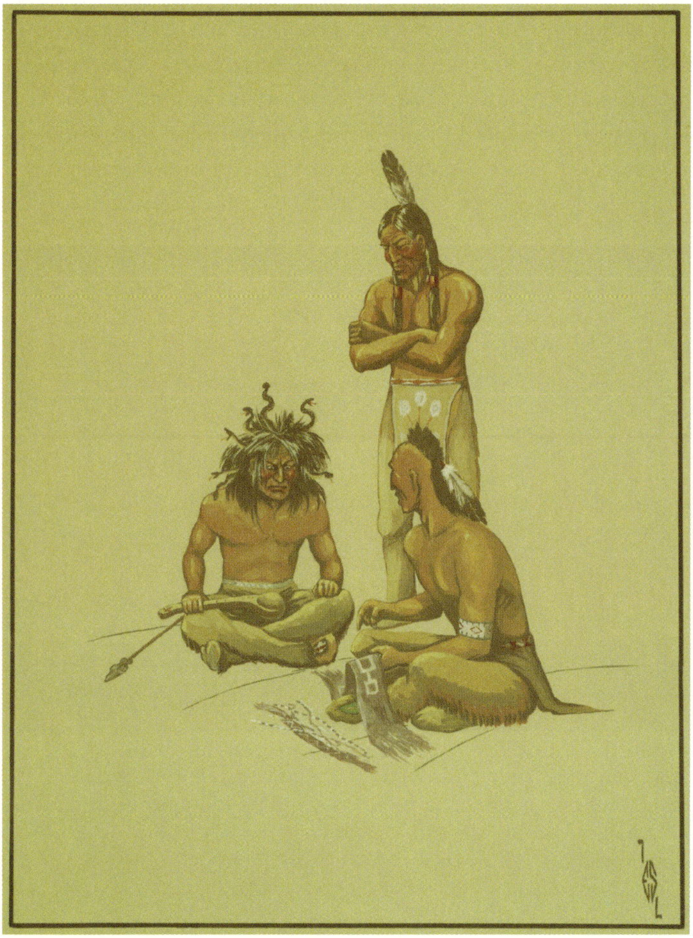

Fig. 4. Ernest Smith, *Council with Tadodaho*, 1936. Watercolor on paper, 18 × 24 in. (45.72 × 60.96 cm). Rochester Museum and Science Center, MR 616. Photo courtesy Rochester Museum and Science Center.

To relate that part of oral tradition, and to remind the chiefs of the responsibility of the title they hold, they made a symbol out of wampum. Two strings on the outside interwoven with one another, representing the chiefs who stood in the circle and locked arms. Their roots going in toward the middle and their branches stretching out to shelter the people on the inside of the circle, reaching toward that tree of peace forever and for all time to come. There's fifty strands of beads, each strand representing the fifty sachems of the Iroquois Confederacy. The Mohawk have nine sachems. The Oneida Nation have nine. The Onondaga Nation raised fourteen men to be sachem chiefs. The Cayuga Nation have ten, and the Seneca Nation ended up with eight, for a total of fifty sachems. The Haye:watha' wampum belt (fig. 5) also represents the formation of the confederacy of the original Five Nations of the Iroquois, with the central Tree of Peace signifying the Onandaga surrounded by and connected to the other four nations (this motif, visible on the belt in fig. 4, is now also used on the Haudenosaunee flag).

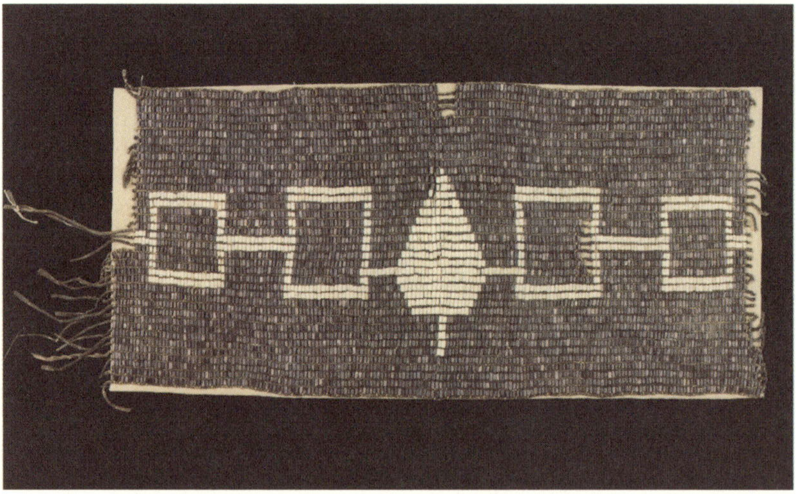

Fig. 5. Haudenosaunee maker, Hiawatha wampum belt representing the original Five Nations Confederacy, n.d. Shell wampum, dimensions unknown. Currently held by the Onondaga Nation, Syracuse, NY. Photo courtesy New York State Museum.

Tales of Loss and Recovery

The first president of the United States recognized the use of wampum among the Haudenosaunee peoples. In 1794, he signed a peace treaty with the Six Nations Iroquois Confederacy. George Washington assigned Timothy Pickering as the active agent to sign this treaty. He also commissioned a belt made from over ten thousand wampum beads to be made by two Oneida women to commemorate the treaty. George Washington sent the belt with Timothy Pickering and ratified the treaty through an exchange of wampum, a custom that has roots within the cultural practices of the Iroquois nations. The relationship between the United States and the Iroquois Confederacy goes back to its roots, before the United States was formed. George Washington adopted an agreement of friendship that had already been put into place almost two hundred years earlier. When the Dutch explorers landed on the shores and entered into the territory of the Iroquois, they made an agreement of friendship. It was based loosely on the concept of a European rope that tied the ship to the land. This was called the Covenant Binding of Friendship. Once the French invaded Haudenosaunee lands, they also expressed their concern for diplomacy with the Haudenosaunee, and they adopted the concept of friendship. But the French were concerned that the "rope" must be fortified to maintain a stronger bond with their Iroquois allies. They insisted that the rope be replaced with an iron chain, and that became known as the Covenant Chain of Friendship.

When the British invaded North America, a man by the name of Sir William Johnson was appointed as superintendent of Indian affairs on their behalf. As part of their efforts to push out the French, they also made allies with the Iroquois. Johnson liked the idea of the Covenant Chain and became well versed in Mohawk and Iroquois cultural practices, especially the use of wampum in diplomatic affairs. On behalf of the British he also made the agreement of joining the Covenant Chain of Friendship. The French and Indian War of the 1750s determined that the British were going to maintain control of the Covenant Chain, but obviously Britain's offspring rose to challenge her for their own independence. George Washington led the colonists to victory and gained their freedom from the Crown in 1783. George Washington, who like Johnson was familiar with Haudenosaunee history, then adopted the Chain of Friendship with the Iroquois. He was insistent, though, that another change be made to the chain itself. An iron chain corrodes quickly over time, so therefore the bond of friendship corrodes with it. He ultimately changed the chain, with the permission of the Iroquois leadership, into a silver chain. Silver does not corrode like iron but needs to be polished every so often to maintain its high luster. George Washington then decreed that the Silver Chain of Friendship be polished every time a new leader was chosen for the new republic. Wampum also carries this metaphor as a Silver Chain of Friendship between the United States and the Haudenosaunee people, maintaining peace and happiness between the two parties.

An exhibit at the National Museum of the American Indian on the National Mall exhibits the treaty made between the United States and the Six Nations, referred to as the George Washington Treaty, the Pickering Treaty, or the Treaty of Canandaigua. It is sometimes called the grandfather of all treaties because it was one of the very first treaties that was made between an Indian nation and the new United States. The wampum belt commissioned in 1794 by George Washington himself to commemorate the Covenant Chain of Friendship is not a "relic" but a living symbol of an agreement between two sovereign states, the United States and the Haudenosaunee.

In the year 1799, the prophet Handsome Lake, a Seneca sachem, had visions and relayed a prophetic message to the Seneca people. Just twenty years earlier, General John Sullivan was sent through Onondaga, Cayuga, and Seneca country to burn and destroy every village that he came upon in what is often referred to as a "slash and burn" campaign. They burned every village, orchard, and cornfield in their path, thereby choking the Haudenosaunee and forcing them from their homelands. This was a tactic by George Washington at the time to draw out the Haudenosaunee warriors who were attacking settlements in the northeast and along the New England coast. Around this time, our people were suffering land loss heavily at the hands of speculators and businessmen. A lot of our traditional communities were being sold off. Hunting grounds were being sectioned off as private property. This had a detrimental effect on the people's morale, and they turned to all these different vices to try to ease the pain of the atrocities they had witnessed on Sullivan's campaign.

Handsome Lake comes to the people and says, "I have this message, and if we follow this message we will survive, but we will have to adapt to the changing times." Handsome Lake claimed that heavenly beings visited him in dreams and in his states of trance they relayed a message from our great Creator. When he stood in front of the people to deliver and preach his message to the Seneca people, he held strings of wampum in his hands. This act of holding wampum gave the people a sense of sacredness and a visual reminder of the ceremonial use of wampum that was at risk of becoming a memory because of the change in cultural practices as a result of colonization. Handsome Lake preached to his Seneca people for fifteen years beginning in 1799 until his death in 1815. His message is still recited in the sanctuaries of the Haudenosaunee communities even today. The bonified preacher of his doctrines still carries on the practice of holding the wampum strings of the different communities of the Six Nations. Wampum is very active matter in this particular ceremony of the recitation of the Code of Handsome Lake.

In the late 1800s, the belts of confederacy were housed by a special person among the Onondaga Nation, who was dubbed Fire Keeper of the Grand Council of the Iroquois Confederacy. One man held the responsibility of keeping sacred wampum belts. These belts ultimately found their way into the collections of educational institutions, museums, and private individuals. It wasn't until the late 1980s that the United States, with the participation of Native communities all around the country, passed the Native American Graves Protection and Repatriation Act. The law states that any institution that receives financial support from the federal government must return items of cultural patrimony, burial items, and sacred human remains to the Native communities with which they are associated. At the time the law was passed, the question being asked among the anthropologists and institutions was, "The objects were purchased by and sold to collectors and museums because they were no longer being used, so why return them? They're just relics." The Haudenosaunee and many Native communities don't look at our precious historic items as being no longer active. We look at them as living beings; because of the stories that they hold, the objects are referred to as grandparents. A grandfather has a lot of experience in life. He can see many ways into things, so when we refer to these belts, we refer to them as our grandfathers, and we treat them with the respect that any grandfather deserves. The stories that they tell are invaluable and remind us of the guiding principles by which our great spirit has meant for us to live. Their loss is detrimental to the survival of our spiritual way of life.

The Orator's Dilemma and the Authenticity of Replicas

The use of replicated (usually glass) wampum beads and items constructed of replica wampum has become an accepted practice today in Haudenosaunee communities. The reduced availability of the large shells that once lined the eastern oceanic coast have made them largely obsolete, along with the loss of skilled people who once worked them. Traditional villages that were once dedicated to producing wampum for trade to inland nations no longer stand in the lands where they once stood. The arrival of Europeans and the ideal that land near the shore was extremely valuable for its easy access to the waterway pushed out the traditional villages and Indigenous people that once lived there. Many of our original belts are still in museums, though we work to identify and repatriate them. What is the orator to do when confronted with such loss? How is he to rely on wampum to deliver the truth of his messages, as our ancestors have done for generations? The contemporary Haudenosaunee have faith in all that we do, and we are able to adapt to such times. Tradition is important, and if a compromise has to be made in order for the practices of our ancestors to continue, then it must be that replica wampum will be used. After all, the practice of two empires sitting down beside the light of a council fire to discuss grievances and agreements has become somewhat obsolete.

When we talk about the orator's dilemma, we ask, "If he is using replica beads or belts, does that mean he is still telling the truth?" The answer is "yes," because the material used is an adaptation of the real thing. European groups around the 1600s took the idea of wampum and the value that it has, and they sent shell beads back to Europe to be manufactured in glass. The resulting glass wampum flooded into the Northeast, and the production of real shell wampum came to an eventual halt. In the absence of abundant shell beads, glass wampum became a suitable if less ideal replacement as a medium for conveying truthful statements. By facilitating the performance of trained oratory, the practice and symbolism of wampum strings and belts survive in this new form.

When we talk about the physicality of wampum—its nature as "active matter"—it has this very strong texture to it. It can survive for many, many generations because of its dense physical properties. The group of calcium molecules and layers in the shell survive through time, which is the reason our ancestors used wampum to carry into the future the special messages that were put into it. Now glass has to serve this same purpose. The material that our mnemonic devices are made from is not the most important thing. What is crucial is that the message gets relayed. What is important is that the medicine we derive from the ceremony in which the wampum is an active participant takes full effect. Because we are an oral tradition culture, the words and the wampum work hand in hand.

I began working in the museum field in 2006, beginning at the Rochester Museum and Science Center as an intern doing data entry tasks for the museum's Native American collections. After a few years, I eventually worked my way up into collections management and would offer conservation advice for the Seneca collections. In the late 1990s, I attended repatriation meetings with tribal leaders as a young teenage observer. My interest at the time was only motivated by the fact that I was able to get out of the house and meet tribal elders from different communities. I listened carefully to the many discussions that took place between the tribal leaders themselves and museum executives. The discussions that went back forth between museum boards and the tribal repatriation committee were interesting and at times somewhat controversial. Each side of the table would present its arguments and concerns for the care and conservation of sacred wampum belts, objects of patrimony, and funerary objects. At that time, as I listened to the museum trustees and board members defend their arguments, it seemed that their approach was a more scientific one and motivated by the monetary value that had been placed on the objects for insurance purposes. It seemed as though they were resistant to Native consultation and participation in conservation because many people lacked the scientific knowledge or degrees to do so. They felt as if all work and any publishing that would take place should be left to official curators, archeologists, and anthropologists. The Native community was in favor of scientific methods for the care of our precious objects because we understood the importance of physically preserving them for the future. However, we argued that the objects need more than just scientific care. They needed to be spoken to with the incense of sacred tobacco and be offered a feast to ensure their spiritual survival.

Since my experience in the late 1990s, things have changed for both the museum and Native communities when it comes to open dialogue. The relationships have improved drastically, and museums today are very open to Native consultation and assistance when caring for objects of special interest, including wampum belts. They're creating intimate spaces for people to burn their sacred tobacco and perform their small feasts to honor the spirits of the objects. Many wampum belts have found their way home, back into the hands of the proper Native stewards. Those that are still in museum collections are being cared for both by museum curators and, most of time, with the advice of the Native communities. We have worked hard to have our voices heard, to bring the proper spiritual care for our living objects, and to ensure that they survive not only physically but spiritually so that they benefit us all now and into the future.

FOURTEEN

Always Becoming Better Stewards: Caring for Collections at the National Museum of the American Indian

Kelly McHugh

Abstract

Conservators of cultural heritage are part of an evolving, active, and interrelated network that includes the people, places, plants, and animals to whom the items we care for belong. Our conservation practice needs to extend beyond a Euro-American basis of thought and a material-centered focus to enact relational conservation care and treatment. Using experiences from the National Museum of the American Indian's Conservation Department, this essay will attempt to describe how a relational conservation methodology is inspired and implemented through collaboration with our Indigenous colleagues.

> Guiding Principles:
> We are part of an organic world in which interrelationships at all levels of life are honored. Our relationship to the place where we live, which includes the land, water, sky, mountains, rocks, animals and plants is tangible and very important. Our sense of social relationships leads us to respect all who have gone before us, our elders, youth and those who will follow us.
>
> We embrace the cycles of our organic world, such as days, seasons and our lives. Our life and death cycles connect us to the cycles of the sky, water and earth. We honor these cycles in our daily tasks, our sustenance patterns and ceremonial activities. Our sense of time is in accord with a natural continuum where past, present and future are interrelated.
>
> —Rina Swentzell, "The Way of the People," 1991

Above are two principles outlined by Pueblo potter and architect Rina Swentzell in "The Way of the People," a foundational report for the creation of the Smithsonian Institution's National Museum of the American Indian. Founding director Rick West committed to the idea that the museum's Native constituency would have a fundamental role in the ideology that guided the museum's creation and its working methodology. This would not only be true for the public-facing sectors of the museum, but it would also inspire an Indigenous concept of stewardship and care for the collection behind the scenes. As it was described in the report, the collection would be professionally housed and cared for in ways carefully designed to meet Indigenous and museological requirements for orientation, environment, access, and conservation and to acknowledge and be respectful for the life that exists within the items housed there. It was, and continues to be, the responsibility of the NMAI collections departments (Conservation, Collections, Archives, and Registration) to realize this guidance through stewardship policies and implementation practices.

Traditionally, museums in their European-American orientation are institutions whose philosophy and purpose may not correlate with Indigenous ontologies and worldviews. The veneration and investigation of inanimate material objects detached from their original context is contrary to these views. Indigenous people see objects as living and as part of a larger environment and museums as places that create boundaries around things. As members of communities, these items are part of ongoing life and derive their importance from their relationships, function, and contact with natural elements. This essay will address ways in which the NMAI, particularly the Conservation Department, tries to rectify traditional Euro-American museum practice by aligning conservation methodologies with Indigenous views. As a conservator at NMAI for over two decades, I will use this discussion to reflect on the recent history and possible future of conservation as it relates to Indigenous cultural heritage more broadly.

Always Becoming

Inspired by "The Way of the People," the *Always Becoming* sculpture project created by Santa Clara Pueblo artist Nora Naranjo-Morse in 2007 stands outside the south entrance of the NMAI's flagship museum on the National Mall in Washington, DC. The group of six hand-built adobe sculptures is rooted in the Pueblo values of community and inclusion and reflects the artist's message of growth, transformation, and Native peoples' relationship to the land. The sculptures are ephemeral and are intended to melt back into the earth with care, much like ceremonial pots are left to do in the community. The ephemerality is intentional and purposeful, and through the structures' material loss, artistically embedded forms are revealed to reinforce the idea of "always becoming." The sculptures have a relationship with the natural world, with the people who visit them, and with the conservators and collections stewardship staff who care for them. Perhaps it is this interrelationship with people, place, animals, and plants that is the most impactful element of *Always Becoming*. The sculptures embody cultural values and knowledge and inspire an organic philosophy of care.

Collectively, the sculptures are an accessioned collection item and as such require the museum's care and stewardship. Nora explains, "Just like your Grandparents age and deteriorate, so do the sculptures. Both require love, attention, care and maintenance."[1] Through tangible care, the intangible aspects of stewardship are also practiced. The ongoing relationship with the artist offers collections and conservation staff, fellows, and interns the opportunity to understand an active, evolving, and interrelated practice of conservation.[2]

This experience with the sculptures and the artist instills a different association with change, in that it requires that we, as members of the stewardship collective, not only accept it but participate in it. It reinforces that embedded Pueblo principles are activated by the community that interacts with and cares for the sculpture family. Deterioration becomes cultural stability and preservation or conservation becomes active and alive. We undertake our work in a connected way, which underscores the need to understand and respect more than just the physical materials from which the components are made.

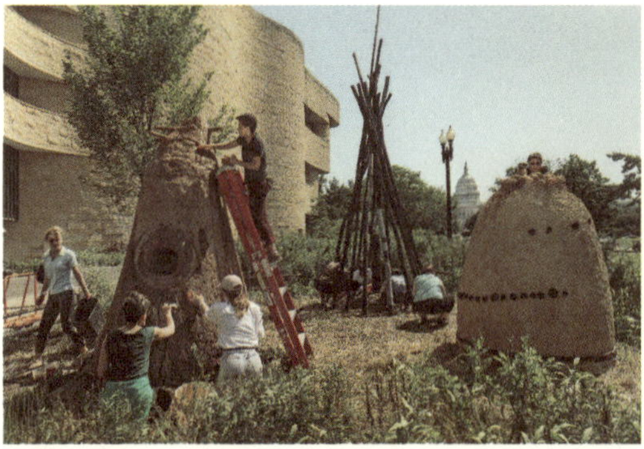

Fig. 1. Nora Naranjo-Morse, *Always Becoming*, 2007. Adobe, stone, wood, ceramic, acrylic paint, mica. NMAI, 265840.000. Photo credit: National Museum of the American Indian.

Always Becoming is a gentle, kind, and generous teacher, and a critical guide. Our training as conservators and collections managers is rooted in the material, the object, and ultimately the collection. However, cultural significance and relational thought are critical components in the care of Indigenous cultural heritage and are often missing in our training and our practice. Conservator Miriam Clavir notes, "The field of conservation deals with the physical aspect of cultural property, and 'cultural significance' is a social construct. Conservation professionals are supposed to refrain from making statements based solely on opinions not rooted in physical evidence. Conservators by their very training and principles attempt to avoid subjective interpretations, yet cultural significance is society's interpretation of what is important. That is always changing."[3] The work on *Always Becoming* requires the acceptance of change as inherent and beautiful. It inspires introspection and reconciliation as we attempt to understand how stewarding the intangible surpasses the preservation of the physical and how this relates to the ethics and principles of our conservation profession. We get down to the essence of what precisely we are preserving and what will stand when the sculptures themselves are no longer here—community, the natural world, and ultimately our relationships.

Implementing this Indigenous practice of "active" or relational conservation becomes more complex when we are dealing with items that were collected within a constructed and fixed period of the past. Historically, it has been the role of the conservator to preserve the items in that "original" state, thus rendering them static. Items were not only decontextualized—removed from their initial time and place—but then their (re)contextualization is fixed. The conservator's purpose then becomes to perpetuate stasis in order to preserve the object as evidence; however, this is counter to the fact that these items belong to and are part of active societies and still have a place and function within those cultures. Art historian and former museum director Ruth Phillips states that "[i]n the past, we thought of an object as a sort of self-contained package of information, from which data could be retrieved. Today objects are viewed as material embodiments of social and cultural perspectives and relationships that are increasingly multi-vocal. Objects are made by people for cultural purposes. Objects have dynamic cultural biographies that do not end but only change when they enter museum collections."[4]

Tlingit weaver and scholar Teri Rofkar shared this philosophy. She described the spruce-root baskets in museum collections as mentors, educators, and elders.[5] Teri was a visionary thinker, and her time spent in museum collections motivated her to advocate for the activation and inclusion of Indigenous science and ways of knowing in the care of cultural heritage. She was interested in utilizing Western analytical tools to illustrate principles long understood by Indigenous people. She saw the baskets' purpose along a continuum, and while their presence in museum collections changed their function, it did not compromise their cultural purpose. Embedded in these objects are a deep understanding of the natural world, a complex understanding of botany, math, and engineering—the knowledge systems of the ancestors. The basket no longer serves as a receptacle for food, so it becomes a teacher, sharing this embedded knowledge with spruce-root basket makers of the present and future. She often said, "We move through the objects—they do not move through us. They will be here long after we are. They simply form different relationships."[6]

Teri's philosophy has been articulated by others in the Indigenous scientific research community as "two-eyed seeing"—the idea of merging different knowledge systems to increase the collective breadth and

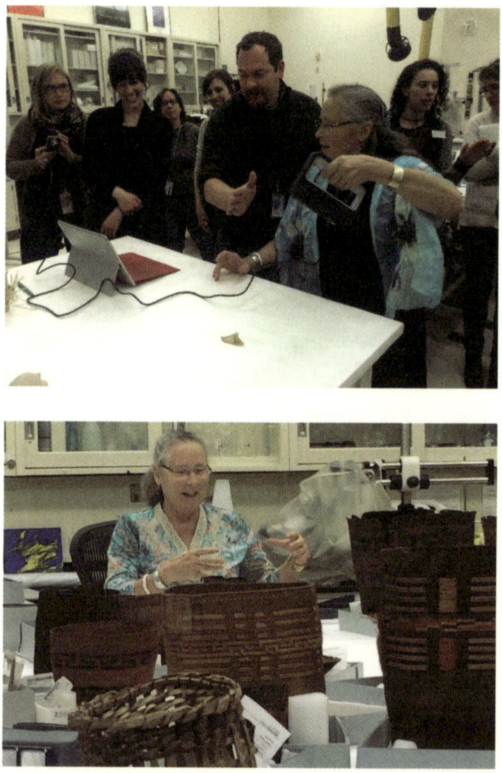

Fig. 2. Tlingit artist and scholar Teri Rofkar, with spruce-root baskets, in the Conservation Department at the NMAI, 2016. Photo credit: National Museum of the American Indian.

depth of understanding. It is a Mi'kmaw concept known as *Etauptmumk*, "To see from one eye with the strengths of Indigenous ways of knowing, and to see from the other eye with the strengths of Western ways of knowing, and to use both eyes together."[7] To responsibly bridge knowledge systems, meaningful engagement with knowledge holders is necessary at all stages of work.

The Collaborative Conservation Model (2000–2020)

The idea of collaborative conservation did not gain the widespread foothold expected from the activity and action of the 1980s and 1990s.[8] Certainly, institutions in the United States and Canada that considered this work essential continued to develop the practice, but there was no centralized movement in professional organizations or training programs. There are numerous attributable reasons for this lack of mainstream progress. One consideration is that given the establishment of the Tribal Museum movement, there was no longer the social pressure to take this on in large museums with diverse collections. Native communities were no longer reliant on European-American museums to tell their story, as they could tell their own. A cultural infrastructure was built, and in the United States, the Association of Tribal Archives, Libraries, and Museums was established.

The United Nations Declaration on the Rights of Indigenous Peoples was passed in 2006, thrusting Indigenous rights involving culture and heritage, and by extension conservation, into the realm of international law. In 2007 the Canadian Conservation Institute held a symposium titled "Preserving Aboriginal Heritage: Technical and Traditional Approaches." The organizers describe that the symposium was developed to meet a recognized need—that Aboriginal people in Canada were concerned about the preservation of their material culture and felt their preservation requirements are not adequately met. It also responded to the growing public awareness of the importance of Aboriginal contributions to collective Canadian culture and identity. The symposium delved into deeper issues surrounding the meaning of "belongings" in museum collections, their role in communities, as issues of access and stewardship.

Sherry Farrell Racette, now professor of art history at the University of Manitoba, stated, "For those of you who have come here from Indigenous communities, we must work harder to connect, listen and reclaim the stories and knowledge within them. Until we do, the objects are suspended in space and time, like an Elder waiting for someone to come and have tea with them."[9] John Moses, now manager of treatment and collections at the Canadian Conservation Institute, wrote about the need for diversity and equity in the conservation profession. He explains, "It is no longer good enough that we simply engage periodically with Aboriginal groups at the community level or at the institutional level on particular finite projects." He advocates for a homegrown conservation profession, "to equip Aboriginal conservation practitioners as expeditiously as possible to assume their rightful place within the ranks of the profession."[10]

NMAI contributed a panel presentation to the symposium on the collaborative work undertaken for the exhibition *Listening to Our Ancestors: The Art of Native Life along the North Pacific Coast.* Barbara Cranmer served as a Kwakwa̱ka̱'wakw community curator for the exhibition and shared her experience on this three-year project. The decision by the Kwakwa̱ka̱'wakw community was to focus on the potlatch for their section of the exhibition. Central to their lives, the potlatch includes the T'seka (Red Cedar Bark) ceremonies, when their people enter the spirit world to gain supernatural power and connect to their history through dance.[11]

The most important part of the T'seka is the Hamat'sa, which tells the story of overcoming Baxbakwalanuksiwe, a great man-eating spirit from the north end of the world. In the third part of the ceremony, the dancer appears wearing a Ha̱msa̱mł mask representing the raven who serves Baxbakwalanuksiwe. He is covered in cedar bark as he dances. The Ha̱msa̱mł mask in NMAI's collection (216534.000) that was selected for the exhibit was made by renowned carver Mungo Martin and included a worn muslin trailer, with tufts of red and white yarn attached to the crown of the raven's head. Eagle feathers were tied to the yarn and hung down the back. Given the importance of cedar bark to this ceremony, and by the visual markings on the mask, it appeared the entire trailer was from an earlier refurbishment made prior to collection. Dancing a mask in such poor condition would be unacceptable to the community and would never happen in practice today.

The community curators felt strongly that the mask in the exhibition should look as if it was ready to be danced at a potlatch ceremony; therefore, another refurbishment was required. Options were discussed in the community, but the one that seemed the most culturally appropriate was to replace all the trailer elements with a new cedar-bark trailer and new feathers. The muslin trailer, feathers, and yarn elements were retained by the museum, and conservation documentation recorded the process through a written report and photographic images. The cedar-bark trailer was fabricated in the community by Donna Cranmer, a relative of Mungo Martin. The work to replace the muslin trailer with the new one at NMAI's Cultural Resources Center was done by her cousin Kevin Cranmer. The results were impactful, and new life was given to this Ha̱msa̱mł. In this case, proper conservation necessitated active material change rather than stabilization to revitalize the relevance of the mask.

While this level of intervention by the community is not a common occurrence at NMAI, the fact that it can and does happen is important. It supports the proclamation that the museum considers itself steward, not owner, of the belongings in the collection and recognizes and respects the community's authority regarding the role and functionality of the mask in contemporary society. If we say these items are "active" or living, we must treat them as such. The functionality of the mask in the Kwakwa̱ka̱'wakw Red Cedar Bark ceremony determined the course of action. The mask in its historic state reflected a difficult period, when the potlatch was outlawed in Canada but still being practiced owing to the

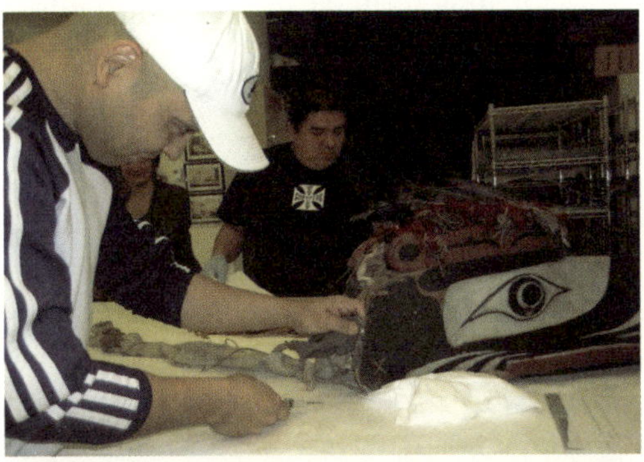

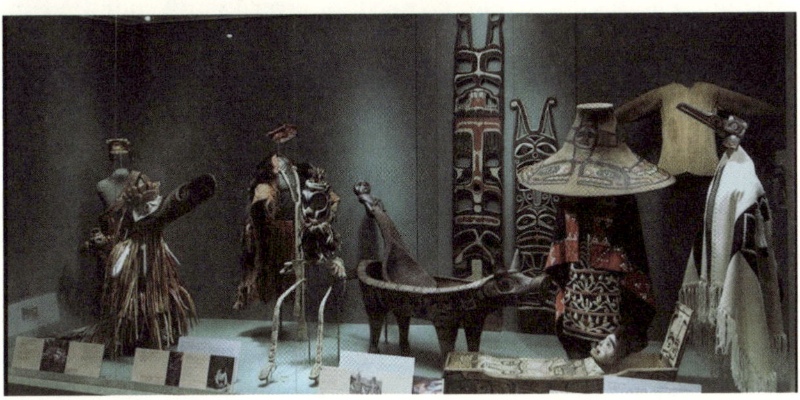

Fig. 3. The Kwakwa̱ka̱'wakw Ha̱msa̱mł mask during and after conservation treatment. NMAI, 216534.000. Photo credit: National Museum of the American Indian.

strength and resilience of these communities. Preserving it in that state did not reflect the Kwakwa̲ka̲'wakw today. NMAI conservation's role was that of documentation, so that the historic evidence was preserved, and the decision to replace the trailer was the right of the Kwakwa̲ka̲'wakw community.

Discrete projects such as this one help balance the inherent asymmetry in the care of museum collections, but ideally we need to move toward more widespread and long-term relational philosophies of care and preservation. Recognizing the need for a resource to help facilitate collaboration between Native communities and museums, the School of Advanced Research–Indian Arts Research Center in Santa Fe formed a group of Native and non-Native museum professionals led by independent conservator Landis Smith; Cynthia Chavez Lamar, now NMAI assistant director for collections; Brian Vallo, now governor of Acoma Pueblo; and Elysia Poon, now director of the Indian Arts Research Center at the SAR to create guidelines for collaboration for both museums and communities.[12] These guidelines serve as a resource for those museums interested in engaging with stakeholder groups and for communities wanting to develop long-term partnerships with museums that are the custodians of their material heritage. The associated website contains case studies, a glossary, and downloadable PDFs of the guidelines in both English and Spanish.

The Future

Conservation must be viewed as a relational process directed by economic, political, religious, social, and cultural dynamics rather than primarily as a process focused only on materials and techniques. In 2007, John Moses pointed to the need for diversity and equity in conservation practice, but that has been slow to happen. This underscores the need to reexamine the conservation education model. While some programs—such as the Andrew W. Mellon–funded opportunity for diversity in conservation at UCLA and the Getty Conservation Institute, or the partnership between the University of Delaware Art Conservation Department and Winterthur Museum, or the Alliance of HBCU Museums and Galleries and Yale University to offer an intensive two-week program of study for four advanced students enrolled in historically Black colleges and universities—are a good start, the training and thinking remain primarily rooted in a Western, materials-based conservation practice.[13] The current conservation education options are inhospitable to people unwilling or unable to dedicate every aspect of their life to conservation. In her 2013 dissertation, Martina Dawley, then at the Arizona State Museum, analyzed the reasons for the low number of Native Americans working in museums and specifically as conservators. Only 0.8 percent of people working in museums identify as Native American, the lowest percentage of any ethnic group.[14] Her research revealed a number of factors: lack of awareness about the field, cost of fulfilling prerequisites and graduate programs, distance from home, familial requirements, limited job market, and low pay, among others.[15] The interviews she conducted with practicing Native American conservators asked about the benefits of having Indigenous people working in the cultural heritage sector. Some responses were knowledge of Indigenous context, values, cultural significance, and relevancy; knowledge of familial lineages and social structures; awareness of the collection items' cultural purpose and function; and knowledge of materials and the way things are made.[16] Additionally, the issue of trust and the fact that Native people can build bridges with Native communities more readily was documented. My experience working in both conservation and collections care and stewardship at NMAI confirms that having Native staff working in these departments creates greater trust and ease among Native people while engaging with their collections at the museum. A comment made by Conrad Fisher from the Northern Cheyenne Nation while visiting collections in 2019 affirms this: "[I]t is so nice to see Native people caring for our heritage—there is a fundamental understanding of what is important."[17]

Increasing diversity will aid in broadening perspectives; however, we also need to create institutional systems that represent a multiplicity of knowledges, cultural structures, and social relationships that ultimately result in more ethical stewardship practices.[18] This needs to happen in the infrastructures that contextualize our collections—the collections information systems. The collections information system utilized at NMAI is a linear, object-focused data management system. A relational database system is required to execute the relational conservation and collections care practice that I am advocating for here. Further, our conservation documentation protocols need to evolve to capture, inspire, or trigger more contextual and relational examination and treatment, including, for instance, the item's relationship to its place, land, water, sky, mountains, rocks, animals, plants, life cycles, and people, as Rina Swentzell articulated.

NMAI is currently utilizing the Rights module within our collections information system, which was originally designed to hold copyright

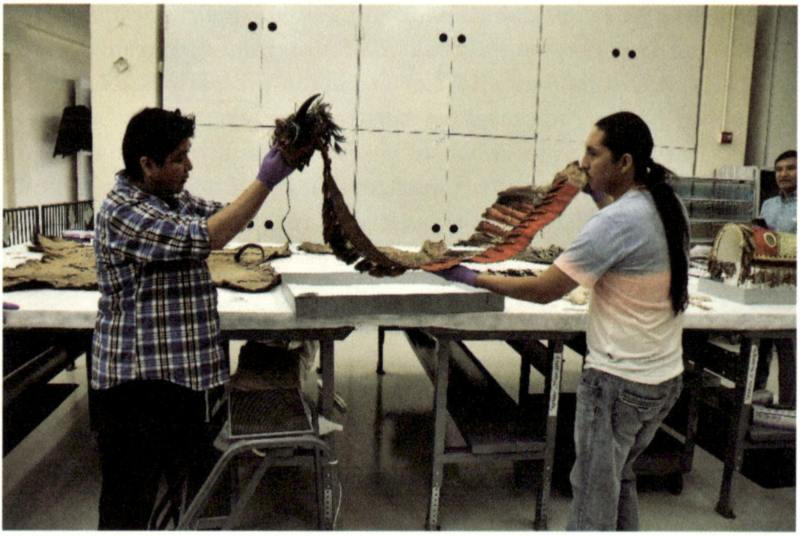

Fig. 4. NMAI collections management staff working with Northern Cheyenne representatives, 2019. Photo credit: National Museum of the American Indian.

information. The Rights module now houses cultural protocol information collected over the past thirty years and will contain cultural protocols shared by Native nations moving forward. The collections departments at NMAI are working with Local Contexts, the organization responsible for the Traditional Knowledge Hub, on the development of Cultural Care Notices (graphic icons). These will align with cultural protocols or "rights" located in the Rights module in our system and will be used both in the database and on the physical housing in collections, making cultural protocol requirements easily accessed and implemented.[19] This adaptation of the collections information system creates the systematic incorporation of Indigenous knowledge into the working practice of collections-based departments at the NMAI, resulting in more ethical and responsible stewardship practices.

A larger digital framework to integrate Indigenous knowledge structures and protocols into knowledge management systems is emerging with examples such as the Mukurtu Content Management System, a free, open-source platform for the management of Indigenous knowledge and heritage that is being utilized by Indigenous organizations globally. In their chapter for a forthcoming volume of the Smithsonian Institution's *Handbook of North American Indians*, Aaron Glass and Kate Hennessy explore the domains created through new digital networks, both online and offline. As they state, "These networks provide a crucial, contemporary means of social reconfiguration and renegotiation of dominant museum values surrounding access to, preservation of, and representation through collections."[20] This has positive implications for the conservation profession, in that contextual models for thinking and acting exist and can provide inspiration for restructuring and rethinking current modes of operation and practice.

Just before leaving NMAI's Conservation Department to become head of collections care and stewardship, I was involved with preparing items selected for display in the 2018 *Americans* exhibition. One item chosen for the Battle of the Little Bighorn gallery is a Sicangu Lakota double eagle-feather headdress. Faced with the care and treatment of numerous headdresses over my career, I understood their significance among the Northern and Southern Plains communities. I also understood that different nations have specific cultural protocols surrounding the handling and care of such revered headdresses and that consultation would be required to ensure we were being respectful and appropriate in our actions. This double-trailer eagle-feather headdress belonged to William Spotted Tail, son of Sinte Gleska, Chief Spotted Tail. It was acquired by the Museum of the American Indian in 1931, and limited information is available on its history.

In Lakota cosmology the eagle is an important and sacred animal whose feathers carry great potency. The eagle feather may be acquired as a result of doing a good deed for the betterment of the community. It represents not only the deed of the individual but also Lakota values of generosity, bravery, persistence, and fortitude.[21] Headdresses are made for warriors or for cultural or spiritual leaders. They symbolize the considerable responsibility that an individual has to his community.

The NMAI Conservation Department works with Sicangu educator and cultural knowledge holder Duane Hollow Horn Bear and has since 2013, when members of the department traveled to the Rosebud Reservation in South Dakota for a place-based educational program. Duane supported the idea of approaching the conservation treatment of the Lakota headdress using relational methodology. We conducted a three-day work session at the Cultural Resources Center in 2017 to discuss all aspects of care and treatment of Lakota eagle-feather headdresses. During the visit, Duane expressed the importance of showing respect and honor to both the eagle and the man who wore the headdress, both beings having served their people together. Showing respect and remembering that these objects are connected to individuals is significant, particularly in museums, where many headdresses became disassociated from the original wearer and community, both physically and often through lost knowledge of their provenance. The ideas Duane shared offered specific guidance for our approach to examining and treating the Lakota headdress. Even ideas that feel intangible, such as showing respect for the eagle and individual, manifested in the practical treatment activities. I supervised the treatment that was executed by Mellon fellow Megan Doxey-Whitfield and preprogram intern Vanessa Ocana-Mayor. It was inspiring to see how these emerging conservators approached the treatment. After understanding the context and the directive from Duane, they proceeded: "We chose to examine the 87 eagle feathers individually and in great detail to express our respect through our care and attention."[22] They acknowledged that each feather represents its own deed and has its own story; therefore, they constructed a condition diagram that considered and reflected this social context (fig. 5).

For Duane, "making the headdress look cared for" and restoring the sheen to the feathers was essential to showing respect to the individual and to the life of the eagle. This mandate directed the treatment to address the surface soiling that had accumulated, particularly the museum dirt from storage and display. The eagle tail feathers were cleaned using vacuum suction with the nozzle covered in Vellux, followed by light passes with a polyurethane foam, and a final pass of the vacuum to remove any foam residue. Duane recommended that the conservators be in the right frame of mind before working on powerful objects such as a headdress, as a measure of personal spiritual protection (a sentiment often shared by other Native American consultants related to their own heritage materials). During this treatment Megan and Vanessa tried to stay mindful that through their actions they were doing more than just reducing surface soiling—their purpose was to bring more life to these feathers and, by extension, to honor the important individual who wore them. Developing and executing this treatment with Duane's technical and cultural expertise ensured the work was carried out in the best interest of the headdress and the community, both materially and culturally. Megan and Vanessa utilized contextual knowledge and cultural awareness to inform their treatment, reinforcing that relational conservation is possible on historic items that have been treated as static in the past.

The treatment followed conventional conservation methods for feather cleaning, but it departed from conventionality in that the treatment was made human. This is not unlike the Pueblo principles

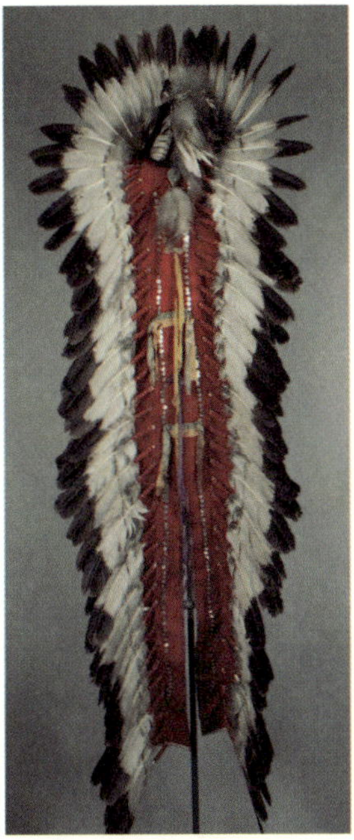

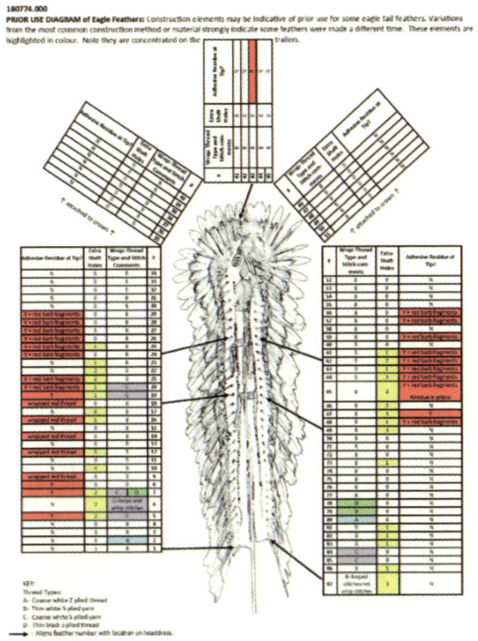

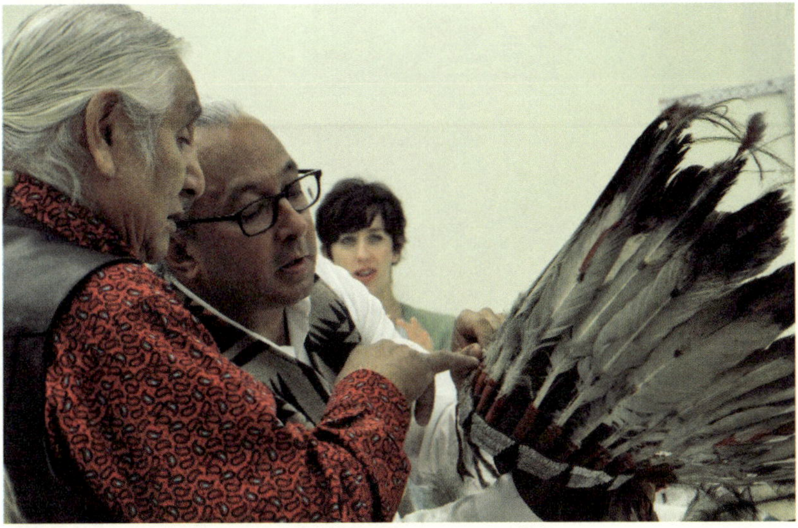

Fig. 5. Sicangu Lakota double eagle-feather trailer headdress. NMAI 180774.000. Conservation condition documentation and Duane Hollow Horn Bear (Sicangu Lakota) working with NMAI staff and fellows. Photo credit: National Museum of the American Indian.

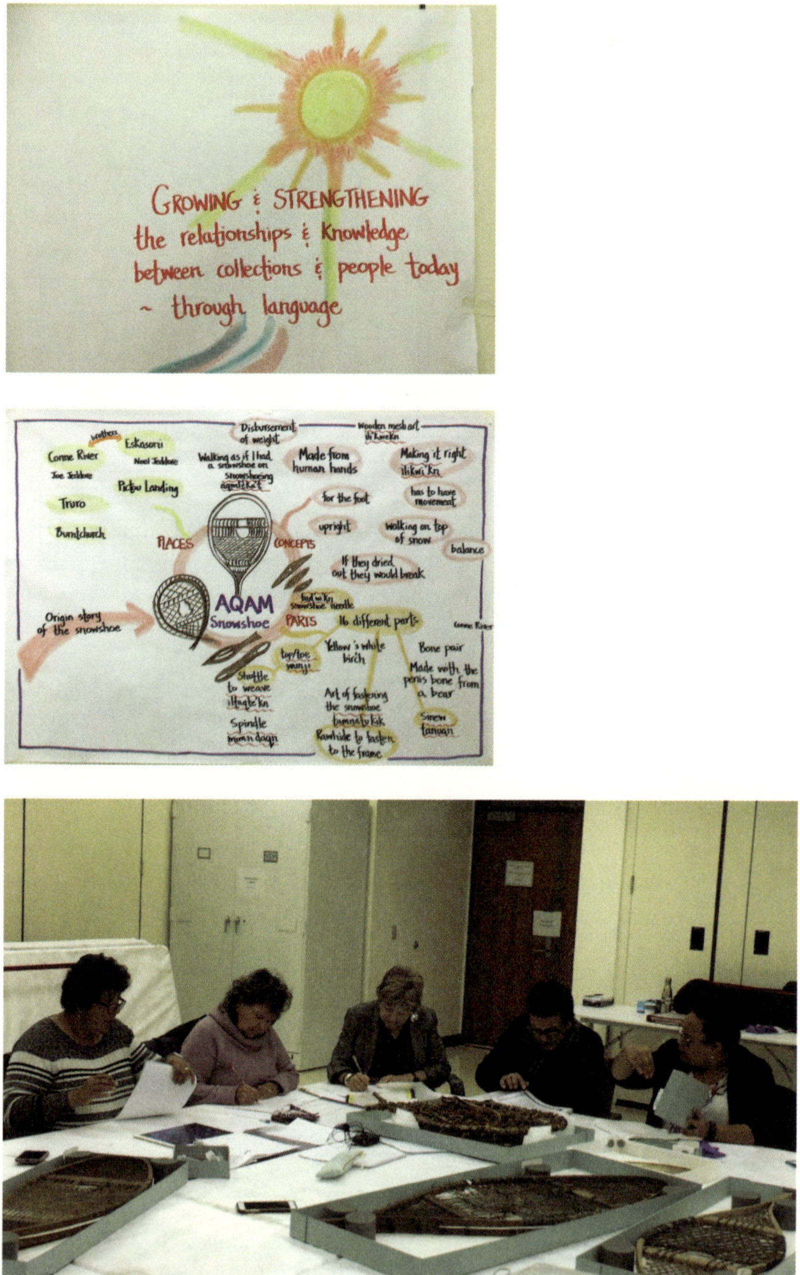

Fig. 6. Mi'kmaq language revitalization work session at the Cultural Resources Center, NMAI, 2018. Graphic recording of language revitalization work session. Graphic recording credit: Corrie Melanson.

embedded in *Always Becoming*, inspiring community, or the outcome of the revitalization of the H̲ams̲amł mask, restoring cultural vitality. Three different conservation treatments that were all inspired by and conducted for people, not materials.[23]

As we look to the future, tribal museums and cultural centers serve as important examples of how to conceive of both exhibitions and collection stewardship in an integrated and relational way. For instance, the NMAI is currently working with the Mi'kmawey Debert Cultural Center, which has spent decades preparing for the current process of establishing a cultural center on traditional Mi'kma'ki homeland in Nova Scotia, Canada.[24] The Mi'kmaq made numerous visits to engage with their belongings at NMAI over the course of this decades-long project. Their working methodology serves as an inspiring model on many fronts. First, they established a memorandum of understanding with NMAI, outlining the conditions of the relationship and establishing equity. Second, they developed and are utilizing a relational collections information system that allows for place, plants, animals, and stories to connect around both archival and cultural material. Third, their work sessions demonstrate Mi'kmaq ways of knowing and thinking that permeate and serve as a foundational structure for everything. An important sentiment was shared during the most recent language revitalization work session: "Our ancestors' work was survival; our work is revival."

This relates back to the Mi'kmaq concept of *Etauptmumk*—two-eyed seeing. As conservators and collections stewards, we play an important role in Indigenous cultural heritage preservation. However, we reside in a small space within an extensive network of people, landscape, plants, animals, and cultural practice that needs to be respected, considered, and well understood. It is not only about opening the door; it is about building relationships that move us in a direction where true and meaningful collaboration happens. It means creating connections. Mostly, it means deconstructing a binary system that leaves us with space "in-between." This allows us to work with great intention and a commitment to "always becoming" better stewards.

NOTES

1. Nora Naranjo-Morse, personal communication, September 10, 2009.
2. Naranjo-Morse, Joice, and McHugh, "Always Becoming."
3. Clavir, "Conservation and Cultural Significance," 139.
4. Ruth Phillips, quoted ibid., 146.

5. This sentiment is also shared by Metis scholar Sherry Farrell Racette, who refers to "visiting the old ladies" when she is researching in collections. For her seminar on this theme, see "Visiting Old Ladies."
6. Teri Rofkar, personal communication, April 23, 2013. Sadly, we lost Teri to cancer in 2016, but her impact and her influence will be long felt.
7. Kutz and Tomasellie, "'Two-Eyed Seeing.'"
8. See Kelly McHugh, contribution on the recent history of "ethnographic" conservation, on the companion website for the exhibition *Conserving Active Matter*, forthcoming, https://exhibitions.bgc.bard.edu/cam/.
9. Racette, "Caring for the Living," 19.
10. Moses, "Preserving Aboriginal Heritage," 24.
11. Kaminitz et al., "Renewal," 75.
12. Indian Arts Research Center, Guidelines for Collaboration (website).
13. Additionally, the Smithsonian Institution provides opportunities through the Andrew W. Mellon–funded ten-week internship program titled Conservation Internship for Broadening Access (CIBA).
14. Dawley, "Analysis of Diversifying Museums," 18.
15. Ibid., 116.
16. Ibid., 118.
17. Conrad Fisher, personal communication, July 2019.
18. Balachandran and McHugh, "Respectful and Responsible Stewardship."
19. For more information on Local Contexts and the Traditional Knowledge Hub, see the organization's website, accessed August 18, 2021, https://localcontexts.org/.
20. Glass and Hennessy, "Museum Collections and Indigenous Knowledge."
21. Duane Hollow Horn Bear, personal communication, September 2018.
22. Doxey-Whitfield and McHugh, "Hold-Off or Hands-On."
23. For more on these issues, see Clavir, "Conservation and Collaboration."
24. For more information, see the website of the Mi'kmawey Debert Cultural Centre, accessed August 18, 2021, https://www.mikmaweydebert.ca/.

BIBLIOGRAPHY

Balachandran, Sanchita, and Kelly McHugh. "Respectful and Responsible Stewardship: Maintaining and Renewing the Cultural Relevance of Museum Collections." In *Preventive Conservation: Collections Storage*, edited by Lisa Elkin and Christopher A. Norris, 3–24. New York: Society for the Preservation of Natural History; American Institute for Conservation of Historic and Artistic Works; Smithsonian Institution; George Washington University Museum Studies Program, 2013.

Clavir, Miriam. "Conservation and Collaboration: A Discussion." In *Heritage Conservation and Social Engagement,* edited by Renata Peters, Iris Den Boer, Jessica Johnson, and Susann Panncaldo, 30–45. London: UCL Press, 2020.

———. "Conservation and Cultural Significance." In *Conservation Principles, Dilemmas and Uncomfortable Truths.* Edited by Alison Richmond and Alison Bracker, 139–49. Oxford: Butterworth-Heinemann, 2009.

———. *Preserving What Is Valued: Museums, Conservation, and First Nations.* Vancouver: UBC Press, 2002.

Dawley, Martina. "An Analysis of Diversifying Museums: American Indians in Conservation." PhD diss., University of Arizona, 2013.

Dignard, Carole, Kate Helwig, Janet Mason, Kathy Nanowin, and Thomas Stone, eds. *Preserving Aboriginal Heritage: Technical and Traditional Approaches.* Ottawa: Minister of Public Works and Government Services, 2008.

Doxey-Whitfield, Megan, and Kelly McHugh. "Hold-Off or Hands-On: Collaborative Engagement as an Integral Part of the Treatment Process." Presentation at the CAC-ACCR 44th Annual Conference. Kinston, Ontario, May 10, 2018.

Glass, Aaron, and Kate Hennessy. "Emergent Digital Networks: Museum Collections and Indigenous Knowledge in the Digital Era." In *Handbook of North American Indians,* vol. 1 (introduction), edited by Igor Krupnik. Washington, DC: Smithsonian Institution, forthcoming.

Indian Arts Research Center. Guidelines for Collaboration (website). Facilitated by Landis Smith, Cynthia Chavez Lamar, and Brian Vallo. Santa Fe, NM: School for Advanced Research, 2019. https://guidelinesforcollaboration.info/.

Kaminitz, Marian, Barbara S. Mogel, Barbara Cranmer, Jessica Johnson, Kevin Cranmer, and Thomas V. Hill. "Renewal of a Kwakawka'wakw Hamsaml Mask: Community Direction and Collaboration for the Treatment of Cultural Heritage at the National Museum of the American Indian." In Dignard et al., *Preserving Aboriginal Heritage,* 75–89.

Kutz, Susan, and Tomasellie, Matilde. "'Two-Eyed Seeing' Supports Wildlife Health." *Science* 364, no. 6446 (2019): 1135–37.

Lonetree, Amy. *Decolonizing Museums: Representing Native America in National and Tribal Museums.* Chapel Hill: University of North Carolina Press, 2012.

Moses, John. "Preserving Aboriginal Heritage: The Need for Diversity and Equity in the Conservation Profession." In Dignard et al., *Preserving Aboriginal Heritage,* 23–27.

Naranjo-Morse, Nora, Gail Joice, and Kelly McHugh. "Always Becoming." *Objects Specialty Group Postprints* 19 (2012): 147–57. Paper presented at the American Institute for Conservation 40th Annual Meeting, Albuquerque, NM.

Racette, Sherry Farrell. "Caring for the Living, Not Embalming the Dead: Storied Objects and Precious Legacy in Museum Collections." In Dignard et al., *Preserving Aboriginal Heritage*, 15–21.

———. "Visiting Old Ladies: Reclaiming Knowledge and Relatives in Museum Collections." Seminar given at the Institute for the Study of Human Rights, Columbia University. March 7, 2019. http://www.humanrightscolumbia.org/events/visiting-old-ladies-reclaiming-knowledge-and-relatives-museum-collections.

FIFTEEN

Hoki Mauri: Bring Back the Life Essence

Rose Evans

Abstract

This paper documents the progress of conservation practice in Aotearoa (New Zealand) through the lens of my thirty-year career as a conservator. It spans a vital period where we have seen the considerable reinvigoration of mātauranga Māori *(Māori knowledge) in our society—conceivably a consequence of Aotearoa addressing its Treaty of Waitangi claims and no doubt intimately connected to the groundbreaking exhibition* Te Maori. *This has helped us as Māori to feel pride in our own culture. I explore why there has been little progress for conservators of Māori descent over that time and connect this directly to Western museums' inability to understand the potency that Māori recognize in the stories that* taonga *(treasure) hold. Activation of taonga is through telling and retelling of these stories by their* whanaunga *(kin). Where this connection is severed, taonga are not active. Two decades ago, my work in developing and enacting the bicultural conservation program at Te Papa (Museum of New Zealand) was considered innovative and progressive. On reflection, I now view this approach as barely tolerated by a mainstream Western ontology. In this chapter I seek to identify the new opportunities to enable Māori-initiated interpretation and conservation of taonga in Aotearoa.*

> *Titiro Whakamuri*
> *Kōkiri Whakamua*
> (Look back and reflect
> So you can move forward)
> —*Whakataukī* (Proverb)

We are on a cusp of change in the museum sector, change that I hope will bring some long-awaited transformation for museums and galleries in Aotearoa (New Zealand). A new *hapū* (subtribe)–based cultural center named Raiatea will open in 2021, representing the first of its kind—independently developed, managed, and operated by a north Hokianga Hapū collective named Ngāti Tamatea. What makes this such a profoundly new development is that Raiatea's vision, and all the policies and processes that flow from this, are driven by hapū values, with *tikanga* (protocol) being the primary driver. As a result, I expect to see, within Raiatea and hopefully beyond, adaptions to standard museum and gallery practice in accordance with the primary objective: articulating tikanga as a set of Indigenous principles and practices to ensure that collections remain vital aspects of living culture in the present and into the future.

I welcome this, as an independent museum conservator of Māori descent looking back over my career to date. Time has given me cause to reflect on my career and its path over the last thirty years and to re-evaluate the museum sector with a thorough, deeper, and more mature lens. Thirty years is a long time to wait for significant change within the sector, and it strangely spans my own career within conservation.

It is not, however, coincidental. On September 10, 1984, the landmark exhibition *Te Maori* opened at the Metropolitan Museum in New York City. The *kaikaranga* (female caller) could be heard along Eighty-Fourth Street and Fifth Avenue, ushering in a ninety-strong contingent of *kaumatua* (elders), cultural performers, and practitioners up the stairs and into the museum galleries. Māori tikanga was upheld, as *whakawātea* (clearing rituals) were undertaken to ensure safe passage for both the Māori and visitors alike.[1] This tikanga continued for each of the following three venues in the United States and no doubt familiarized museum staff and audiences about the close dynamic relationship Māori have with their own *taonga* (ancestral treasures).

Te Maori's international success initiated a paradigm shift in how Māori material culture was valued and viewed within its own country, ultimately driving cultural policy changes throughout Aotearoa. It is widely acknowledged that the exhibition instigated transformation toward what are now the more consultative and codevelopment models prevalent in many of the modern museums in Aotearoa.[2]

Sir Pita Sharples, a *manutaki* (warrior) at the Metropolitan Museum opening, recalls the opposition by some *iwi* (tribes) to their taonga traveling offshore, further emphasizing the dynamic relationship with taonga, so deeply rooted in *whakapapa* (genealogical ties).[3] "The opposition to it going was the fact that these are artifact forms of living people. So, they carry the *mana* [authority] of that whakapapa and that tribe . . . but of course we go and we are the living taonga if you like, so I didn't see a difference."[4] It is important to note that a Māori subcommittee, chaired by Kaumatua Kara Puketapu, the New Zealand government's secretary for Māori affairs, guided this protocol, ensuring what roles Māori would play concerning the selection and interpretation of loans, copyright controls, and conducting of traditional opening ceremonies. This was perhaps the first time that Māori participated in and drove the process for exhibiting their taonga within a Western context.

On the return of *Te Maori* to Aotearoa, there was vigorous discussion and debate among staff in cultural institutions and the wider Māori communities. This included notions of the ownership, meaning, and value of taonga Māori within institutions. Tipene O'Regan's article "Taonga Maori Mana Maori" raised points that consolidated some crucial issues that remain relevant today:[5]

> My own view is that the most productive approach could be in the effective penetration of the administrative structures which control our heritage resource. How many museums, for example, have specific provision for Maori tribal representation in amongst their Trustees? What criteria are applied in the appointment of staff with responsibilities for Maori collections? How many insist that staff for whom they are responsible refer to appropriate Maori tribal sources prior to publication?[6]

In the same issue of *Agmanz Journal*, Mina McKenzie, Aotearoa's first Māori museum director, revealed her own concerns about Aotearoa museum standards: "Hopefully, Te Maori will hasten the redevelopment of our national institutions. Hopefully, the Government will take the lead and make it possible to increase the numbers of staff necessary to do justice to the Maori collections."[7]

In response to the debates and the concerns about museum and gallery standards that *Te Maori*'s return to Aotearoa triggered, educational initiatives were launched. One was the introduction of the Interim (later formalized) Committee for the Conservation of Cultural Property. This government committee was formed to guide and facilitate collection conservation, training, funding, and policy. The majority of New Zealand museum conservators were trained with funding from the committee from this period. It was during the committee's active period that conservators of Māori descent were advocated for and trained. I was one of the five Māori conservators trained under this scheme. In 1991 the committee was reviewed, and its activities were discontinued.

My recent rereading of Sir Hirini Moko Mead's reflections and thoughts about the mana of taonga Māori in the 1984 publication produced to accompany the exhibition *Te Maori* has had an impact on my thinking.[8] After decades of *mahi* (work) within some major museums in New Zealand and overseas, his words made me appreciate issues of the greatest importance to my work. They are the deep connection and closeness I feel with taonga Māori and the whakapapa that connect and protect me within museum and *marae* (communal meeting place) environments.

Whakapapa perfectly articulates my place and connection within the natural world—a vast cosmic ecosystem of birds, fish, people, and planet within a layered world of humans, gods, and ancestors. Within this ontology, Māori states of being—*tapu* (realm of dead, requiring respect) and *noa* (everyday and living)—define the disciplines and restrictions of the "sacred" and "profane" that shape our behavior and beliefs to a greater or lesser extent.[9]

As a conservator trained within a Western scientific framework, with knowledge and experience of my mother's Māori worldview, there were times in my early career where I felt my beliefs were regularly compromised. I found ways to quietly listen and potentially solve many conservation issues while still holding fast to tikanga. I knew that every cultural challenge would be different; after all, there are many tribal stories, perspectives, and *kōrero* (conversations). The challenge was to constantly learn, build on knowledge gained, and remain flexible and respectful at all times. A swift movement back and forth between both worlds seemed at the time the best option. For me, the *mana* (prestige) and active power of taonga Māori was locked within the tribal stories they chose to tell, and I was aware that their potency was not always known within the museum fraternity and perhaps even among many iwi communities.

In fact, my *kaupapa* (purpose) in the 1990s was to embed Māori values within the dominant *Pākehā* (Western) system, to create a more integrated model that recognized Māori community needs and demands in areas such as treatment, storage, and access to taonga. Hirini Moko Mead's reflections on the inherent potency and active power of taonga clearly define the close connection between stories, *tūpuna* (ancestors), and maker: "The mana of the personalities together with the power of words and of the form created by the artist conspire to produce a powerful taonga—a taonga with mana, a taonga with immanent power, an object that has within it a hidden force.[10]

As my career developed, Mina McKenzie would repeatedly tell us that the potency of taonga without known whakapapa often needed more attention and care. She warned many of us of our own need for cultural safety and protection when working with them. Mead further accentuates this Māori ontology of collections with his kōrero about the often-unknown history of taonga within museum care, frequently exacerbated by institutional collecting habits that often recorded a collector's identity, mode of manufacture, material, and sometimes function—but not often the stories embedded within the taonga themselves. "Just as a *koroua* (elderly man) and a *kuia* (elderly female) are venerated for their knowledge and wisdom, so are old taonga, respected for the knowledge and wisdom that lie buried in the korero which has been revealed. But great age brings uncertainties about the nature of the korero, for who can possibly know what the *tohunga* (priest) of the early period said, what spells he recited, what curses he may have placed on certain items."[11]

In 1990, after training in Australia and on beginning my internship at the New Zealand National Museum in Wellington, I was fortunate to be mentored by Maui Pomare, the chair of the National Museum Board of Trustees (fig. 1). Maui was politically astute, diplomatic, well educated, and of *Rangatira* (chiefly) status. He sat comfortably in both Māori and Pākehā worlds and was one of the few Māori at that time to have a museum leadership role of such influence. Although he exercised authority in governmental circles, his role and impact now seem invisible, as is often the case with many pioneering Māori in our politically charged museum sector. Maui was my *whanaunga* (relative) and was willing to transfer his knowledge and provide the cultural protection necessary for me to work with tapu (sacred) taonga.

While acknowledging previous academic research, Maui's primary kaupapa was to focus on observation and to develop familiarity with taonga Māori. His calm, practical, and observation-based thinking was underpinned by a rigorous adherence to tikanga. This has shaped my work and approach to this day.

I was very privileged to experience this customary *tuākana-tēina* relationship. Tuākana-tēina is a reciprocal and balanced learning approach drawn from *Te Ao Māori* (the Māori world), where experience is shared and knowledge is exchanged through a supportive rapport between the tuākana (older person or mentor) and the tēina (younger person or mentee). This customary learning relationship was used in

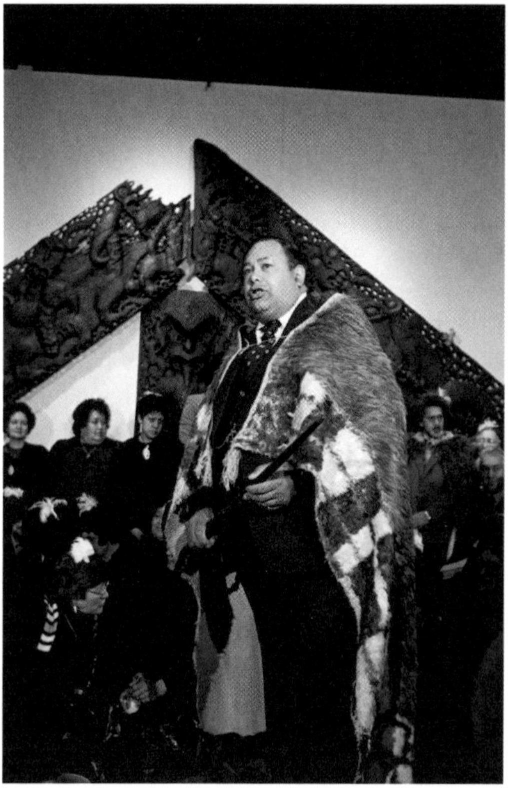

Fig. 1. Maui Pomare providing *whaikōrero* (formal welcome speech) at the National Museum's opening ceremony of *Te Maori: Te Hokinga Mai* (the return New Zealand leg of the *Te Maori* touring exhibition, 1986). Photo: *Te Maori* exhibition, ca. 1986, New Zealand, by Brian Brake. Museum of New Zealand Te Papa Tongarewa (E.005587/11).

the traditional *wananga* (schools of learning) and was enhanced by exchange and transferal of knowledge by both *kaiako* (teacher) and *ākonga* (student).

Most of our mahi was concentrated on *Toi moko* (human remains), which required careful, reflective observation and shared, pooled opinion. I offered my conservation perspective, complementing Maui's depth of knowledge and the cultural support and protection he offered. I quickly learned that any well-meaning, conventional conservation training is a secondary support skill when working with Toi moko and other tapu taonga. My ability to offer open and flexible opinions that provided agile options and outcomes was central to my learning from this relationship.[12] Maui's vision for the care, research, and eventual repatriation of Toi moko to their descendant communities has now been enacted through government statute.[13] Now Te Papa, as a crown agency, is expected to implement the customary care required by Māori communities until their eventual return home.

Conservation care for Toi moko is preventive in approach and is performed in accordance with tikanga that respects the tapu nature and significance of human remains. From my own perspective, I view Toi moko as tūpuna; they are an active connection into my own whakapapa as well as a tangible connective thread to an earlier precolonial time. This perspective was and still is at loggerheads with the traditional conservation profession, often unable to nimbly respond to problems that require knowledge and understanding of the tangible and intangible stories connected to taonga Māori.

The conservation of a prestigious *pātaka* (storehouse) named Te Tākinga from the Ngāti Pikiao iwi further demonstrates this learning outcome. This project started in 1994 and spanned a three-year duration. I worked closely with the *Iwi rūnanga* (tribal committee) with a clearly developed project plan that allowed for resource contingencies, ensuring ample time for discussion and negotiation, resource-gathering workshops, iwi intern conservation training, and ongoing *hui* (tribal meetings). Pivotal to its success was the recognition that many stories of this taonga had been lost; as a result, Te Tākinga had become the focus for enthusiastic tribal discussion and a source of tribal mana (power and authority) and ancestral strength.[14]

The three-year process entailed a high level of liaison and informed discussion over each step in the treatment. Throughout this treatment, tikanga was observed, and treatment outcomes remained open and flexible. Some conservation outcomes varied from conventional approaches, incorporating customary practice, ensuring what was culturally crucial for Ngāti Pikiao to *manaaki* (care) for their taonga. For example, iwi were encouraged to undertake the gathering of traditional materials—such as *muka* (flax-fiber) binding and *raupō* (swamp-grass) thatching—and engage in ongoing customary practice—such as feather dressing of the pātaka—themselves, rather than have conservators perform all the physical treatments (see the chapter by Kelly McHugh, in this volume). All these supportive activities exhibited the *kaitiakitanga* (guardianship) for a living taonga, which was no longer a static "object" in storage, or on display, in a context entirely defined by a Western perspective. Conservation treatments varied, in that they included cosmetic repairs for tūpuna features to ensure mana through their intact nature and form (required by Ngāti Pikiao). The incorporation of two Ngāti Pikiao interns for the treatment duration, and the inclusion of a leading Ngāti Pikiao carver to undertake recarving for areas of large *whakairo* (carving) losses within the treatment program, ensured both iwi capability and a continuation of kaitiakitanga (care) and tikanga.

These treatments and forms of iwi participation were calmly folded into the project plans.[15] This approach was an outcome of my time working quietly alongside Maui Pomare, early in my career (fig. 2).

As much as this type of process appears to have been incorporated into some contemporary conservation approaches in museums worldwide, this adaptation, innovative at the time, is today one step short of the current expectations of many iwi and hapū.[16]

Looking forward, it is salutary to survey the state of conservator training in New Zealand. After *Te Maori*, training and employment of Māori conservators within the museum sector was an important imperative. It must remain so. While the New Zealand Government's Ministry for Culture and Heritage is now providing some relevant project funding, there is no dedicated conservation training funding available for New Zealanders, let alone conservators of Māori descent, to train internationally.[17] Additionally, there is no proactive recruitment policy in any of our regional or national institutions for conservators of Māori descent.

With the imminent prospect that iwi and hapū cultural centers will develop throughout New Zealand in varied formats, it is critical that the absence of training opportunities is addressed. The prospect that the *whare taonga* will form part of these cultural centers is high in my opinion. Significantly, taonga still shape the basis for our stories and cultural knowledge and in many cases represent the single ancestral link and dialogue to a time when these stories thrived and knowledge was widely held.

When I was speaking with Raiatea's principal curator, Whina Te Whiu, about a hapū-based perspective for potential conservation treatment of taonga, Whina commented, "Collection care of taonga is firstly underpinned by *Te Kaupapa o te Marae,* the foundation principles which benefit hapū members in the context of *whanaungatanga* (their relationships) and founded on the principle of addressing, enhancing and restoring tapu (sacred being and dignity) and the rightful exercise of their mana; their spiritual power and authority."[18] Another Māori ontological principle applied to assess taonga and its care is *te wā*—the cycles of time (past, present, and future), life, and death. Has this taonga served its purpose in this time? Does it have another purpose in the future? How do we address, enhance, or restore its mana and tapu?

One example Whina shared was of a particular *korowai,* the hapū's prized feather cloak. Most days the cloak hangs in the ancestral house of the hapū, but a tribal member can request to borrow the cloak for special occasions because it is understood this practice is imbued with

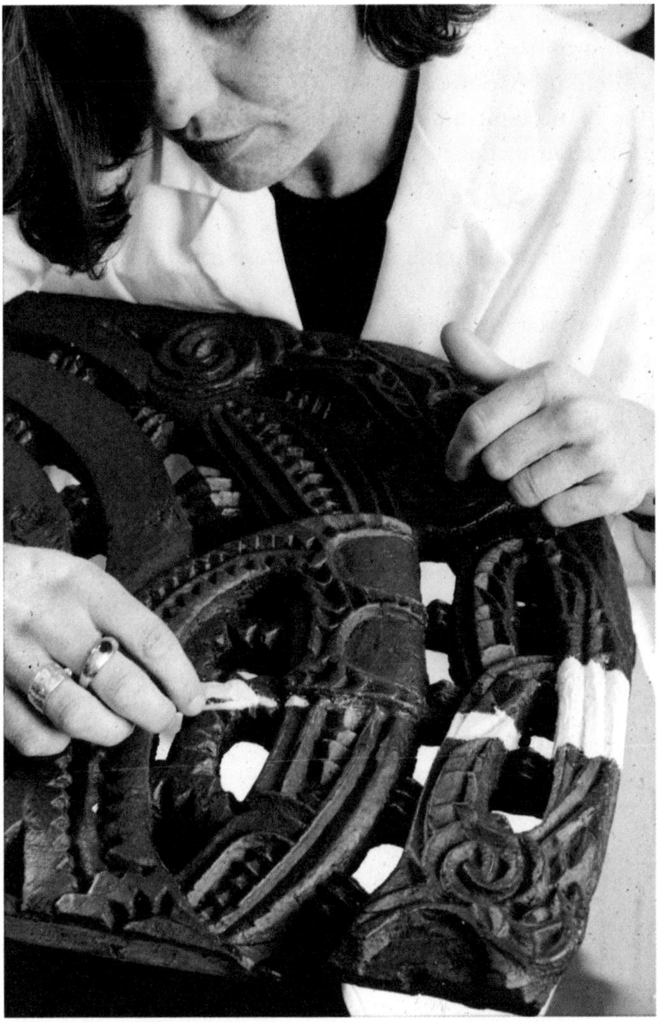

Fig. 2. Rose Evans conserving one of the large *maihi* (barge boards) of Te Tākinga pātaka. Photo: Rose Evans.

Te Kaupapa o te Marae (principles of marae), by which the taonga and hapū members are in a reciprocal mana-enhancing relationship. However, there is physical cost. The physical deterioration of the cloak means its life span is shorter, and it will no longer be fit to be worn in thirty-five years or so.

The hapū decides that this is acceptable, as the remnants of the original cloak will be woven into a new cloak sometime in the future.

To preserve the spiritual and temporal aspects of the taonga is the priority, with the whakapapa (lineage and provenance) of the original taonga becoming the foundation knowledge of the new cloak; this ensures that the mana (spiritual power) and tapu of the original are transferred. In the meantime, it is decided that remedial treatment not be considered, and only preventive conservation is utilized to reduce further deterioration.[19]

This practice of readaption is not a new one for the hapū; 1,500 years ago, their eponymous ancestor, Kupe, came to Aotearoa on his *waka* (canoe) called Matawhaorua. He established a settlement in the Hokianga and then returned to Tahiti. Several generations later his descendant Nukutawhiti refashioned his grandfather's waka and renamed it Nga Toki Matawhaorua. Following in Kupe's footsteps, Nukutawhiti traveled to the hapū's new home, Aotearoa.

This perspective of te wā, the cycle of time, was a prevalent perspective at the start of my career among many iwi I met with. This was a time when much knowledge and skills were almost completely severed and lost from our cultural memory. These taonga remained as a last remnant of customary practice in some cases, and whakapapa in others. Decisions were made by many kaumatua at this time to conserve these taonga for future generations and so provide a portal back to this knowledge and connection. Much expertise of the old ways has been gathered since then, and new materials, designs, and formats have been incorporated to create new and innovative ways to express and connect with whakapapa.[20] The concept of fluidity and change is very apparent in patterns of production, with shifts in fabrication, significance, and use from the precolonial times. This asserts the evident dynamic and innovative nature of Māori customary practice, which continues to incorporate change and disruption while retaining some traditional features and use. So it would seem that more discussions at this current point in our history might be timely and appropriate to readdress these conversations and reconsider the notion of te wā.

For me, this demonstrates how hapū and iwi can exert their own unique form of *tino rangatiratanga* (self-determination) in how they engage with tikanga and how this determines their relationship with taonga. The 1991 lodging of a Treaty of Waitangi Wai 262 claim to investigate the Crown's existing laws, policies, and practices is significant in this respect and to current and future conservation practice.[21] Many iwi and hapū are at a moment where they are seeking to address new ways to interpret and connect with their own stories and histories. Cultural centers and whare taonga offer an opportunity to provide a path for iwi and hapū to define an independent future and identity through their taonga.

A 2021 scan of the museum and gallery sector shows only two full-time Māori conservators are employed, with one at the National Library and the other at Auckland Library.[22] Does the museum sector's pattern of employing international (non-Māori) conservators suggest a lack of commitment to or understanding of the critical importance of the specific worldviews and skills Māori bring into the conservation profession? Regrettably, internships and mentoring programs are currently also absent, and there seems to be no future scoping of approaches to cultivate young people of Māori descent into the sector.[23]

As the approach at Raiatea shows, and based on my experience, this is a significant risk in my opinion. Institutions will simply miss out on the culturally appropriate and responsive approaches and strategies that a Māori conservator can offer. Museums need to value these skills as part of a conservator's expertise and capabilities—but under the current situation, opportunities to develop cultural networks and pathways that build trust with Maori communities will not evolve.[24]

There have been recent innovative museum approaches such as the collections-based Pacific Collections Access Project and Te Awe, a *Māori raranga* (textile) collections project. Both these Auckland War Memorial Museum schemes utilize expert practitioners in the wider Māori and Pacific communities to inform and enrich collections knowledge, help develop future collecting strategies, and build relationships and networks.

These are examples of good initiatives that function within a conventional museum framework—their objective is to significantly improve institutional knowledge. But the taonga remain separate, isolated from their whakapapa. Initiatives designed to ensure collective, cooperative project exhibition development ultimately do not change these cultural dynamics. Perhaps in time we will look back and see these approaches simply as a "holding pattern"—prior to the development of more whare taonga within tribal *rohe* (boundaries). The outcomes of the Wai 262 claim mentioned above may well shed light on the Crown's responsibility to demonstrate enablement of kaitiaki to fully exercise kaitiakitanga over their taonga and *mātauranga Māori* (Māori knowledge).

Māori in governance and staffing positions can have the capacity and potential to innovate and disrupt conventional approaches to care and interpretation, not just for taonga Māori but for the benefit of the entire collections. The opportunity to advocate for Māori as conservators to work in museums, galleries, and libraries as well as whare taonga is now overdue.

My hope is that whare taonga will multiply and specific taonga currently held within museums, galleries, and libraries will be repatriated back into the care of their whanau, hapū, or iwi (fig. 3).[25] If tino rangatiratanga was recognized in our museums, then so would tikanga and the differing care and kaitiaki outcomes that might come from that. Essentially, conservation as currently practiced in Aotearoa is intended primarily to halt decay and preserve physical qualities of an "authentic" object defined by its past status and condition. In doing so, it ceases to recognize the living tikanga and customs that we hold vital to our taonga and culture. Institutions can facilitate and enable contact and collaboration between taonga and community, but this denies the very essence of active matter: the revitalization and activation of a community through the ability to express kaitiakitanga over their own taonga, and in their own way, into the future.

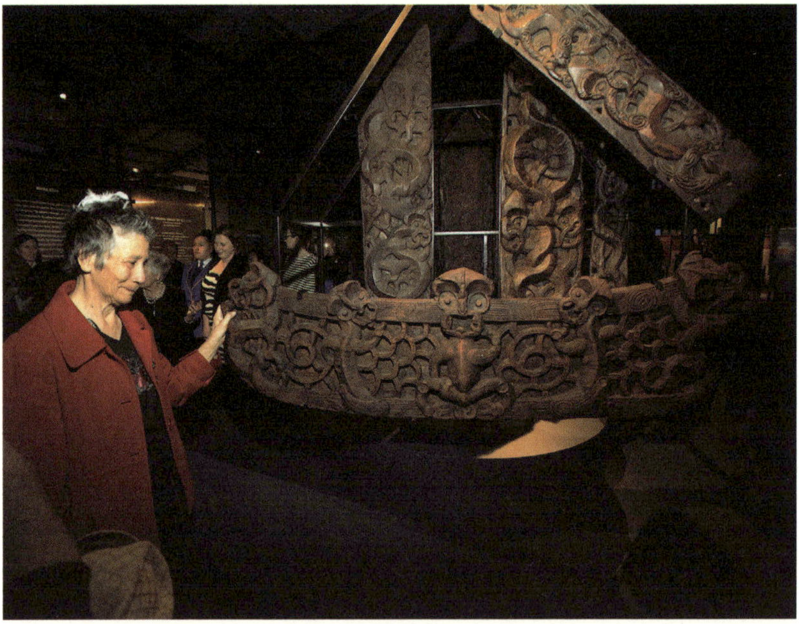

Fig. 3. Ngāti Rāhiri kuia (elder) welcoming back the carved Motunui panels. These whakairo (carvings) were exported illegally in 1972 and sold to private collector George Ortiz. They were returned in 2014 as part of Waitangi Treaty settlements to the custodianship of Puke Ariki regional museum for Te Āti Awa. It is widely anticipated that these carvings will initiate the recovery of mātauranga Māori for iwi. It is also expected that their return to the rohe will ensure further repatriations from other national institutions. Image permission granted by Ngāti Rāhiri whanau.

Hirini Moko Mead's comments from 1984 are still pertinent, and his kōrero about the many ways that iwi, hapū, and whanau create words in multiple and meaningful ways around their taonga merely stresses that the need to connect through whakapapa is vital and persistent:

> That the korero can be rediscovered and brought back into the oral traditions of the group is certainly possible, but this is often conditional upon the object being brought home where it can be seen and talked about. Although the object itself can be appreciated as a unique thing, for many people that is not sufficient. They must know the korero and because of this some individuals go to lengths to find the right stories that go with a particular object. If the korero cannot be found, the owners can resort to several alternatives that are well known to scholars; they can create a new story; they can give it a temporary, speculative story; or they can leave out the unknown part of its history and commence the korero from the moment and circumstances of the rediscovery of the object. Quite an exciting story can be told in each case. What is important for the owning group is that the object is "brought home" so it can be slotted into the art style, the history, and the oral traditions of the people. When taonga is not brought home into the tribal territory but rests instead in some museum hundreds of miles away, the object and its associated korero remain lost to the owning group.[26]

> *E tu ki te kei o te waka, kia pākia koe e nga ngaru o te wā*
> (Stand at the stern of the canoe and feel the spray of the future biting at your face)
> —Whakataukī, attributed to Api Mahuika

NOTES

1. A whakawātea is a ritual to lift *tapu* off persons so they are no longer under restrictions and are able to move freely and undertake their duties.
2. In 1997, Te Papa Tongarewa initiated a revolving Iwi Exhibition Programme as an expression of the *Mana Taonga* (principle that recognizes the right of Māori to their material culture). These exhibitions usually run for two and a half years. The process involves iwi

working collaboratively with Te Papa staff to develop their exhibition. *Kaumatua* (tribal elders) are chosen by that iwi to remain in residence at Te Papa for the duration of the exhibition.

3. Sharples is an academic and politician (former leader of the Māori Party).
4. Sharples, "Te Māori."
5. O'Regan is an academic, company director, and former longstanding chairman of Ngāi Tahu Māori Trust Board.
6. O'Regan, "Taonga Māori Mana Māori," 16.
7. McKenzie, "Te Ao Hau o nga Kaitiaki o nga Taonga," 28.
8. Mead, *Te Maori.* Mead is a writer, anthropologist, teacher, and Māori leader and was cocurator of the *Te Maori* exhibition.
9. Further academic writing exploring whakapapa as "practical ontology" is included in a special issue of the *Journal of the Polynesian Society.* Lythberg, McCarthy, and Salmond, "Transforming Worlds."
10. Mead, *Te Maori,* 22.
11. Ibid., 23.
12. This focus was central to my presentation, "Making Our Past Visible," at the BGC Conserving Active Matter symposium, April 18, 2018, https://www.bgc.bard.edu/research-forum/articles/286/conserving-active-matter-indigenous-ontologies.
13. The care of Toi moko is now undertaken by a government-mandated authority that negotiates the repatriation on behalf of Māori. This government agency, Karanga Aotearoa Repatriation Programme, takes its direction from Māori communities through an expert panel comprising Māori academics and practitioners. This program has been government resourced since 2003, to repatriate, care for, and research Toi moko, with the goal of returning them to their descendant communities throughout Aotearoa. The program supports these descendant communities to define their own management practices, which will ultimately determine a final resting place for their tūpuna.
14. In this instance, the resources to undertake a project that could compile and examine relevant oral histories, analyze museum collections and inventories, and assess current and traditional customary practice provided Ngāti Pikiao the ability to further cultivate and develop their own mātauranga Māori and in this way regenerate and nurture the mana of Te Tākinga.
15. Evans, "Tribal Involvement."
16. Johnson et al., "Practical Aspects of Consultation." A few examples of consultation in conservation practice would be both the National Museum of the American Indian, Washington, DC, and the Museum

of Anthropology, University of British Columbia, Vancouver. Clavir, *Preserving What Is Valued*; Johnson et al., "Practical Aspects of Consultation"; Chang, Documenting and Implementing; Heald and Ash-Milby, "Woven by Their Grandmothers."

17. This national funding is to support work programs whose scope might include exploration into crucial issues such as kaitiakitanga (care), partnership, stewardship and governance options, and approaches to taonga and *mātauranga Māori* (Māori knowledge). This, in combination with ongoing treaty settlements between Iwi and Crown may provide even further potential and incentive for iwi and hapū to explore independent ways to define their own futures. The ministry has recently added *whare taonga* (storehouses) to its intended recipient category for major funding. Inclusion of iwi or Māori as key stakeholders is required within its criteria for funding. There is no domestic training program within New Zealand for conservation. All budding conservators need to train internationally. National Library and Auckland Library have previously supported international training qualifications undertaken by two of their Māori conservators.
18. Whina Te Whiu, personal communication, September 1, 2020.
19. According to Whina, any small taonga remnants remaining would be returned to their respective *Atua kaitiaki* (spiritual guardian)—therefore, a waka would be buried in the sand close to *Tangaroa* (sea), and the korowai would return to *Tane* (forest).
20. Rueben Friend discusses in "He Whare Toi, He Whare Whakaruruhau" that artistic practice can be very reflective of place and society. Current use of new digital technologies reflects both retention of mātauranga Māori (cultural knowledge) and an extension of artistic practice.
21. The Treaty of Waitangi was signed in 1840, on behalf of two signatories: the Crown and Iwi. This treaty underpins legislation and represents an exchange of promises between these parties. The Waitangi Tribunal was formed in 1975 as a permanent commission of inquiry to consider claims by Māori that the Crown had breached principles of the treaty. In 1985 the tribunal's jurisdiction was extended to cover Crown acts and omissions since the signing of the treaty in 1840. Further amendments to the Treaty of Waitangi Act expanded the tribunal's membership and extended its capacity for research, hearings, and report writing. The Waitangi Tribunal released its report into the Wai 262 claim in 2011 and specifically recommends a move beyond settlement of historical grievances to consider the place of Māori culture, identity, and traditional knowledge in New Zealand's laws, and in government policies and practices today. The Wai 262 claim considered who is entitled to make or participate in

decisions affecting indigenous flora and fauna, the environment, Māori culture, and the products of Māori culture. It relates to "indigenous flora and fauna me ō rātou taonga katoa (and all their treasures) including but not limited to *mātauranga*' (knowledge), *whakairo* (carving), *wāhi tapu* (sacred sites), biodiversity, genetics, *Māori* symbols and designs and their use and development and associated indigenous cultural and customary heritage rights in relation to such taonga. (*Taonga* being defined as tangible/intangible)." For the tribunal's strategic goals, see "Waitangi Tribunal."

22. At Te Papa's opening in 1997, there were three Māori conservators in a team of eight conservators employed in a full-time capacity. There are now no conservators of Maori descent employed at Te Papa.
23. Five conservators of Māori descent were funded by the Interim Committee for the Conservation of Cultural Property: Dean Whiting, Rangi Te Kanawa, Vicki Heikell, Nick Tupara, and Rose Evans, who trained in the late 1980s. All remain working in conservation in varying private and public capacities. In the 1990s two conservators, Erina McCann and Jade Hadfield, were funded by Te Māori Manaaki Taonga Trust to train in Australia, and they remain there, working specifically within Indigenous conservation practice. Two paper conservators, Damon Joe and Tharron Blomfield, were fully supported by Auckland and National Libraries. Damon Joe is still working at Auckland Library, and Tharron is working in a related field. Two further conservators, Kararaina Te Ira and Karel Kaeo, were self-funded, and both work in related fields in Aotearoa.
24. It is widely believed that Te Papa's early proactive recruitment of Māori in all fields resulted in considerable reinvigoration of mātauranga nationwide and contributed to significant positive cultural developments in the heritage sector. McCarthy, *Te Papa*.
25. Two significant repatriations have been undertaken recently: Mataatua wharenui (ancestral meetinghouse) was repatriated from Otago Museum to Ngāti Awa in 2015 and the Motunui panels (ancestor figures) were returned from international private ownership in 2014 to Te Āti Awa. Both these repatriations were part of Waitangi Treaty settlements.
26. Mead, *Te Maori*, 29.

BIBLIOGRAPHY

Chang, L. H. "Documenting and Implementing Collections Care and Treatment Requests from Native Communities." In *American Institute for Conservation 33rd Meeting*. Textile Specialty Group postprints. Minneapolis, 2005.

Clavir, Miriam. *Preserving What Is Valued: Museums, Conservation, and First Nations*. Vancouver: UBC Press, 2002.

Evans, R. "Tribal Involvement in Exhibition Planning and Conservation Voices: A New Institutional Approach." *ICOM Ethnographic Newsletter*, no. 19 (April 1999): 10–12.

Friend, R. "He Whare Toi, He Whare Whakaruruhau: Creating Space for Maori Curatorial Practice." In *Becoming Our Future: Global Indigenous Curatorial Practice*, edited by Julie Nagam, Carly Lane, and Megan Tamati-Quennell, 135–52. Winnipeg: Arbeiter Ring, 2020.

Heald, Susan, and Kathleen E. Ash-Milby. "Woven by Their Grandmothers: 24 Blankets Travel to the Navajo Nation." *Journal of the American Institute for Conservation* 37, no. 3 (1998): 334–45.

Johnson, Jessica S., Susan Heald, Kelly McHugh, Elizabeth Brown, and Marian Kaminitz. "Practical Aspects of Consultation with Communities." *Journal of the American Institute for Conservation* 44, no. 3 (2005): 203–15.

Lythberg, Billie, Conal McCarthy, and Amiria J. M. Salmond. "Transforming Worlds: Kinship as Practical Ontology." In "Te Ao Hou: Whakapapa as Practical Ontology." Special issue, *Journal of the Polynesian Society* 128, no. 1 (2019): 7–18.

McCarthy, Conal. *Te Papa: Reinventing New Zealand's National Museum 1998–2018*. Wellington: Te Papa Press, 2018.

McKenzie, Mina. "Te Ao Hau o nga Kaitiaki o nga Taonga." *Agmanz Journal* 15, no. 4 (1984): 28–29.

Mead, Sidney Moko. *Te Maori: Maori Art from New Zealand Collections*. New Zealand: Heinemann/AFA, 1984.

O'Regan, Tipene. "Taonga Maori Mana Maori." *Agmanz Journal* 15, no. 4 (1984): 15–18.

Sharples, Pita. "Te Māori." Radio New Zealand (RNZ). August, 28, 2015. https://www.rnz.co.nz/national/programmes/nz-society/audio/201768261/te-maori.

"Waitangi Tribunal." Te Puni Kokiri (website). Last updated June 9, 2021. https://waitangitribunal.govt.nz/.

PART 4

Materials

SIXTEEN

Introduction: Developing Informed and Sustainable Responses to the Alteration of Cultural Artifacts; Materials Engineering Meets Material Culture

Jennifer L. Mass

Abstract

Our collective understanding of the physical, chemical, and biological reactivities of cultural materials has long guided our preservation of these objects. Even seemingly static objects are in a constant state of chemical change, interacting with their environment or even reacting with each other within a single museum case. This introduction provides a current overview of the field, through the lens of the four materials science chapters that follow. These projects include investigations of metal soap formation in paintings, the degradation mechanisms of holograms, biologically active time-based media conservation issues, and the use of materials analysis to undo long-held misconceptions about Meiji-era woodblock prints. Our understanding and acceptance of material activity are, like the objects entrusted to our care, in a constant state of change. In response to urgent environmental concerns, we are working toward a relaxation of traditionally strict museum climate standards to achieve a delicate balance between environmental and object preservation. We also must constantly adapt to the challenges of contemporary artists' media and confront the rapidly changing requirements of objects that perform in new ways for larger and more inclusive audiences.

The restiveness of the materials that constitute our global cultural heritage is something that the chemical, physical, and biological scientists who work in museum laboratories encounter daily. Historically, we have endeavored to monitor, retard, and in certain cases even arrest or reverse the chemical and physical changes that artworks undergo.

We also elucidate the mechanisms of interaction and degradation of the diverse range of materials (and their combinations) that one encounters in an encyclopedic museum collection. These materials can span the plant fibers used to make Polynesian bark cloth or tapa, the intricate castings of depletion-gilded pre-Colombian alloys, and the late nineteenth-century semisynthetic plastics such as cellulose nitrate. In the past we have carried out these efforts (always in conjunction with our conservation and art history colleagues) to render objects' appearances and functionalities as close as possible to our perceptions of the intentions of the artist, craftsperson, or collective who produced the work. However, as the disciplines involved in the preservation and conservation of cultural heritage mature, we have passed through multiple eras that have demanded reevaluations of our work, in terms of both how it is practiced and the new values that may be assigned to the objects that we transform. These reevaluations can be internally driven (as in Andrew Oddy's publication of the British Museum's treatment of the Parthenon marbles) or they can come about through external stakeholders.[1] These stakeholders range from art historians (consider John Richardson's "Crimes against the Cubists") to environmental activists, social justice activists, Indigenous communities, the general public, and scholars in numerous adjacent disciplines.[2] Today the fundamental tenets of art conservation and cultural heritage science, especially those surrounding stabilization, are being tested in the face of new and increasingly urgent demands that are being placed on cultural materials. These include both climate change and a growing international desire for museums to be more than warehouses for objects of colonial conquest.[3] Cultural heritage is being stored, displayed, and accessed in new ways to tell larger and more inclusive narratives. Our improved understanding of how the materials that constitute these objects will react with their environments is a crucial component of our discussions regarding these new uses.

The inorganic materials used to make cultural heritage objects (glasses, minerals, and alloys) provide an excellent example of how this understanding has evolved over time. The ancient Romans started what would become an over-two-thousand-year inquiry into historic bronze surfaces when they first hypothesized about bronze sculptures' activity with respect to their surface color and texture changes. The nineteenth-century archaeologists who excavated these objects continued to postulate explanations for the phenomenon, and in the twenty-first century conservators and scientists are still exploring the scientific basis of this blue patina through technical studies of the

Delphi Charioteer.[4] The legacy of millennia of study into these subjects is an unusual nomenclature for the physical and chemical changes that occur in inorganic works of art. This nomenclature includes such evocative terms such as "bronze disease," "caries," "métaux malades," "malignant patina," "glass sickness," "glass disease," "vitrum putrefactibilium," "ultramarine sickness," and "ultramarine disease." Given that the scientists or natural historians who first used these terms were working without knowledge of microorganisms or information about their reaction with copper alloys or silicates, these terms make sense both from a phenomenological perspective and within their historical contexts. Scientists today work within rigid frameworks for what constitutes an inorganic versus an organic material and the possible realms of activity for these objects. However, by looking more deeply into the genesis of this ancient medical language for inorganic activity and alteration, we have an opportunity to explore the foundational concepts of cultural heritage science and how they changed to keep pace with the new discoveries and worldviews of scientists from the seventeenth through the nineteenth centuries. Such research also allows us to explore scientists' evolving understanding of the range of agents of degradation responsible for the activity of inorganic materials. What emerges is a portrait of how scientists, artists, and craftspeople understood (and now understand) the natural world and its realm of possibilities. This portrait allows us to identify the different moments in history in which certain types of activity were allowed or disallowed under the paradigms of the day.[5]

An additional emerging challenge for cultural heritage science is artists' exploration with novel materials, which are active owing to their biological basis or intentionally biodegradable nature, or as in the case of YInMn blue and Vantablack, because their alteration pathways are incompletely understood.[6] The materials science essays included here showcase a variety of objects that are actively changing—both chemically and in our understanding and interpretations of them. These objects include artworks as seemingly inert as oil paintings to ones as seemingly unstable as a multimedia installation with decaying organic matter at the center of its medium and message. We also explore here the documentation, alteration, and preservation of holograms and holographic images, raising important questions about conservation prioritization, decision-making, and where the art in imaging technologies truly lies. Finally, we consider how the technological innovation of surface-enhanced Raman spectroscopy has allowed for the identification of the brightly colored organic dyes in Meiji-era woodblock prints. This research uncovers long-held Western misunderstandings about Japanese artists' materials and their significance during this period of rapid technological change.

Francesca Casadio's essay reveals critical misperceptions about the fundamental nature of oil paintings, a category of objects that includes some of Western art's most treasured cultural properties. Even after an oil painting appears to have "dried," fatty acids from the oil binder and metal ions from the pigments can combine to form new molecules that, because of their limited solubilities, slowly self-aggregate or agglomerate in ways that can have remarkably diverse and damaging consequences.[7] Manifestations of this phenomenon can include the formation of translucent regions of paint, the rise and eruption of spherical pustules the size of a grain of sand, and in the case of mid-twentieth-century paintings such as abstract expressionist works, the spalling of paint flakes that can be the size of potato chips. Even amid our growing realization of how destructive these undead molecules moving through our paintings can be, we have to balance this with the knowledge that removing them from a painting entirely would leave the painting less flexible and vulnerable to other types of mechanical damage. As a result, art conservators and scientists take a minimalist approach of limiting the mobility of these molecules using carefully controlled visual monitoring and relative humidity levels. Such an approach is a far cry from well-intended but aggressively interventionist nineteenth-through mid-twentieth-century treatments regarding the structural stability of paintings. These might include the "cradling" of paintings on wooden panels by adding a secondary network of wooden supports that would restrain the expansion and contraction of the wooden panel.[8] While the intent was to limit the interlayer cleavage and paint flaking that is caused by different degrees of expansion with changes in relative humidity (for the paint versus the wood), restraining the wood from movement simply led to increased internal stress and cracking. A less-is-more approach, accepting some degree of change but managing its rate, aligns Western cultural heritage conservation with more sustainable and global attitudes toward preservation and change. One notable example of this is the Japanese concept of wabi-sabi, which ascribes value to both imperfection and transience. This approach may also put Western conservators and scientists more in line with the original thinking of artists such as Vincent van Gogh, who was well aware of the fugitive and changeable nature of his pigment and who constructed the vibrancy of his compositions to accommodate this.[9] The idea that his compositions would reach a harmonious balance with age is in particular borne out by the Metropolitan Museum of Art's *Van Gogh: Irises and Roses* exhibit, where reconstructed images of how his works would have appeared when they were first painted can be compared with their current appearance.

Marco Leona and Henry D. Smith's essay on the materiality of Meiji-era woodblock prints examines how a recent technological innovation, surface-enhanced Raman spectroscopy, allows us to dismantle some of the long-held myths about the impact of Western synthetic organic dyes on the *nishiki-e* (ukiyo-e) art form. Western art history has long viewed these dyes as synonymous with the precipitous decline of the art form, resulting in prints that were considered unpleasantly bright, garish, and decadent in the twentieth century. Among Leona and Smith's revelations, however, is that the Meiji-era *nishiki-e* incorporated very few synthetic dyes, employing only eosin red (to obtain a delicate pink), crystal violet (toned down with Prussian blue), and the naphthol reds. The objectionable fields of bright red color that were viewed as unacceptably vivid and tawdry (and thus prepared with cheap synthetic materials that represented a decline of the medium) are in fact made with cochineal carmine, a traditional natural red dye similar to lac dye. This reveals how a greater understanding of the materiality of cultural heritage can challenge long-held misunderstandings about an art form and, perhaps more importantly, deeply held cultural prejudices. The dominant twentieth-century paradigm that any brightly colored art was cheap and gaudy was an outgrowth of an unstated but widely agreed-upon Protestant aesthetic that predominated in Europe and the United States. This paradigm has long been used to relegate Asian and other forms of non-Western art to the category of "lesser" or "primitive" art on the basis of the brightness of the media used. In doing so, status-conscious twentieth-century scholars and art historians revealed their ignorance of the recent past, ignoring the blazingly bright American Fancy period of the 1830s and the aesthetic-era interiors of the 1870s to the 1890s.

Chris McGlinchey's essay on working together with contemporary Cuban artist Tania Bruguera and his colleagues to reproduce Bruguera's multisensorial installation *Untitled (Havana 2000)* at the Museum of Modern Art explores the difficulties that arise when the decay of certain components of an artwork is an essential element of the work (in this case bagasse, sugarcane waste, used to protest the violence of Cuba's colonial past). This project is emblematic of a larger problem in the conservation of contemporary art installations, which commonly incorporate intentionally ephemeral materials that are meant to undergo dramatic transformations during the lifetime of the installation, such as the bananas in Zoe Leonard's *Strange Fruit* (1992–97). While the rotting fruit and sugar cane are used in these works to deliver powerful messages about brutal regimes and social orders, their inclusion in a museum environment is problematic from the perspective of the preservation of the rest of the museum's collections. Rotting organic materials release a number of gasses that can impact other collection objects and cause microbial growth that can be damaging to both museum objects and visitors. The successful installation of *Untitled (Havana 2000)* using a floor of kiln-dried bagasse combined with a simulated nontoxic rotting scent reveals how scientists in contemporary art museums must constantly innovate and collaborate across a variety of disciplines to bring an artist's vision to life without sacrificing the safety of other collection objects. The US Visual Artists Rights Act (1990) allows artists to disavow works that they contend were installed or conserved in such a manner that the work no longer accurately represents their vision. As more and more artists use this legislation to protect their moral rights, McGlinchey's work reveals a best-case scenario of creative problem solving that simultaneously protects the artist's vision, MoMA's collections, the live performers participating in the artwork, and the museum visitor.

Marc Walton and his colleagues' essay explores the role of the cultural heritage scientist in the preservation of a specific form of mid-twentieth-century innovative imaging technology—the hologram. This essay incorporates a number of themes relevant to conservators and scientists working not only on holographic materials but also in time-based media preservation and digital-born artworks. In this field of conservation, one of the most pressing questions to be addressed is which components, exactly, should be preserved. If the holographic image can be reproduced successfully through migration (by creating a new image plate), as has been done at MoMA with Louise Bourgeois's spectacular holograms, then what level of preservation does the exceedingly delicate and photosensitive silver salt-coated plate that originally generated the image require? Walton and his colleagues make the choice here that the technology that generates these images, in particular the technological innovations of holographic pioneer Stephen Benton, are well worth preserving given his mastery of the medium and role in growing it to produce dynamic, polychrome, and exquisitely detailed three-dimensional images. The concept of the medium being (part of) the message has a long history in cultural heritage preservation, from our preservation of Thomas Edison's original metal recording masters to national cellulose nitrate film vaults and repositories. Ironically, the reactive nature of these materials enables them to capture images, movement, and sound but works against their long-term preservation. Walton and his coauthors' development of strategies for diagnosing condition issues, retrieving images, and uncovering the working methods of a master holographer point the way to a rigorous approach for the preservation of a medium that captivated twentieth-century viewers with a startlingly present yet frustratingly noncorporeal image that always remains tantalizingly out of reach.

We often speak in materials science about our search for high-performance materials that push the boundaries of what constitutes a useful lifetime, whether their purposes are optical, electronic, or aesthetic. Museum objects are asked to perform far outside their expected lifetimes, in some instances for millennia, and in ways that their creators could never have anticipated. Our increasing understanding of the mechanisms of degradation of these materials is taking place in a shifting landscape of our expectations for them, and this will bring about a new set of concerns and objectives for the generation of scientists that succeed us. Will some of the forms of alteration and degradation that we now consider unacceptable become regarded as beautiful or at least intrinsic to artists' media? Will we learn to relax our visual "performance standards" for cultural heritage objects for the sake of the environment and become more forgiving of a certain degree of visual change? Andrew Juniper notes, "The tides of time should be able to imprint the passing of the years on an object. . . . It is the changes of texture and colour that provide the space for the imagination to enter."[10] By generating a clearer understanding of how exactly cultural materials are reacting with their environments and each other, we can perhaps continue to allow our environmental and aesthetic standards to evolve and keep pace with the emerging demands that are being made of global cultural materials.

NOTES

1. Oddy, "Conservation of Marble Sculptures," 145.
2. Richardson, "Crimes against the Cubists."
3. Both of these issues are addressed in Janes and Sandell, *Museum Activism.*
4. For the nineteenth century, see Villenoisy, "Patine du bronze antique." For contemporary investigations, see Franke and Mircea, "Plutarch's Report."

5. A full discussion of the origins and implications of the glass disease, ultramarine disease, and bronze disease as cultural heritage science terminologies can be found in "Historical Perspectives on Inorganic Activity through the Lens of Glass, Bronze, and Ultramarine Diseases" on the *Conserving Active Matter* website, 2022, https://exhibitions.bgc.bard.edu/cam/.
6. For biological materials, see Antonelli et al., *Neri Oxman Material Ecology Catalogue.* For biodegradability, see Li and Subramanian, "Inorganic Pigments." For new artists' pigments with incompletely determined alteration pathways such as YInMn blue and Vantablack, see Adams et al., "Vantablack Properties in Commercial Thermal Imaging Systems."
7. There are numerous important publications on this topic, but a comprehensive review can be found in Casadio et al., *Metal Soaps in Art.*
8. See, for example, Dardes and Rothe, *Structural Conservation of Panel Paintings.*
9. Brack, "Refreshing Van Gogh's Faded Flowers."
10. Juniper, *Wabi Sabi.*

BIBLIOGRAPHY

Adams, Arnold, Fred Nicol, Steve McHugh, John Moore, Gregory Matis, and Gabriel Amparan. "Vantablack Properties in Commercial Thermal Imaging Systems." In *Infrared Imaging Systems: Design, Analysis, Modeling, and Testing XXX,* edited by Gerald C. Holst and Keith A. Krapels, 110010W-1–110010W. Proceedings of the Society of Photo-Optical Instrumentation Engineers 11001. Bellingham, WA: SPIE, The International Society for Optics and Photonics, 2019. https://doi.org/10.1117/12/2518768.

Antonelli, Paola, Anna Burckhardt, Emily Hall, Jennifer Liese, and Neri Oxman. *The Neri Oxman Material Ecology Catalogue.* New York: Museum of Modern Art, 2020.

Brack, Paul. "Refreshing Van Gogh's Faded Flowers." Chemistry World (website). Royal Society of Chemistry. October 1, 2015. https://www.chemistryworld.com/news/refreshing-van-goghs-faded-flowers/9005.article.

Casadio, Francesca, Katrien Keune, Petria Noble, Annelies Van Loon, Ella Hendriks, Silvia A. Centeno, and Gillian Osmond. *Metal Soaps in Art: Conservation and Research.* Cultural Heritage Science Series. Cham: Springer, 2019. https://doi.org/10.1007/978-3-319-90617-1.

Dardes, Kathleen, and Andrea Rothe, eds. *The Structural Conservation of Panel Paintings: Proceedings of a Symposium at the J. Paul Getty Museum, 24–28 April 1995.* Los Angeles: Getty Conservation Institute, 1998.

Franke, Walter A., and Magda Mircea. "Plutarch's Report on the Blue Patina of Bronze Statues at Delphi: A Scientific Explanation." *Journal of the American Institute for Conservation* 44, no. 2 (Summer 2005): 103–16.

Janes, Robert R., and Richard Sandell, eds. *Museum Activism.* Museum Meanings Series. New York: Routledge, Taylor & Francis Group, 2019.

Juniper, Andrew. *Wabi Sabi: The Japanese Art of Impermanence.* Tokyo: Tuttle, 2019.

Li, Jun, and M. A. Subramanian. "Inorganic Pigments with Transition Metal Chromophores at Trigonal Bipyramidal Coordination: Y(In,Mn)O3 Blues and Beyond." *Journal of Solid State Chemistry* 272 (2019): 9–20. https://doi.org/10.1016/j/jssc/2019.01.019.

Oddy, Andrew. "The Conservation of Marble Sculptures in the British Museum before 1975." *Studies in Conservation* 47, no. 3 (2002): 145–54.

Richardson, John. "Crimes against the Cubists (Protecting Cubist Paintings)." *New York Review of Books,* June 16, 1983.

Villenoisy, F. "La patine du bronze antique." *Revue archéologique* 1 (1896): 69–71.

SEVENTEEN

Contextualizing the Installation of Tania Bruguera's *Untitled (Havana 2000)*

Chris McGlinchey

Abstract

The various scales of time on which artistic media remain stable are explored in this essay. Working in the field of cultural heritage science for thirty-five years, the author draws from his experience in traditional artistic media in both fine and decorative art and thinks through how this experience might inform the installation and performance of Tania Bruguera's Untitled (Havana 2000). *This work of performance art was acquired by the Museum of Modern Art, and in order to exhibit it successfully staff worked closely with the artist. Working together made it possible to exhibit Bruguera's work and, in the process, to learn and document how to install and perform the work in the future. In reflecting on the challenges of this piece, thoughts are offered on how to reframe issues related to the preservation and exhibition of traditional artistic media.*

The term "active matter" suggests that energy is being absorbed and somehow transformed. Having worked in the area of cultural heritage science for three and a half decades, I can say that this is decidedly a scientist's approach to the phrase. For example, colors are observed when visible light is absorbed and selectively released on the basis of a molecule's architecture. Some molecules have a vulnerable structure and can be degraded in the same light that enables them to be seen. Similarly, artist's liquid oil paint solidifies when exposed to light and air. When all goes well, the paint becomes a robust solid by the time a painter considers the work finished. This, however, is not always the case. Museums, and in particular those that collect modern and contemporary art, deal with active matter for good or ill. If they are fortunate enough to have a scientist, then that person—or group—helps provide technical support to conservation and curatorial staff typically to aid decisions that prevent or slow deterioration. This dynamic process leads one to observe, as I did when I switched from being a scientist primarily caring for Old Master paintings at the Metropolitan Museum of Art to being responsible for the entire collection at the Museum of Modern Art, one fundamental fact: change is greatest for things that are young.

In actuality, during my fifteen years at the Met, I never fully appreciated a few examples where painting conservators pointed out changes in condition or composition that likely occurred within a few decades after the work was created, some plausibly within the artist's lifetime and possibly observed—or even made—by the artist him- or herself. Shortly after I joined MoMA, this became more concrete. Some works contain light-sensitive pigments that, if not sheltered from light, would be lost, while others cracked not too long after they were made, despite being housed in a stable climate. The few small, slightly dodgy-looking retouches on Henri Matisse's oversized work *The Dance*, which have chemical similarities with the original paint, were likely in-painted by Matisse himself after the work was first installed. As a result, the imperfect yet easily overlooked marks remain present for all to see in the ultramarine background if one looks closely enough. This is nothing like the complete posthumous repainting that has occurred with some of Ad Reinhardt's paintings from his Black Square series. The beauty in Reinhardt's matte velvet minimalist surface is achieved by removing much of the oil medium in the paint, making the dried paint vulnerable to scuffs or cracking from a fluctuating climate. One such example was donated for research purposes in a collaboration between MoMA and the Solomon R. Guggenheim Museum with the goal to explore ways of separating the overpaint from the fragile surface below. It's worth contemplating whether or not small imperfections communicate a message of vulnerability and register on some level. In seeing a perfect flawless surface, it is sometimes easy to forget that it may be incredibly fragile.

It is not uncommon to repaint decorative art objects, but while at MoMA I played a small role in helping to peel away overpaint to reveal the original paint of white cabinets from what was then a recent acquisition of an apartment kitchen designed by Charlotte Perriand and Le Corbusier from Unité d'Habitation, Marseille, France, from 1952. The restored surface of the original paint is again beautiful and in harmony with the work's other aged but otherwise unaltered elements. Part of what made that a successful treatment was the fact that the original paint was durable. A more typical example of historic surfaces being repainted might be the Met's Lansdowne Dining Room, a late eighteenth-century interior by Robert Adam, where tiny paint cores were analyzed and small areas adjacent to some of these sampled sites were cleared of overpaint so that original paint could reveal the exact hue to the colors cited in the original invoice. Interestingly, paint schemes in layers above the original appeared to be more evocative of a palette thought to be from Adam's *earlier* and more colorful interior paint schemes. The tolerance for authenticity can be variable, and arguments for what exactly is authentic, complex.

While one understandably laments the loss of freshness in young paintings and sculpture, modern media such as installation and performance-based art require consideration of an important set of dynamic issues in order for the work to become realized. One can't simply open up a crate, install the work in a gallery, and shine a light on it. *What* is needed to make the installation of such a work authentic, and who determines that? MoMA's 2018 exhibition of Tania Bruguera's (Cuban, b. 1968) *Untitled (Havana 2000)* is an excellent case study to demonstrate the ways in which change *activates* the work and, in fact, may provide further inspiration to the artist. The many factors that needed to be considered in bringing this work to life can help us appreciate how much we might be missing when it comes to preserving and exhibiting more traditional media.

For the installation, MoMA *transformed* a "white box" gallery space into a convincing re-creation of a prison chamber of a Spanish colonial fortress in Havana, Cuba, where *Untitled (Havana 2000)* was first performed. This prison had a long history of torture, murder, and imprisonment of those who were deemed political threats; this darkness was not lost on Bruguera, whose literal use of it for her piece would compel the viewer to gather as much information from other senses as possible. This was not the traditional way of "looking" at art; rather, the multisensorial experience with a bare minimum of illumination was critical to Bruguera. She wanted visitors to avoid forming preconceived ideas of what they would eventually see by forcing them to experience the art first.[1] Another critical active element was the four nude male performers standing on a floor bedded with four to eight inches of sugarcane bagasse. Bagasse is the structural detritus left after the valuable sugar has been removed from the cane. The performers acted out a choreographed set of movements of bowing, rubbing at their skin, and putting their hand in their mouth. The bagasse waste and the nude performers bear reference to how Cuba's colonial past generated wealth for the country and the very human cost it extracted.[2] The only illumination in the space emanated from a television facing downward that ran a video loop of footage from previously aired Cuban television programing showing Fidel Castro acting with bravado in both his public and private life.

The original installation was part of the 7th Havana Biennial, themed "Closer to One Another." The Cuban officials initially approved what they thought was the work—a cavernous space only filled with the bagasse—but once the work was activated by installing the television with the video loop and the performers began to execute their routine of choreographed suffering, the work was censored. The quickest way for the authorities to do this was to shut off the electricity. Fortunately, Bruguera was able to retrieve the monitor and video before they were confiscated. (In 2010 when MoMA acquired this work, this was the sole physical, and worth noting authentic, medium obtained.) Only the bagasse remained in the prison. Within a few days the sugarcane had begun to rot, and the stench became overwhelming.

Returning to the notion that change is greatest for things that are young, the bagasse is inherently better suited for compost than a medium found in museums; it breaks down that rapidly because the sugars and the trapped moisture will rapidly ferment and promote mold growth. When the rotting medium is a component of the message, it can be difficult, if not impossible, to perform the work on the time scale typical for a museum exhibition. Installing the work at MoMA was going to require a balancing act. Finding a way to slow the decomposition meant the concomitant smell could also be reduced. Curatorial, conservation, and exhibition design and planning worked closely together with Bruguera to ensure that the artistic merit of the work was not compromised in any way as research was being carried out. This process began internally more than a year before the installation was scheduled to open, and the artist's opinion was sought frequently as planning and research progressed.

It is important to understand the magnitude of the proposed space for the MoMA installation, approximately 164 feet long by 39 feet wide, with an arched ceiling reaching about 13 feet high; it was a very close approximation of the original prison chamber. The amount of bagasse needed to cover the floor was more than thirty cubic yards. A large truckload of potentially rotting matter is not a museum's ideal material to install since biodeterioration is always a concern in museums; living and rotting matter causes mold and can host insects and vermin that scatter once the installation is over, endangering the preservation of the collection and even presenting potential health problems to people. Pulling mold spores into the museum's HVAC system would wreak havoc on objects and people. As serious as these issues were, they were not the only reason for investigating ways to slow down the rotting of the bagasse; it was a matter of sourcing. In New York it had to be trucked in from the southern United States. Samples sent to New York from the closest source, in Florida, arrived moldy. This was not necessarily a new issue, as for the 2015 Venice Biennial, where reinstallation of the work was first attempted, the bagasse arrived too rotted to use. Unfortunately, Bruguera couldn't assist with that installation; she was under house arrest in Cuba for a staging of her piece *Tatlin's Whisper*, a public performance work that amplified the voice of anyone who wished to have their voice heard. In Venice, organizers used gravel as a substitute, and when the artist finally was able to visit the work, she deemed it a significant and undesirable departure from the original. The multisensorial aspect critical to the work was absent. Significantly, Bruguera does not include this installation as part of her oeuvre despite some critics' having written (positively) about it.

Another critical challenge that was quickly identified in preparation for the MoMA installation was the type of bagasse to use. Bruguera originally used a type processed with machinery from the nineteenth century that crushed the cane in a way that broke it open but left the original length of the cane, approximately two meters, intact. Modern extraction methods grind the sugarcane into a pulp biomass to make sugar extraction more efficient and permit the fibrous bagasse to be converted into fuel pellets for extraction equipment: a closed-loop system. In sourcing bagasse, this pulped form was initially the type of material that was received as a moldy fibrous mass. Fortunately, the MoMA production designer working on the installation and former Floridian, Matthew Cox, knew of the Pioneer Florida Museum, a working museum in Dade County that harvests its own sugarcane and processes it using antique farm equipment. As research for the exhibition was underway, he made inquiries and established a critical connection who was willing to help crush the cane to Bruguera's specifications. A first sample of the properly crushed bagasse, the way it is still extracted at some farms in Cuba, arrived at MoMA; however, it, too, was covered in mold.

With a source established, it was necessary to find a way to get the material to New York in good condition. One possibility was to treat the bagasse with a blend of fungicides and mildewcides immediately after pressing to stop biodeterioration. However, the closed confines of the space meant that the performers would be standing barefoot on the material, and additional physical contact for performers—whose part included flagellating themselves with the cane—meant that the hazards these additives posed to their health and safety had to be taken seriously, especially since the two teams of performers, who rotated, would each be in this space four hours each day, six days a week for six weeks. The classes of chemicals that would prevent mold growth on woody, cellulosic matter all came with the warning that they were skin, eye, and respiratory irritants and that direct physical contact should be avoided. During one of the early organizational meetings when this news was shared and concerns about the performers' safety was brought up by MoMA's scientists, the group recognized other options were worth investigating.

MoMA's scientists decided to explore the possibility of getting the moisture content low enough to prevent mold growth. The installation team's luck continued when a lumber mill near the farm museum was willing to kiln-dry the bagasse after the sugarcane was harvested and pressed. There was a lag of about six months between harvest season and the exhibition's opening, and through Matthew's perseverance, he was able to locate a fruit storage facility nearby, where the dried bagasse could be kept cool and dry until it was needed. Removing the moisture quickly through the circulated hot, dry air of the kiln followed by keeping it cool and dry are critical factors in slowing the decay of organic matter. As conservation received samples of the dried bagasse for evaluation—without mold—it was clear this would be the best method to ensure not only that the bagasse arrived in New York in good condition but also that it would be able to last the length of the exhibition.

While the drying procedure effectively stopped the rotting process, it also reduced the stench that intensified over time. In consultation with the artist, it was decided that stabilizing the crushed bagasse was a tremendous success but that, to maintain integrity of the installation, a scent had to added. The multisensorial quality critical to Bruguera meant it was essential to reestablish the rotting scent. During one of her visits Bruguera evaluated fabricated scents that were formulated to smell like the rotting bagasse. Theatrical performance utilizes manufactured scent in a variety of ways, and profilers can provide just about any scent imaginable. The scent was aerosolized during the exhibition as part of the performance exactly the same way it might be used in a theatrical setting. The chemical compounds in it were volatile organic compounds of neutral pH, neither acidic nor alkaline, an important factor to consider when thinking of the other art in adjacent galleries and the safety of visitors.

While we relied on the theater arts community for some components of this installation, like the manufactured scent, it was decided to work with staff plasterers to create the appearance of the Havana space; it was more important to create the feel of a Cuban prison than it was to create its replica. The rough plaster was hand-troweled and tinted in patches to make it appear, in the artist's words, "as if it had been repaired."[3] There was no need for paint; the tinted plaster captured the rough appearance of the original site effectively. As per Bruguera's instructions, stains of rust occasionally streaked down the wall. These elements were more easily observed in the stray light scattering in from the entrance and into the antechamber that managed to trap most of the light before entering.

Following an open casting call and series of interviews, nine performers were selected by the artist, curators, and MoMA's performance production team; all nine were men and either born in Cuba or of Cuban ancestry. One of the most documented components of the work, to assist with installing and performing it again, centered on Bruguera working with performers on the gestures they would make during the performances. Bruguera trained the performers over several days at the museum, specifying four physical positions in the installation. The performers cycled through four movements: bowing, wiping the arms and body, placing the hands on the head, and placing a hand in the mouth. Given the durational nature of the performance, which started when the museum opened at 10:30 a.m. and ended upon closure at 5:30 p.m., a schedule was devised that would allow performers to rotate out in shifts. Many hours of video document Tania and the performers working through these movements as part of their training, and documentation continued during the installation via performers' journals and weekly debriefs. The video files and written documentation are now part of the museum's database and are an archived digital asset for the work. The process and the artist's intent have been so carefully documented that it will be much easier for the work to be reinstalled or lent by the museum.

As the exhibition date neared, the dried bagasse was shipped to New York a few weeks prior to the opening. At this time, as an application to the New York City Fire Department was requested to forego an illuminated exit sign, it was learned from the fire department that a salt-based fire retardant had to be applied to the bagasse. This process was carried out offsite by a company certified to treat theatrical stage sets that had flammable components. The waiver was granted once the fire department received certification the process had been carried out. Fortunately, this procedure happened during the winter when the humidity was low and the bagasse dried quickly, reducing the risk of mold growth. Final preparations centered on a variety of details: securing a wheelchair with oversize wheels that could roll over the bagasse if a visitor required it; making sure the "greenroom" for the performers had what they wanted; reviewing how security would work to ensure their safety while performing; reviewing how conservation would monitor air quality conditions with exhibition design and planning.

As all the components came together, museum staff from visitor services began deploying their plans that had started many months earlier, including the establishment of a method to bag cell phones in signal- and light-blocking pouches. Educators helped guide the conversation with visitors who were to wait in a line adjacent to the long side of the installation; they would engage and raise issues about life in Cuba and remind them, as Bruguera herself would point out, that in Cuba "people have to wait in line for everything." The *line in* was part of the artist's intent. When visitors reached the corner near the entrance, they would be given the signal to enter by a security guard standing "sentry," who was instructed by Bruguera to look serious and kind.[4]

Controlling the number of visitors inside the space was critical in two ways: it helped maintained the potency of the work, and it also assured the cue integral to the work, something easily accomplished at both venues. Experiencing the work is a better way to describe it because one's vision struggled in the barely lit space—this disorientation was a critical component of the work. Since this was a multisensorial work, Bruguera wanted visitors to utilize senses they didn't traditionally use in the presence of art. The lengths of bagasse and its depth made a stable footing difficult to achieve; it was necessary to lift and drop feet to avoid tripping. Once inside, visitors might imagine feeling Francisco Goya's *Disasters of War* being played out live when they finally got close enough to see the performers bathed in the light from the television.

The challenges of this work may very well have impacted Bruguera's thinking and ideas about related works. Instead of re-creating *Untitled (Havana 2000)* in other cities, she would find ways to address their own unique social issues. For instance, she used local individuals and sites of host cites to create *Untitled (Kassel, 2002)*, *Untitled (Moscow, 2007)*, and *Untitled (Bogota, 2009)*. These installations dealt with topics related to complicity, trust, and the drug trade, respectively. They were very different installations and realized before MoMA's acquisition of *Untitled (Havana 2000)*. In fact, *Untitled (Kassel, 2002)* employed the near-blinding light of an interrogation room. The original title she proposed for *Untitled (Havana 2000)* was "Engineers of the Soul," a reference to a remark Stalin made about the artist's role in crafting propaganda, but it was vetoed by officials. Consequently, she renamed the works to pay homage not only to fellow Cuban-born artist Félix González-Torres and his approach to naming many of his works, but also to the intent he had behind that nomenclature: that the viewer be needed to participate in the work to activate it.

Are there any lessons to be learned from installing Bruguera's work that could be applied to more traditional media found in paintings, sculpture, and works on paper? When a physical object is "finished," in an ideal world, that object has ceased to change, provided it is well cared for. But even with the best care, physical works will continue to change, if even on a timescale imperceptible to humans. Under the rubric of "change being greatest for things that are young," it is important to be reminded that a work that we are seeing for the first time may somehow have been transformed by a variety of factors that may or may not keep it within a window of authenticity. And for this reason, research and documentation of what is acceptable change are essential for all forms of media. Direct research on an artist's work, regardless of medium, requires that one's resources interact with the work and, once documented, become incorporated into the body of knowledge future generations have to process.

NOTES

Martha Joseph and Kate Lewis, The Museum of Modern Art, provided critical information regarding curatorial issues and matters related to performance and documentation. Jim Coddington, formerly MoMA, for his thought provoking comments and edits, and Sarah Scaturro, The Cleveland Museum of Art, for her encouragement, thoughtful observations, and edits.

1. Fuentes, "Sugarcane, Fidel Castro."
2. Bruguera, *Stuart Comer in Conversation.*
3. Bruguera, "*Untitled (Havana 2000)*."
4. Ibid.

BIBLIOGRAPHY

Bruguera, Tania. *Stuart Comer in Conversation with Tania Bruguera.* December 2017.

———. "*Untitled (Havana 2000).*" *Boundary 2* 29, no. 3 (August 2002): 43–46.

Fuentes, Elvis. "Sugarcane, Fidel Castro, and Performance Art in Cuba: Tania Bruguera's *Untitled (Havana, 2000).*" Museum of Modern Art, 2018.

EIGHTEEN

Moving beyond the Binaries: Exploring the Active Matter of Metal Soaps in Paint

Francesca Casadio

Abstract

This essay explores active matter in oil paint through the science of metal soaps, the product of chemical reactions of organic and inorganic paint components that continue to evolve, recombine, and decay long after the brush has left the artist's palette. The activity of metal soaps in paintings exposes their intersectional identities: as deceivingly static painted images that are brimful with microscopic active matter, as accumulations and erasures of conservation states and societal values, as artistic and scientific objects. Reframing the paint medium as active through the lens of metal soaps provides the opportunity to advance an activist view of collaborative research. By using a vectorized approach to visualize the entanglement of conservation, scientific research, and art historical practice in the study of objects, it is possible to chart a new path toward the authentic co-creation of knowledge.

> Tὰ πάντα ῥεῖ (everything flows).
> —Heraclitus of Ephesos

In Plato's dialogue *Cratylus* Socrates cites a saying—"ta panta rhei"—attributed to the Greek philosopher Heraclitus of Ephesos (sixth century BCE) expressing how, like the stream of a river, all things are in motion and nothing is at rest, so that one can never step in the same river twice.[1]

The ancient Greek verb rhéō (ῥέω)—or "to flow"—is also at the root of rheology: the property of how wet paint flows and reacts to external pressure upon application. Wet/dry is a constraining binary when we want to accurately describe the matter of paint, as the chemistry of oils does not cease to be in flux after the painted surface has hardened. Even centuries-old paint is subject to constant transformation.

This essay will explore one area of active change in oil paints, the evolution of metal soaps, and situate the impact of this phenomenon in the variable landscape that can be traced around the poles of conservation, scientific research, and art historical interpretation. Metal soaps are the product of chemical reactions that happen when fatty acids in the oil binding medium combine with lead, zinc, copper, manganese, and other heavy metals from common pigments, paint extenders, and driers to form organic salts called soaps.[2]

Recent advances in our knowledge of metal soaps provide an opportunity to reconsider some of the binaries that have characterized perceptions and misconceptions of oil paintings until the present moment.

Immutable/Dynamic

A painting's evolving values in society and throughout its history are widely understood. In October 1888 Vincent van Gogh made a painting of his bedroom to send to his brother Theo. In September 1889, returning from hospitalization after the infamous ear incident and finding that the first bedroom painting had been damaged by humidity and efflorescence, he painted a second version for himself and one for his mother and sister.[3] All three were born of a very personal experience, starting with the anticipation of Paul Gauguin's arrival at the Yellow House in Arles, continuing through the excitement and exhaustion of painting feverishly in preparation, and culminating with the artist falling asleep on that same now-immortal bed and sleeping for sixteen hours straight.[4] None of the *Bedrooms* sold during the artist's lifetime. Now the paintings are among the most reproduced images in the world, and though not for sale, they are worth a fortune. The fixity of their ubiquitous digital reproductions conceals their material activity: metal soaps have been forming and growing under the surface of the paint film for over a century (fig. 1).

Most people consider paintings as immutable objects when thinking of their materiality, in stark contrast with their evolving societal values. Barring fire, water, and other catastrophic natural and anthropogenic events (including vandalism, destruction, and theft) oil paintings are deemed to be stable. Instead, they are incredibly dynamic: their timescales flow like Heraclitus's river. Oil paintings beat to the rhythm of more than one relevant clock, evolving over historical,

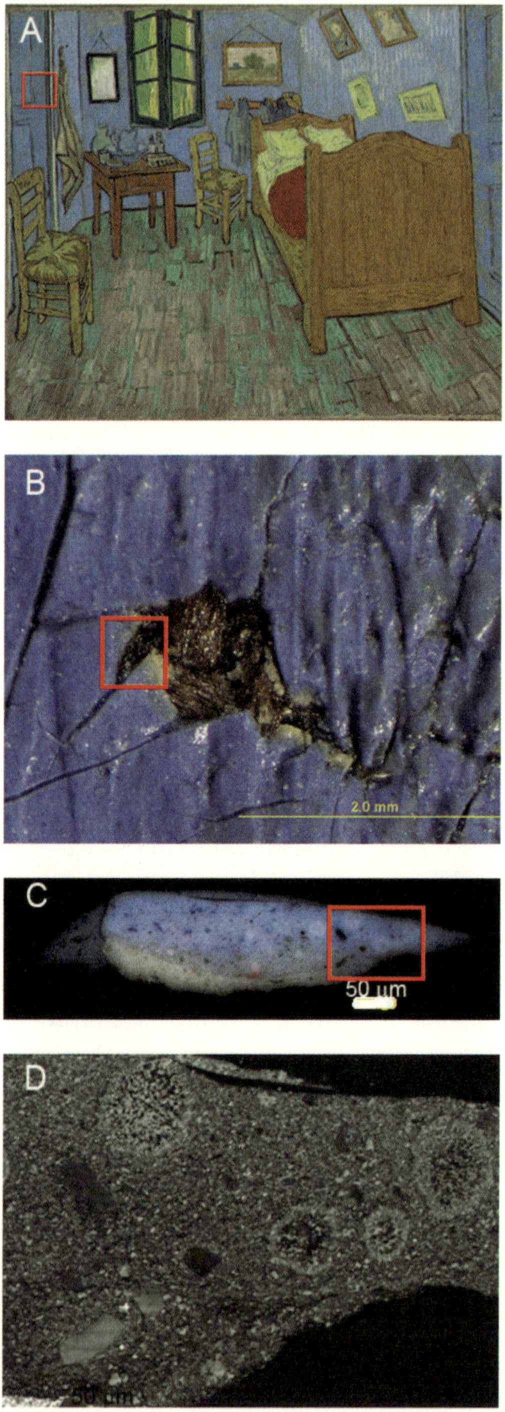
A
B
2.0 mm
C
50 µm
D
50 µm

scientific, and collecting timescales. Historical timescales of oil paints used by artists are measured in centuries. While the technology was known in the ancient world and documented for some medieval paintings alongside the more common egg tempera, the use of drying oil as a widespread medium for easel paintings took hold in Europe only in the fourteenth and fifteenth centuries.[5] Since then, the media of linseed, walnut, and poppy oils were pervasive until the industrialized making of paints in the nineteenth and twentieth centuries, when different types of oil including castor, rapeseed, safflower, and many others, as well as myriad paint modifiers, were marketed for both artists' and house paints.[6]

There are scientific timescales of oil paint: once the paint is applied to the canvas, solvent evaporation takes minutes; then the film starts to harden through a process called autoxidative curing that develops over hours, weeks, and ultimately years, leading to a biopolymer that is tough yet flexible. Subsequently, under the surface of the painting, the oil continues to evolve: through hydrolysis some of the chains in the network are broken, with chemical reactions that can take months and continue over the years. Some of these chain fragments interlock with the metals in pigments and paint additives to form metal soaps at timescales as fast as a few days after mixing or as slow as months and years, sometimes with a stabilizing effect, most times leading to problematic deterioration. There are also collecting timescales to consider, as paintings will require cleanings, treatments, and other interventions throughout their history.

Scientists working with conservators on issues of metal soaps are compelled to understand how endogenous aging mechanisms will influence the painting's response to these types of exogenous inputs.

Fig. 1. *Left: A*, Vincent van Gogh, *The Bedroom*, 1889. Oil on canvas, 29 × 36⅜ in. (73.6 × 92.3 cm). The Art Institute of Chicago, Helen Birch Bartlett Memorial Collection, 1926.417. *B*, a detail from the door on the left (outlined in red in *A*) shows the macroscopic manifestation of zinc soaps as small pinhole protrusions scattered over the surface. *C*, a cross-section sample taken from an area of loss (outlined in red in *B*) showing whitish zinc soaps in the top light blue layer. *D*, a detail of the cross-section (outlined in red in *C*) magnified with a scanning electron microscope and imaged with backscattered electrons clearly visualizes the globular forms that are typical of zinc soap formations at the micro-scale, one of which is almost breaking up the surface of the paint as a protrusion (as visible at top left).

2D/3D

Reducing easel paintings to two-dimensional (2D) works of art is a fallacy, as they are actually objects in three dimensions (3D). It is simply a matter of scale. The paint itself is a complex, heterogeneous, multilayer, and multiscale system of organic and inorganic components: below the surface myriad chemical reactions take place at every single moment. Simulation of the early stage curing of oils, hours after they have left the artist's palette, has calculated 1,087,360 possible unique reactions for ethyl linoleate only, a simple molecule often used as a model for the more complex biopolymer linseed oil.[7]

Our ability to understand and even simulate these dynamic processes of chemical evolution of oil paint has advanced dramatically in the past ten years. Researchers have also mapped transport mechanisms of fluids and humidity throughout the 3D paint structure, to cast light on causal correlations of metal soap formation with changes in the environment or the application of solvent-based conservation treatments.[8] Advances in this area will lead to the formulation of sophisticated models with predictive capabilities for a variety of scenarios of display, storage, and care.

Good/Bad

The values we attribute to transformations in oil paints change over time. At the beginning, the dynamic qualities of oil paint in combination with pigments are "good": paint properties can be modified and oils made to dry faster or slower, to render thicker texture or smooth, translucent thin glazes.[9] The artist is a demiurge, transforming raw materials into something else, a beautiful image. When Raphael painted the *Mond Crucifixion* around 1502–3, he not only used his hands to model the wet paint, as revealed in the fingerprints now preserved on the surface and visible under high magnification; he also used his empirical knowledge of the changing properties of the oil medium to his aesthetic advantage.[10] He employed walnut oil—which darkens less upon aging—for the sky and linseed oil for the darker passages, knowing all too well that the darkening of linseed oil would enhance the contrast in the painted scene over time.[11] In this case the artist, his chosen materials, and time contribute to making meaning.

Throughout their history, paintings can accrue a desirable patina, but, conversely, their surfaces can also darken excessively or develop degradation phenomena, including metal soaps, that may lead to undesirable damage and loss.[12] At multiple points along this temporal arc, conservators are called to intervene, and it is often in these moments that art historians, and sometimes the public opinion, take notice of their work.[13] In the twentieth century it was the cleaning of paintings that came to the fore in the art historical discourse and public opinion, with prominent cases such as the National Gallery cleaning controversy in London in the 1930s and 1940s, debates over the radical cleaning of early Italian paintings at the Yale University Art Gallery in the 1950s and 1960s, and the cleaning of the Sistine Chapel in the 1980s and 1990s causing the loudest uproars.[14] In most of those cases, conservation treatments were highly criticized and labeled "bad" by prominent critics, while their proponents championed the "good" use of science to support treatment decisions.

These controversies and other forces have led to the formulation in the Western conservation profession of the concept of minimal intervention. However, minimal intervention as a construct only works within a fixed set of values, mostly held by a homogenous group. It becomes fuzzy and fluid when multiple groups are involved, with different sets of values and different thresholds for what should be considered minimal.[15]

Metal Soaps in Paints

In Rembrandt van Rijn's *The Anatomy Lesson of Dr. Nicolaes Tulp*, 1632, the Amsterdam surgeons cluster around the corpse of the robber Aris Kindt, but when we look closely, the lifeless surface of his skin appears blemished with dark spots.[16] The cause? Clusters of lead soaps.[17] Since that initial observation and subsequent survey in the early 2000s, thousands of paintings affected by metal soaps across centuries and geographical regions have been revealed and inventoried.[18] The collaboration with scientists and the availability of advanced scientific techniques were instrumental in the identification of a phenomenon that had been observed before but was never understood.

One of the most notable examples of such misinterpretation has repeatedly occurred in relation to Johannes Vermeer's *View of Delft*, ca. 1660–61.[19] In describing the painting, Martin Bailey noted the artist's meticulous addition of grains of sand.[20] In "Vermeer's *View of Delft* and His Vision of Reality" Arthur Wheelock and Kees Kandelbach wrote eloquently about the master's skills and how his technique endows the paint surface with the ability to evoke nearly tactile sensations.

> One of his most successful effects are the red tiled roofs along the left side of the painting. [. . .] Augmenting this effect is the manner in which Vermeer enhanced the grainy quality of the tile roof with small bumps, which protrude from the surface. These bumps are large particles of lead white paint that Vermeer either applied to the ground in this area or mixed with the thin reddish layer that established the base color for the roof.[21]

These scholars' observations of the painted surface are incredibly perceptive, though their proposed explanation proved to be an alternative reality, as the grains of sand, or large particles of lead white, are in fact lead soap aggregates. When looking at a small, detailed scale, the active matter of oil paint has actually enhanced the visual impact of the painting, adding a textural effect to the roofs of the city of Delft that goes beyond the artist's initial rendering (fig. 2).

The Macroscopic Ugliness and Microscopic Beauty of Metal Soaps' Degradation

When Marten Soolmans and Oopjen Coppit commissioned Rembrandt to make their imposing portraits in 1634, they thought they would commit their timeless image to posterity.[22] Yet what the paintings have come to mean has drastically changed in the subsequent four centuries, just as the paint layers below the portraits' surface have continued to evolve. When the newlyweds sat for Rembrandt, they intended to seal their status and wealth as marks of respectability in the Dutch elite of Amsterdam. Today, as we reconsider the colonial legacy of Western art, we see Marten as the offspring of a family that accumulated riches through sugar refineries relying on imports of raw cane sugar from the Caribbean, an enterprise built on the exploitation of Black slave labor.[23] At over two meters tall, Oopjen's life-size marriage portrait is large by the standards of art history, yet it shrinks to a micro-manifestation of beauty when juxtaposed to the societal macro-ugliness of her husband's trade. At an even smaller scale, below the painted surface and the many layers of successive historical restorations and varnishing campaigns, the most recent treatment has uncovered a lunar landscape of soap protrusions.[24]

In addition to protrusions, other visible manifestations of metal soaps in paintings include increased transparency and associated dark-

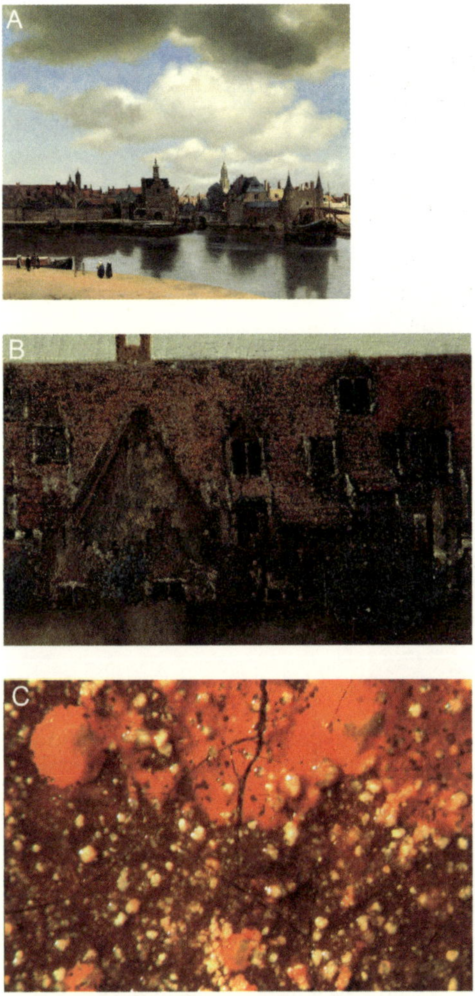

Fig. 2. *A*, Johannes Vermeer, *View of Delft*, ca. 1660–61. Oil on canvas, 38 × 45⅝ in. (96.5 × 115.7 cm). Mauritshuis, The Hague, inv. n. 92. *B*, a detail showing the red roofs at center left; *C*, a microscopic detail showing translucent, whitish lead soap protrusions. Details reproduced with permission from Noble, "Brief History of Metal Soaps," 9.

ening of paint layers due to the dissolution of lead white, hazes, pinpoint paint losses, cracks, crusts, or delamination. At the microscopic scale, they compose a captivating visual atlas of self-directed creation of form in paints, where matter breaks apart and recombines beyond the intention of its makers (fig. 3). Though initially the visual impact of metal soap formation at the macro-scale may be considered negligible, over time soaps can cause substantial deterioration and loss of original paint. Some of these effects can be visually mitigated, but because instability generates from inside the paintings, metal soaps will continue to evolve even after treatment.[25]

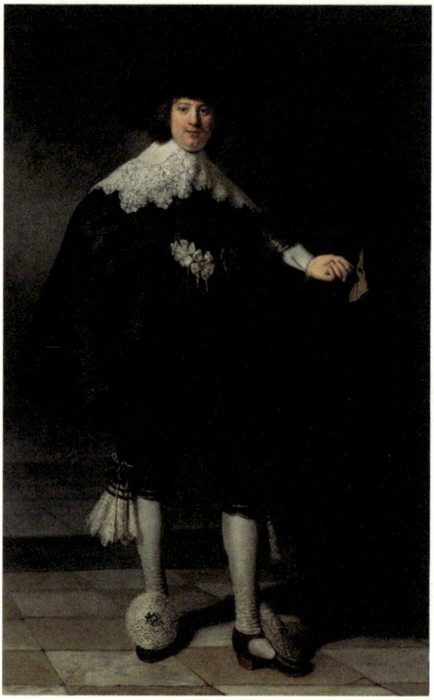

Fig. 3. Macro-micro manifestations of metal soaps in art. *Left*: Rembrandt van Rijn, *Portrait of Marten Soolmans* (1613–41), 1634. Oil on canvas, 82¾ × 53⅝ in. (210 × 136 cm). Rijksmuseum, inv. sk-a-5033. *Right*: *A*, microscopic detail of the blackish background paint before treatment showing cracks and whitish protrusions of lead soaps. *B*, a lead soap protrusion is evident in the paint cross-section taken from this area. Detail and paint cross-sections reproduced with permission from Eumelen et al., "Computational Modelling," 2. *C*, Meindert Hobbema, *Wooded Landscape with Cottage and Horseman*, 1663. Oil on canvas, 40⅛ × 51¾ in. (101.9 × 131.4 cm). The Art Institute of Chicago, Mr. and Mrs. W. W. Kimball Collection, 1922.4460. *D*, a detail of the top left corner of the painting during treatment shows how the saponified lead white in the sky has become transparent, showing the dark ground underneath. *E*, Pierre Soulages, *Peinture, 12 mars 1960*, 1960. Oil on canvas. Sara Hildén Foundation/Sara Hildén Art Museum, Tampere, Finland. A detail of the painted surfaces shows severe delamination occurring between ground and paint layer because of the presence of soaps. Reproduced with permission from Hélou-de La Grandière, "Montparnasse Disease?" 394. *F*, László Moholy-Nagy, B-10 *Space Modulator*, 1942. Oil and incised lines on Plexiglas, in original frame, 16⅞ × 11½ × 2⅜ in. (42.9 × 29.2 × 6 cm), framed, 32⅝ × 26⅝ in. (82.9 × 67.6 cm). Solomon R. Guggenheim Museum, New York, Solomon R. Guggenheim Founding Collection, 47.1063. Detail of scanning electron microscopy image of cross-section of the gray paint imaged with backscattered electrons showing a superficial crust of zinc and lead soaps. For more, see Salvant et al., "László Moholy-Nagy's Painting Materials," 314–18.

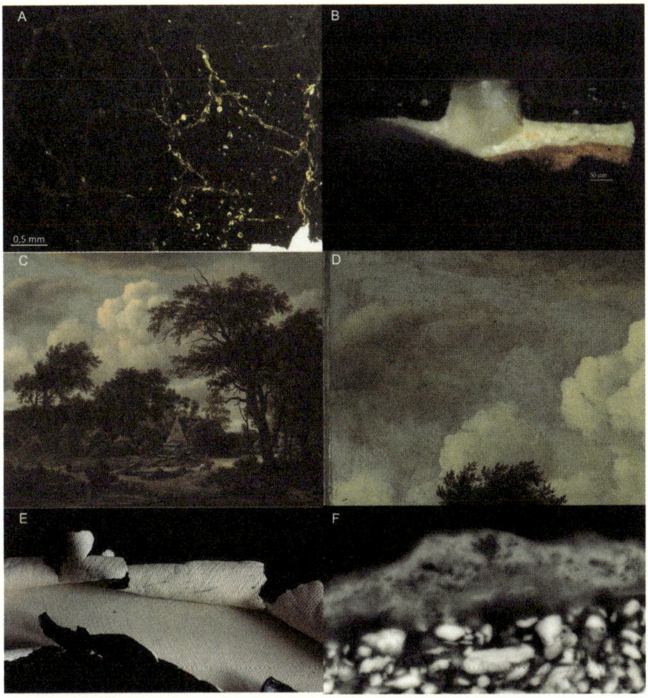

Metal Soaps: A Leading Scientific Research Problem in Paintings Conservation of the Twenty-First Century

Rarely has a single topic of research catalyzed the sustained interest of so many conservation and scientific research groups around the globe, with hundreds of publications, a dedicated symposium (Rijksmuseum, Amsterdam, March 14–15, 2016), and a book.[26] Recent advances of science allow us to interrogate the dynamic processes of metal soaps at infinitesimal scales thanks to synchrotron facilities and scientific equipment that can analyze ultramicroscopic samples with enhanced sensitivity. When combined with sophisticated noninvasive mapping techniques, these insights into micro- and nanoscopic processes can be extrapolated to macroscopic phenomena observed at and below painted surfaces.[27] Actively traveling through multiple-length scales we can now penetrate the surface of van Gogh's 1889 *The Bedroom* painting to advance our understanding of zinc soaps (fig. 1) and make predictions about the appearance of objects in the future, extrapolating current degradation processes.[28]

Research in this area is exemplary as it leverages a three-pronged approach that combines empirical case studies of historic paintings with the investigation of modern paint reconstructions and mechanistic studies by computational modeling. Yet, while the "hows" of metal soaps have been extensively explored, the "whys" are still not fully understood. Why do metal soaps have a stabilizing effect on some paints, while in other cases they self-assemble to form disfiguring protrusions or cause delamination by organizing in arrays of newly formed soaps at the interfaces of paint layers? These and other mysteries remain unresolved and are currently the subject of intense research.

New answers and knowledge will continue to emerge thanks to massively increased computing power for modeling. For instance, the whitish lead soap globules sprinkled across the shadowy background of Rembrandt's *Portrait of Marten Soolmans* have been used to propose a numerical simulation of how paint composition can influence nucleation, growth, and induced fracture of paint surfaces by metal soaps.[29] New developments in surface metrology with machine learning offer enhanced vision tools to monitor the active evolution of soap protrusion formation and cracks over time and in 3D (fig. 4), opening the possibility of epidemiological studies in art.[30]

Moving beyond Minimal Intervention toward a "Vectorized" System of Values

Taco Dibbits, general director of the Rijksmuseum, states in the foreword to the volume on metal soaps,[31]

> As a museum, it is our responsibility to present paintings in optimal condition and to ensure their preservation for future generations. We strive to treat and maintain paintings in the best possible manner. Metal soap-related degradation is a good example of the changes that paintings can undergo over time. It is therefore important for conservators to know how to deal with this. Should you treat it? Can you stabilize it? *Can we ignore it?*[32]

Increasingly, as conservators and allied professionals reflect on the theory and practice of conservation, we ask ourselves, What values or meanings of the artwork are undermined if we do nothing? As our ability to correctly identify and diagnose metal soaps evolves, so do approaches to their treatment. Today prevailing conservation practices center around managing change rather than trying to halt these dynamic processes of alteration.

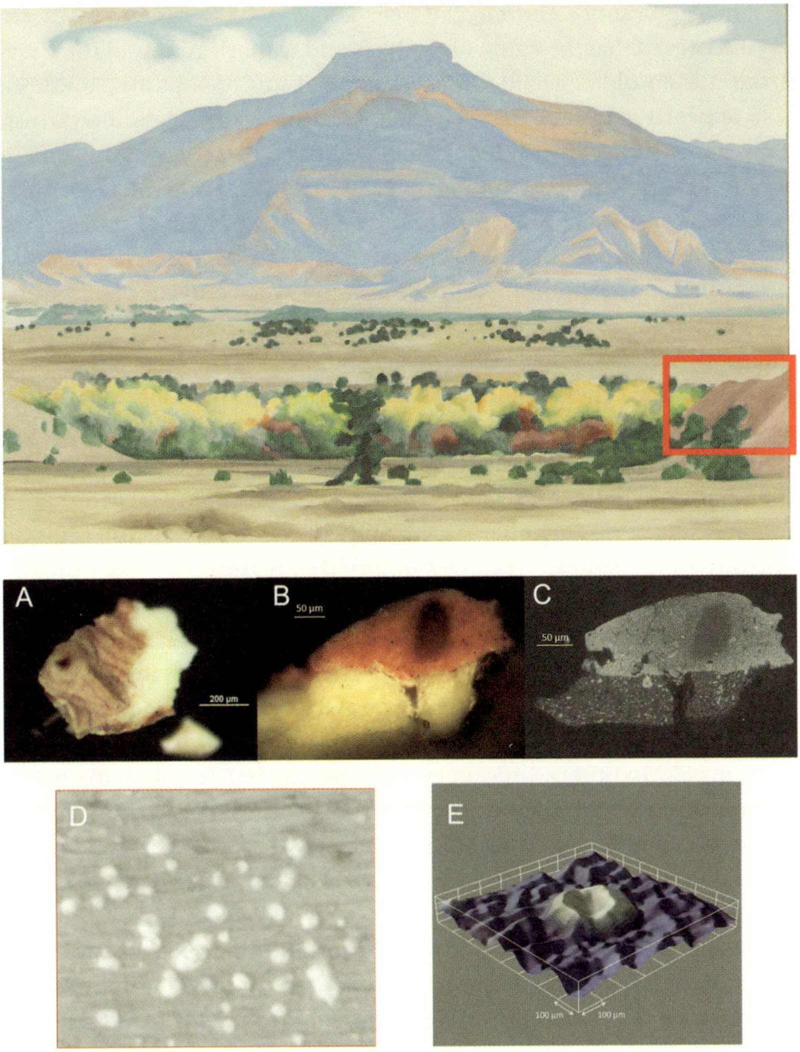

Fig. 4. *Top*: Georgia O'Keeffe, *Pedernal*, 1941–42. Oil on canvas, 20⅛ × 30¼ in. (51.1 × 76.8 cm). Georgia O'Keeffe Museum, Gift of The Burnett Foundation and The Georgia O'Keeffe Foundation, 1997.5.12. *A*, a sample of lead white protrusion from the landscape outlined in red at bottom right: when the sample was mounted as a cross-section, the round lead soap protrusion was clearly visible as a medium-rich area in both visible light, *B*, and with a scanning electron microscope, *C*. *D* shows computational imaging maps of soap protrusions in the area sampled, and *E*, a three-dimensional rendering of a lead soap protrusion. For more, see Plisson et al., "Photometric Stereo."

In Rembrandt's *Portrait of Marten Soolmans*, Rijksmuseum researchers discovered that the cause of the haze and scarce legibility of the background is an old egg-white varnish, applied during a past restoration, now appearing cracked and yellowed. Over time, this egg-white varnish caused micro-cupping of the original paint and reacted with the lead in the underlying paint layers to form soaps, making its complete removal nearly impossible. The decision of leaving the varnish on or taking it off therefore required a risk-management approach. Models exist to guide preventive conservation measures for entire collections but need to be customized for cleaning individual paintings.[33] As important as the tangible risk of material loss is for the painting, it is useful here to consider other risks and other actors. Salvador Muñoz Viñas has long argued that conservators need to shift from a supposedly objective goal of minimal intervention (an impossible and inherently self-contradictory concept: Minimal for whom? And with respect to which thresholds?) to that of minimizing the loss of potential meanings. Decisions on varnish removal and other interventions "can be supported with greater or lesser quantities of scientific data, and implemented with more or less scientifically monitored techniques, but this does not make the *decision* to perform these operations anymore objective."[34] The goal of the treatment should therefore be to "enhance or preserve the preferred meanings of the object while impairing as little as possible its ability to convey any other meanings."[35] In other words, in considering the deterioration caused by metal soaps, we should move, as Miriam Clavir proposes, "from a strictly scientific definition of deterioration to include a social definition of deterioration, such as 'We describe deterioration as those changes that we [regard as] undesirable.'"[36] These undesirable changes can and should go beyond aesthetics and appearances and into the social meaning of the work, as well as its potential to convey further scientific information to the researchers of the future. A simple yet elegant solution to visualizing these changes is offered by Jonathan Kemp, who proposes to plot these values along three *x*, *y*, *z* axes: change in function, change in interpretation, and change in material. All axes originate at a point zero of intersection that Kemp maintains as the original, "authentic" state of the work of art (fig. 5).[37] This "vectorized" approach effectively allows us to start delineating the topologies of impact on artifacts not only of conservation treatments but also of other events such as installation, scholarship, public interpretation, scientific analysis, storage, and care.

The realization that the treatment of a painting cannot be accurately described on a linear trajectory along a before- and after-treatment binary that is supposed to bring us closer to an authentic ideal for the

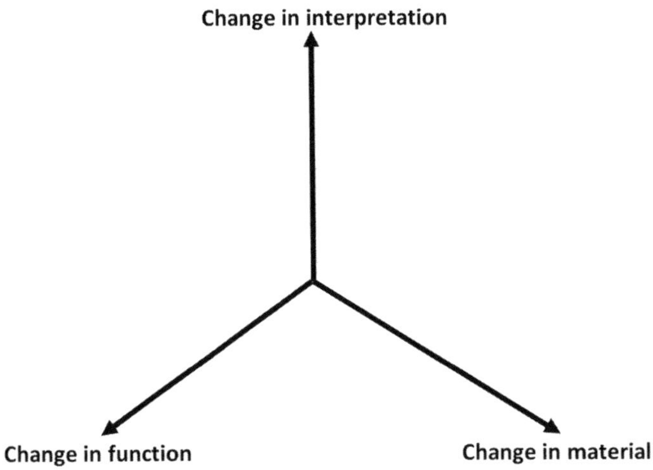

Fig. 5. Diagram after Jonathan Kemp's vectorized axes of authenticity. Kemp, "Practical Ethics v2.0."

object represents a giant conceptual leap. It allows us to comprehend how the impact of a conservation treatment or deterioration phenomenon resonates along multiple dimensions of value. It also questions an object's singular identity, opening up the opportunity to recognize multiple identities in a single artifact. To capture this complexity the graph should be expanded to three dimensions, or visualized with a radar chart, as the perception of change along Kemp's three axes will likely be different for different observers (conservators, scientists, art historians, the media, and the general public).

As a case in point concerning metal soaps, though the media has occasionally underlined the importance of this topic with sensational headlines, and conservators and scientists have intensely researched this subject for several decades, the art historical community has hardly taken notice.[38] This gap between scientific research in conservation and art historical scholarship exists despite the fact that their subject of investigation is the same: oil paintings.

A keen observer of this dichotomy was Giovanni Urbani, the director of Italy's Istituto Centrale del Restauro in Rome from 1973 to 1983. Urbani, who was trained as an art historian, was early to recognize that if the primary function of a work of art is to stimulate our aesthetic sensibility, there can be a tremendous dissonance between the determination on the part of the scientist of the severe state of deterioration of the artwork's materials and the actual impact of this deterioration on the ability of the object to perform its primary aesthetic function.[39] Still, Urbani identified a meaningful role for science—instead of simply leading to a superficially cosmetic conservation operation, the new knowledge could allow the slowing down of those processes of deterioration in a rigorous and measurable way. Almost three decades after Urbani's lectures, as science becomes increasingly adept not only at dealing with measurable quantities at minuscule scales but also at unraveling complexity in paint systems through computational modeling tools, there is promise for advancement.

A vectorized approach to examining the impacts of deterioration, treatment, or other processes will also highlight how different types of objects will tolerate different degrees of damage before they can no longer fulfill their function as relevant to different actors.

An Activist View of Collaborative Research

Ultimately, considerations on the active matter of metal soaps may provide us with a chance to reframe the entanglement of conservation, scientific, and humanistic efforts around the study of objects.

Decay opens up opportunities by unlocking instability, unleashing unstable or meta-stable states of being that can challenge established paradigms of knowledge creation. As researchers in the field of object-based studies, can we shift our thinking from binary descriptions to a flow, across temporal scales but also dimensional scales, and will this shift our perceptions of what is valuable and important? What happens when instead of thinking of a linear and universally understood scale of values, we vectorize values for different actors, to create topographical contours of impact? Would a vectorized approach be a more meaningful tool to describe the networked domains of art history, conservation, and science, so as to orient our efforts toward research practices that truly sit at the intersection of disciplines? How could a more inclusive concept of intersectionality, rather than the prevalent view of interdisciplinarity, better inform our practice? Would openly recognizing the interaction of various social, aesthetic, and material identities for historical artifacts, and acknowledging their effect on power hierarchies of scholarship, allow a more expansive practice of object-based research?

Typically, research that has involved the collaboration of scientists, conservators, and art historians, curators, or other humanists has been labeled cross-disciplinary or interdisciplinary at best. I see cross-disciplinary studies of objects as describing the case when sometimes curiosity leads us to hop over to another discipline, though staying within conventional boundaries remains the norm, even when multiple disciplines are concerned with the same object. In the study of oil paintings, this is the current status of research on the active matter of metal soaps. A more evolved approach involves interdisciplinarity. I see this as the case when a project brings together different specialists who nonetheless advance and publish primarily within their own disciplines. At times they compare findings and cross-pollinate ideas to move forward, on separate though occasionally intertwined tracks. Research focused on active deterioration in paints that lead to color change is almost at this point.

Embracing an activist direction for object-based studies, I here propose the more radical concept of intersectionality.[40] Intersectionality affects both the subject and the method. Using intersectionality as a lens means recognizing an object's intersection of identities: as accumulations of conservation states and accrued or erased societal values; as aesthetic object, scientific object, and historic object; and as a macroscopic painted image that appears fixed over time while being composed of microscopic active matter that continues to evolve, recombine, decay. Used as a framework for a shared research practice, intersectionality encourages us to truly meet and remain at the center where our respective disciplines cross, fighting the discomfort of not knowing all the answers or methods, enhancing and informing each other's knowledge with a multiplying rather than additive effect.

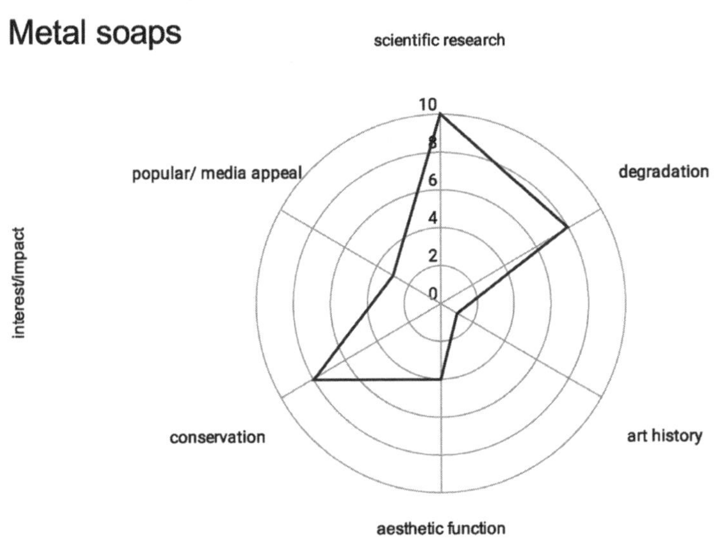

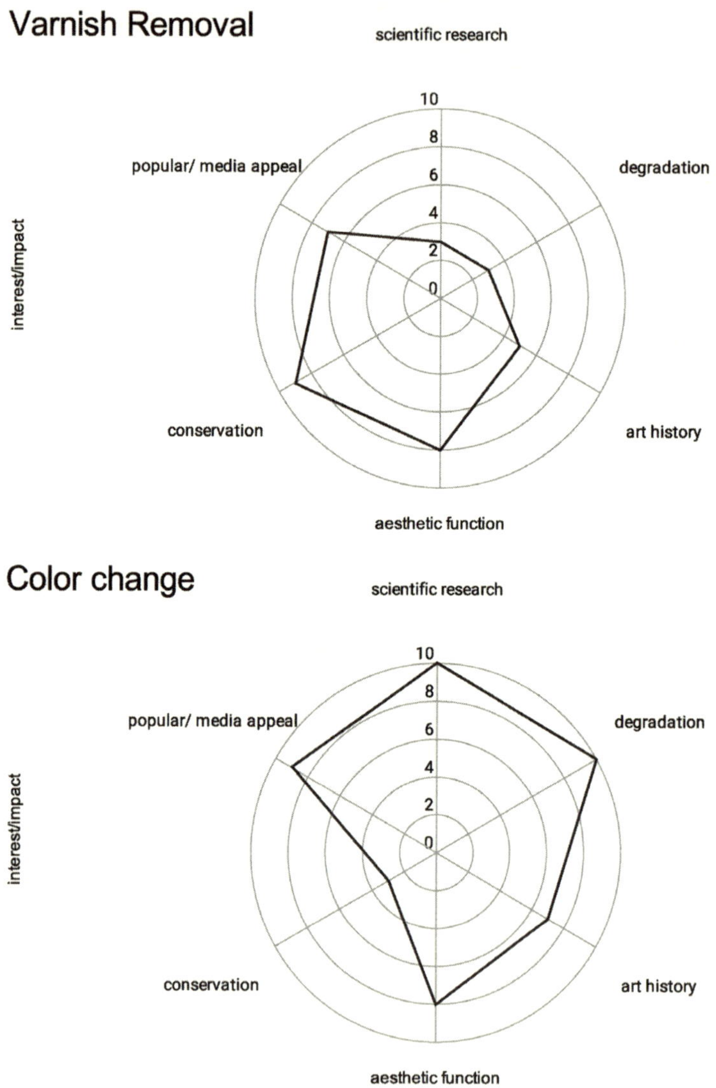

Fig. 6. Examples of radar charts visually representing a vectorized approach to mapping the topological contours of research interests in various topics (metal soaps, varnish removal, color change) for different disciplinary areas (conservation, scientific research, and art history) and their impact on different qualities (degradation, aesthetic function, and popular or media appeal). Values on a 0–10 scale have been arbitrarily assigned by the author as a matter of example for this thought exercise and go from 0 (no interest and/or no impact) to 10 (highest interest and/or highest impact).

Intersectionality has the potential to act as a true knowledge amplifier if combined with a vectorized approach to mapping the topographical contours of research value to various actors. Researchers become explorers of this multidimensional space of materials and meanings. The largest area of impact will focus our attention on questions where the application of this approach of co-creation of new knowledge can be most effective (fig. 6).

Conclusions

In 1949, in his opinion on the National Gallery cleaning controversy, Cesare Brandi wrote in the pages of the *Burlington Magazine* that "the materials in a work of art, must be induced to serve, in a subordinate capacity, the image itself."[41] Metal soaps are an example of how painting materials resist this definition: rather, they are insubordinate and active. When we are able to see paintings as complex, multilayer systems with intersectional identities, "pictures resist the quick technical fix and for that matter aesthetic or historical dematerialization."[42] It is therefore critical to understand that "technical, aesthetic and historical choices, far from being mutually exclusive, are interdependent," and that the important topics to discuss are "what is known and not known about the complex systems that paintings are, and the values and choices that are brought to caring for them."[43]

NOTES

1. Plato, *Dialogues* 1:344.
2. This essay concerns itself only with metal soaps that are understood as the cause and manifestation of paint decay. Their history of use as cosmetics and additives intentionally manufactured to improve the properties of paints is extensively covered elsewhere—see, for example, Cotte et al., "Lead Soaps in Paintings."
3. Groom and Hoyle, *Van Gogh's Bedrooms.*
4. Vincent van Gogh to Theo van Gogh, Arles, October 15, 1888.
5. For egg tempera, see Nadolny, "Problem of Methodology." For drying oil, see Eastlake, *Methods and Materials*; Higgitt and White, "Analysis of Paint Media."
6. Berg et al., *Issues in Contemporary Oil Paint.*
7. Oakley, "Mechanistic Studies"; H. Oakley et al., "Modeling the Evolution."

8. Baij et al., "Review of Solvent Action," 43; Keune et al., "Impact of Lead Soaps."
9. For drying rate, see Carlyle, *Artist's Assistant*; Carlyle, "Exploring the Grammar." For texture, see Walter and Viguerie, "Materials Science Challenges."
10. Oil on panel, 111⅝ × 65⅞ in. (283.3 × 167.3 cm), National Gallery, London, NG3943.
11. Roy, Spring, and Plazzotta, "Raphael's Early Work."
12. Starn, "Three Ages of 'Patina.'"
13. Talley, "Old Fiddle."
14. For the National Gallery, London, see Gombrich, "Controversial Methods"; Brandi and Gombrich, "Cleaning of Pictures," 296–97; Mahon, "Miscellanea"; Maclaren and Werner, "Some Factual Observations." For Yale, see Garland and Mention, "Loss and Restoration"; "Cleaning Early Pictures"; Kanter, "Some Early Sienese Paintings." For the Sistine Chapel, see Colalucci, "Michelangelo Buonarroti"; Beck, "Final Layer."
15. Richmond and Bracker, *Conservation Principles.*
16. Oil on canvas, 66¾ × 85¼ in. (169.5 × 216.5 cm), Mauritshuis, The Hague, inv. N. 146.
17. Noble et al., "Aspects of 17th Century"; Noble, Van Loon, and Boon, "Chemical Changes."
18. The narrative of soaps is mostly dominated by Dutch paintings, with Rembrandt and Dutch golden age paintings being the poster children for lead soaps, Van Gogh paintings for zinc soaps. This is correlated to the tremendous efforts that Dutch conservators and scientists have put into understanding and documenting these phenomena. Many more efforts in various other countries exist and have joined in this investigation.
19. Petria Noble, head of paintings conservation, Rijksmuseum, Amsterdam, has written about this in her essay "Brief History of Metal Soaps." Noble also included this example in her talk at the symposium "Conserving Active Matter: Materials Science," held at Bard Graduate Center on March 28–29, 2019.
20. Bailey, *Vermeer*, 60–62.
21. Wheelock and Kaldenbach, "Vermeer's *View of Delft.*"
22. For the *Portrait of Marten Soolmans* (1613–1641), see fig. 3; *Portrait of Oopjen Coppit (1611–1689),* 1634, oil on canvas, 82⅞ × 53 in. (210.5 × 134.7 cm), Musée du Louvre, inv. sk-c-1768.
23. Bikker and Richards, *Marten and Oopjen*; Sint Nicolaas et al., *Slavery.*
24. Noble et al., "Exceptional Commission," esp. 315, 324–25.

25. Centeno and Mahon, "Chemistry of Aging."
26. Casadio et al., *Metal Soaps in Art.*
27. Ma et al., "Revealing the Distribution"; Liu et al., "Nano- to Microscale."
28. In the case of Van Gogh's *Bedroom*, though this simulation has not yet been done for the effects of metal soap degradation, elegant predictions have been made with respect to past, present, and future color change. See, for example, Hendriks, Brokerhof, and Meiracker, "Valuing Van Gogh's Colours"; Geldof et al., "Reconstructing Van Gogh's Palette," 17.
29. Eumelen et al., "Computational Modelling," 1310.
30. Xu et al., "3D Surface Measurement"; Plisson et al., "Photometric Stereo."
31. Casadio et al., *Metal Soaps in Art.*
32. Taco Dibbits, foreword to Casadio et al., *Metal Soaps in Art*, v–vi. The italics are mine.
33. For generalized risk-management models, see Waller and Michalski, "Paradigm Shift"; Cannon et al., "Towards the 'Collectively Subjective.'" Petria Noble shared considerations concerning customized risk management during her talk at the symposium "Conserving Active Matter: Materials Science."
34. Muñoz Viñas, "Minimal Intervention Revisited," 51.
35. Ibid., 56.
36. Clavir, "Conservation and Cultural Significance."
37. Kemp, "Practical Ethics v2.0." I am using the term "authentic" in quotation marks, as increasingly the field recognizes not one but multiple "authentic" states of works of art, a debate that is particularly lively for painted surfaces in the context of polychrome sculptures. See, for example, Marincola and Kargère, *Conservation of Medieval Polychrome*; Tom et al., "Many Faces of Guanyin." These discussions are also very relevant for contemporary art, especially time-based media art. See, for example, Laurenson, "Authenticity, Change and Loss."
38. Mass, "Paintings Suffer"; Cascone, "There Is a Secret Villain."
39. Urbani and Zanardi, *Intorno al restauro.*
40. This is a term that is widely used in gender, race, and class studies and was first introduced in Crenshaw, "Mapping the Margins." For an overview of its expanded footprint as a framework for research, practice, and activism in the following two decades, see Al-Faham, Davis, and Ernst, "Intersectionality."
41. Brandi, "Cleaning of Pictures."
42. Starn, "Three Ages of '"Patina,"'" 103.
43. Ibid., 108.

BIBLIOGRAPHY

Al-Faham, Hajer, Angelique M. Davis, and Rose Ernst. "Intersectionality: From Theory to Practice." *Annual Review of Law and Social Science* 15, no. 1 (October 10, 2019): 247–65. https://doi.org/10.1146/annurev-lawsocsci-101518-042942.

Baij, Lambert, Joen Hermans, Bronwyn Ormsby, Petria Noble, Piet Iedema, and Katrien Keune. "A Review of Solvent Action on Oil Paint." *Heritage Science* 8, no. 1 (May 6, 2020): article 43. https://doi.org/10.1186/s40494-020-00388-x.

Bailey, Martin. *Vermeer.* London: Phaidon, 1995.

Beck, James. "The Final Layer: 'L'ultima mano' on Michelangelo's Sistine Ceiling." *Art Bulletin* 70, no. 3 (1988): 502–3. https://doi.org/10.2307/3051180.

Berg, Klaas Jan van den, Aviva Burnstock, Matthijs de Keijzer, Jay Krueger, Tom Learner, Alberto de Tagle, and Gunnar Heydenreich, eds. *Issues in Contemporary Oil Paint.* New York: Springer, 2014.

Bikker, Jonathan, and Lynne Richards. *Marten and Oopjen: Two Monumental Portraits by Rembrandt.* Amsterdam: Rijksmuseum, 2016.

Brandi, Cesare. "The Cleaning of Pictures in Relation to Patina, Varnish, and Glazes." *Burlington Magazine* 91, no. 556 (1949): 183–88.

Brandi, Cesare, and E. H. Gombrich. "The Cleaning of Pictures in Relation to Patina, Varnish and Glazes." *Burlington Magazine* 92, no. 571 (1950): 296–98.

Cannon, A., H. Privett, M. McCubbin, and R. Waller. "Towards the 'Collectively Subjective': The Challenges of Estimating Loss in Value during Collection Risk Assessments." In *ICOM-CC 18th Triennial Conference Preprints, Copenhagen, 4–8 September 2017,* edited by J. Bridgland, article 1504. Paris: International Council of Museums, 2017.

Carlyle, Leslie. *The Artist's Assistant: Oil Painting Instruction Manuals and Handbooks in Britain, 1800–1900, with Reference to Selected Eighteenth-Century Sources.* London: Archetype, 2001.

———. "Exploring the Grammar of Oil Paint through the Use of Historically Accurate Reconstructions." In *Conservation of Easel Paintings,* edited by J. H. Stoner and R. Rushfield, 33–38. London: Routledge, 2012.

Casadio, Francesca, Katrien Keune, Petria Noble, Annelies van Loon, Ella Hendriks, Silvia A. Centeno, Gillian Osmond, et al., eds., *Metal Soaps in Art—Conservation & Research.* Cham: Springer Nature, 2019.

Cascone, Sarah. "There Is a Secret Villain Threatening the World's Masterpieces. It's Soap." Artnet News, May 24, 2017. https://news.artnet.com/art-world/scientists-lead-soaps-oil-paintings-936128.

Centeno, Silvia A., and Dorothy Mahon. "The Chemistry of Aging in Oil Paintings: Metal Soaps and Visual Changes." *Metropolitan Museum of Art Bulletin* 67, no. 1 (2009): 12–19.

Clavir, Miriam. "Conservation and Cultural Significance." In Richmond and Bracker, *Conservation Principles*, 139–49.

"Cleaning Early Pictures: Seeking the Surface." *Burlington Magazine* 144, no. 1191 (2002): 331. https://www.jstor.org/stable/889576.

Colalucci, Gianluigi. "Michelangelo Buonarroti: Restoration of the Frescoes on the Vaulted Ceiling and the Last Judgment in the Sistine Chapel." *Conservation Science in Cultural Heritage* 16 (2016): 89–126. https://doi.org/10.6092/issn.1973-9494/7166.

Cotte, Marine, Emilie Checroun, Wout De Nolf, Yoko Taniguchi, Laurence De Viguerie, Manfred Burghammer, Philippe Walter, et al. "Lead Soaps in Paintings: Friends or Foes?" *Studies in Conservation* 62, no. 1 (January 2, 2017): 2–23. https://doi.org/10.1080/00393630.2016.1232529.

Crenshaw, Kimberle. "Mapping the Margins: Intersectionality, Identity Politics, and Violence against Women of Color." *Stanford Law Review* 43, no. 6 (1991): 1241–99. https://doi.org/10.2307/1229039.

Eastlake, Charles Lock. *Methods and Materials of Painting of the Great Schools and Masters: Two Volumes Bound as One.* Mincola, NY: Dover, 2001.

Eumelen, G. J. A. M., E. Bosco, A. S. J. Suiker, J. J. Hermans, A. van Loon, K. Keune, and P. D. Iedema. "Computational Modelling of Metal Soap Formation in Historical Oil Paintings: The Influence of Fatty Acid Concentration and Nucleus Geometry on the Induced Chemo-Mechanical Damage." *SN Applied Sciences* 2, no. 7 (June 30, 2020): article 1310. https://doi.org/10.1007/s42452-020-3038-z.

Garland, Patricia Sherwin, and Elisabeth Mention. "Loss and Restoration: Yale's Early Italian Paintings Reconsidered." *Yale University Art Gallery Bulletin* (1999): 32–43.

Geldof, Muriel, Art Ness Proaño Gaibor, Frank Ligterink, Ella Hendriks, and Eric Kirchner. "Reconstructing Van Gogh's Palette to Determine the Optical Characteristics of His Paints." *Heritage Science* 6, no. 1 (March 20, 2018): article 17. https://doi.org/10.1186/s40494-018-0181-6.

Gogh, Vincent van. Letter to Theo van Gogh. Arles, October 15, 1888. Vincent van Gogh: The Letters (website). Van Gogh Museum and Huygens ING. http://vangoghletters.org/vg/letters/let704/letter.html.

Gombrich, E. H. "Controversial Methods and Methods of Controversy." *Burlington Magazine* 105, no. 720 (1963): 90–93.

Groom, Gloria, and Michael Hoyle. *Van Gogh's Bedrooms.* Chicago: Art Institute of Chicago, 2016.

Hélou-de La Grandière, Pauline. "A Montparnasse Disease?

Severe Manifestations of Metal Soaps in Paintings by Pierre Soulages from around 1959 to 1960 (Delaminating Oil Paint Layers, Medium Exudates, Discolorations)." In Casadio et al., *Metal Soaps in Art*, 393–412.

Hendriks, E., A. W. Brokerhof, and K. van den Meiracker. "Valuing Van Gogh's Colours: From Past to Future." In *ICOM-CC 18th Triennial Conference Preprints, Copenhagen, 4–8 September 2017*, edited by J. Bridgland, article 1306. Paris: International Council of Museums, 2017.

Higgitt, Catherine, and Raymond White. "Analysis of Paint Media: New Studies of Italian Paintings of the Fifteenth and Sixteenth Centuries." *National Gallery Technical Bulletin* 26 (2005): 88–104.

Kanter, Laurence. "Some Early Sienese Paintings: Cleaned, Uncleaned, Restored, Unrestored; What Have We Learned?" *Yale University Art Gallery Bulletin* (2010): 46–65.

Kemp, Jonathan. "Practical Ethics v2.0." In Richmond and Bracker, *Conservation Principles*, 60–72.

Keune, Katrien, Rick P. Kramer, Suzanne Stangier, and Margriet H. van Eikema Hommes. "Impact of Lead Soaps on the Formation of Age Craquelure." In Casadio et al., *Metal Soaps in Art*, 107–21.

Laurenson, Pip. "Authenticity, Change and Loss in the Conservation of Time-Based Media Installations." *Tate Papers* 6 (Autumn 2006): https://www.tate.org.uk/research/publications/tate-papers/06/authenticity-change-and-loss-conservation-of-time-based-media-installations.

Liu, Xiaoyang, Valeria Di Tullio, Yu-Chung Lin, Vincent De Andrade, Chonghang Zhao, Cheng-Hung Lin, Molly Wagner, et al. "Nano- to Microscale Three-Dimensional Morphology Relevant to Transport Properties in Reactive Porous Composite Paint Films." *Scientific Reports* 10 (October 27, 2020): https://doi.org/10.1038/s41598-020-75040-6.

Ma, Xiao, Victoria Beltran, Georg Ramer, Georges Pavlidis, Dilworth Y. Parkinson, Mathieu Thoury, Tyler Meldrum, Andrea Centrone, and Barbara H. Berrie. "Revealing the Distribution of Metal Carboxylates in Oil Paint from the Micro- to Nanoscale." *Angewandte Chemie-International Edition* 58, no. 34 (August 19, 2019): 11652–56. https://doi.org/10.1002/anie.201903553.

Maclaren, Neil, and Anthony Werner. "Some Factual Observations about Varnishes and Glazes." *Burlington Magazine* 92, no. 568 (1950): 189–92.

Mahon, Denis. "Miscellanea for the Cleaning Controversy." *Burlington Magazine* 104, no. 716 (1962): 460–70.

Marincola, Michele D., and Lucretia Kargère. *The Conservation of Medieval Polychrome Wood Sculpture: History, Theory, Practice.* Los Angeles: Getty Trust, 2020.

Mass, Jennifer. "Paintings Suffer from Breakouts, Too—but What Is 'Art Acne'?" *Apollo Magazine*, February 22, 2019. https://www.apollo-magazine.com/art-acne-paintings-conservation/.

Muñoz Viñas, Salvador. "Minimal Intervention Revisited." In Richmond and Bracker, *Conservation Principles*, 47–59.

Nadolny, Jilleen. "A Problem of Methodology: Merrifield, Eastlake and the Use of Oil-Based Media by Medieval English Painters." In *ICOM Committee for Conservation 14th Triennial Meeting The Hague 12–16 September 2005: Preprints*, edited by Isabelle Sourbès-Verger, 1028–33. London: James & James; Earthscan, 2005.

Noble, Petria. "A Brief History of Metal Soaps in Paintings from a Conservation Perspective." In Casadio et al., *Metal Soaps in Art*, 1–22.

Noble, P., J. Wadum, K. Groen, R. Heeren, and K. J. van den Berg. "Aspects of 17th Century Binding Medium: Inclusions in Rembrandt's *Anatomy Lesson of Dr Nicolaes Tulp*." In *Art et chimie, la couleur: Actes du Congrès*, edited by Jacques Goupy, 126–29. Paris: CNRS Éditions, 2000.

Noble, Petria, Annelies van Loon, and J. J. Boon. "Chemical Changes in Old Master Paintings II: Darkening Due to Increased Transparency as a Result of Metal Soap Formation." In *ICOM Committee for Conservation 14th Triennial Meeting The Hague 12–16 September 2005: Preprints,* edited by Isabelle Sourbès-Verger, 496–503. London: James & James; Earthscan, 2005.

Noble, Petria, E. Duijn, Erma Hermens, Katrien Keune, Annelies van Loon, S. Smelt, Gwen Tauber, and R. Erdmann. "An Exceptional Commission: Conservation History, Treatment and Painting Technique of Rembrandt's Marten and Oopjen, 1634." *Rijksmuseum Bulletin* 66 (December 1, 2018): 309–43.

Oakley, L. "Mechanistic Studies of the Autoxidative Curing of an Oil-Based Paint Model System." PhD diss., Northwestern University, 2017.

Oakley, Lindsay H., Francesca Casadio, Kenneth R. Shull, and Linda J. Broadbelt. "Modeling the Evolution of Crosslinked and Extractable Material in an Oil-Based Paint Model System." *Angewandte Chemie-International Edition* 57, no. 25 (June 18, 2018): 7413–17. https://doi.org/10.1002/anie.201801332.

Plato. *Dialogues of Plato.* Translated by Benjamin Jowett. Vol. 1. New York: Cambridge University Press, 2010.

Plisson, Johanna Salvant, Marc Walton, Dale Kronkright, C. K. Yeh, Oliver Cossairt, F. Li, and Aggelos Katsaggelos. "Photometric Stereo by UV-Induced Fluorescence to Detect Protrusions on Georgia O'Keefe's Paintings." In Casadio et al., *Metal Soaps in Art*, 375–91.

Richmond, Alison, and Alison Bracker, eds. *Conservation Principles, Dilemmas and Uncomfortable Truths.* London: Routledge and Victoria and Albert Museum, 2020.

Roy, Ashok, Marika Spring, and Carol Plazzotta. "Raphael's Early Work in the National Gallery: Paintings before Rome." *National Gallery Technical Bulletin* 25 (2004): 4–35. http://www.nationalgallery.org.uk/technical-bulletin/roy_spring_plazzotta2004.

Salvant, J., J. Barten, F. Casadio, M. Kokkori, F. Pozzi, C. Stringari, K. Sutherland, and M. Walton. "László Moholy-Nagy's Painting Materials: From Substance to Light." *Leonardo* 50, no. 3 (2017): 314–18. https://doi.org/10.1162/LEON_a_01430.

Sint Nicolaas, Eveline, Pierre Bouvier, and Geri Klazema. *Slavery: The Story of João, Wally, Oopjen, Paulus, Van Bengalen, Surapati, Sapali, Tula, Dirk, Lohkay.* Amsterdam: Rijksmuseum; Atlas Contact, 2021.

Starn, Randolph. "Three Ages of 'Patina' in Painting." *Representations* 78, no. 1 (2002): 86–115. https://doi.org/10.1525/rep.2002.78.1.86.

Talley, M. Kirby. "An Old Fiddle on a Green Lawn: The Perverse Infatuation with Dirty Pictures." *Metropolitan Museum Journal* 40 (2005): 37–50.

Tom, Cybele, Ken Sutherland, Clara Granzotto, and Inge Fiedler. "The Many Faces of Guanyin: Conservation and Context of a Polychrome Bodhisattva." In *Transcending Boundaries: Integrated Approaches to Conservation; ICOM-CC 19th Triennial Conference Preprints, Beijing, 17–21 May 2021,* edited by J. Bridgland. Paris: International Council of Museums, 2021.

Urbani, Giovanni, and Bruno Zanardi. *Intorno al restauro.* Milano: Skira, 2000.

Waller, R., and S. Michalski. "A Paradigm Shift for Preventive Conservation, and a Software Tool to Facilitate the Transition." In *ICOM Committee for Conservation 14th Triennial Meeting The Hague 12–16 September 2005: Preprints.* Vol. 2, edited by Isabelle Sourbès-Verger, 733–38. London: James & James; Earthscan, 2005.

Walter, Philippe, and Laurence de Viguerie. "Materials Science Challenges in Paintings." *Nature Materials* 17, no. 2 (February 2018): 106–9. https://doi.org/10.1038/nmat5070.

Wheelock, Arthur K., and C. J. Kaldenbach. "Vermeer's *View of Delft* and His Vision of Reality." *Artibus et Historiae* 3, no. 6 (1982): 9–35. https://doi.org/10.2307/1483203.

Xu, Bingjie, Florian Willomitzer, Chia-Kai Yeh, Fengqiang Li, Vikas Gupta, Jack Tumblin, Marc Walton, and Oliver Cossairt. "3D Surface Measurement and Analysis of Works of Art." In *Conference Record of the 2019 Fifty-Third Asilomar Conference on Signals, Systems & Computers,* edited by M. B. Matthews, 1779–82. New York: IEEE, 2019.

NINETEEN

Characterizing the Immaterial: Noninvasive Imaging and Analysis of Stephen Benton's *Engine no. 9*

Marc Walton, Pengxiao Hao, Marc Vermeulen, Florian Willomitzer, and Oliver Cossairt

Abstract

Invented in 1962, holography is a unique merging of art and technology. It persisted at the scientific cutting edge through the 1990s, when digital imaging emerged and supplanted film. Today, holography is experiencing new interest as analog holograms enter major museum collections as bona fide works of art. In this essay, we articulate our initial steps at Northwestern's Center for Scientific Studies in the Arts to describe the technological challenges in the conservation of holograms, emphasizing their nature as an active material—a holographic image requires user interaction to be viewed, and the materials are delicate and prone to deterioration. Specifically, we outline our methods for creating digital preservation copies of holographic artworks by documenting the wavefront of propagating light. In so doing, we demonstrate why it remains challenging to faithfully capture their high spatial resolution, the full parallax, and deep depths of field without terabytes of data. In addition, we use noninvasive analytical techniques such as spectral imaging, X-ray fluorescence, and optical coherence tomography, to provide insights on holograms' material properties. Through these studies we hope to address current concerns about the long-term preservation of holograms while translating this art form into a digital format to entice new audiences.

Introduction

> The fragment breaks loose with a snap, sending Luke tumbling head over heels. He sits up and sees a twelve-inch three-dimensional hologram of Leia Organa, the Rebel senator, being projected from the face of little Artoo. The image is a rainbow of colors as it flickers and jiggles in the dimly lit garage. Luke's mouth hangs open in awe.[1]

This scene of the Princess Leia hologram from George Lucas's film *Star Wars* perfectly embodies the futuristic zeitgeist of the mid-1970s. Indeed, anyone who has seen a large-format analog hologram in person is familiar with the beguiling realism it offers. In the early days when the technology was still new and fresh, the impression of a complete three-dimensional (3D) picture frozen in time could be unnerving, as illustrated by the apocryphal story of Cartier's 1972 Fifth Avenue display of a hologram depicting a hand holding jewelry (fig. 1) causing an elderly woman to attack the virtual arm floating in thin air.[2] Since this heyday of holography, many of these works have been collected by museums. Notably, the MIT Museum holds one of the largest archives with over two thousand holograms originating from the now defunct Museum of Holography (1976–92).[3] More recently, in 2018 the Art Institute of Chicago obtained *Huddle* by Simone Forti, a two-hundred-degree multiplex hologram.[4] In 2019, the J. Paul Getty Museum acquired 105 hologram pieces related to the *C-Project,* a collection of holograms produced in the 1990s by major contemporary artists to promote holography as an artistic medium.[5] Yet, while scientists, engineers, and artists have been creating holograms for over sixty years, very little work has been done to document the wide variety of imagery stored within them, nor to study how to preserve the large body of holograms housed at museums and galleries.

Holography, though closely related to its cousin photography, is a vastly more complicated medium, with image information recorded in a different way than conventional photographic processes. Conventional photography relies on the projection of light rays incident onto a photosensitive surface. This optical arrangement results in a direct, and easily understandable, relationship between the density of the silver grains and the observed contrast of a photographic image at the macroscopic level. A hologram, on the other hand, relies on recording the interference produced by the wavefront reflected from an object or scene. Also, because holograms rely on the wave nature of light, they can only be recorded using coherent light sources, such as a laser (light

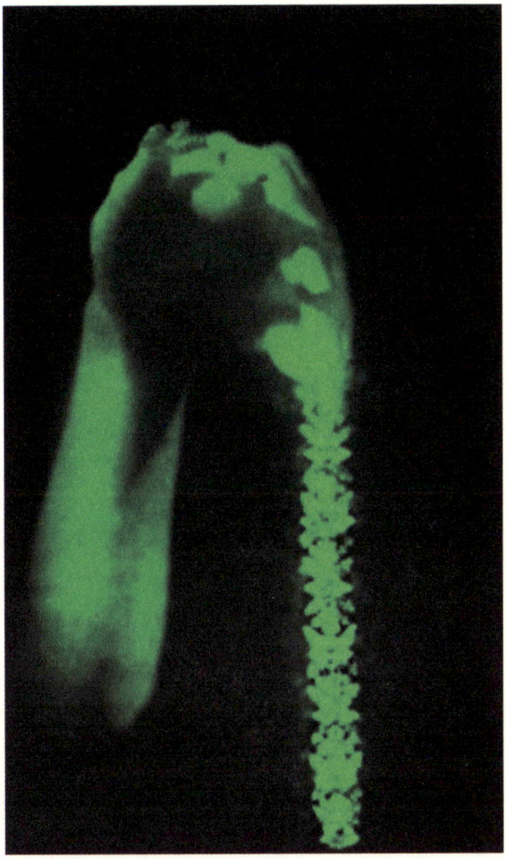

Fig. 1. Robert Schinella, *Hand and Jewels*, 1980s–90s. Holograms, 12 × 16 in. to 18 × 24 in. (30.48 × 40.64 cm to 45.72 × 60.96 cm). MIT Museum collection. Photo adapted from "[Cartier] Hand and Jewels."

amplification by stimulated emission of radiation). The 3D image of a hologram is therefore encoded in the emulsion as nanometer-spaced interference fringes, analogous to a diffraction grating. These fringes bear no physical resemblance to the wavefront image that is formed only once the correct angle and wavelength of light are reflected off its surface or transmitted through its thickness. Perhaps it is the ambiguous nature of where the image forms in the physical material that accounts for the lack of conservation efforts and scientific study of holographic materials. It is intimidating to even suggest where one should begin when considering the material aspects of an object formed from light. Also, what exactly is one attempting to preserve with something so immaterial? Whatever the reasons, knowledge about holographic materials and the technical processes of making them remains ambiguous, and conservation approaches lack clarity and precedence.[6]

Holograms As Active Matter

One of the primary problems of the material study of holograms is that a hologram itself is not matter at all. Put another way, a hologram is a 3D object entirely sculpted by light interference. The only physical material in a hologram is the photographic diffraction grating, and arguably this encoded structure is no more than a vehicle of delivery for the image content. Therefore, there is only an incidental relationship between what one observes in a hologram and the physical materials constituting a holographic artwork. This is a very different situation than one would find in most any other artistic medium wherein appearance is the direct result of light being absorbed, scattered, or reflected in or off the materials constituting the object itself. Rather, a hologram is akin to the effect of light in a gothic cathedral, where the material properties of colored glass are subservient to its primary function of filtering wavelengths of natural light, thus synthesizing an illusionistic environment that elicits a metaphysical response—what Abbot Suger referred to as the *lux nova* in the twelfth century.

So, on the face of it, the conservation of holograms might seem fairly straightforward. Since the carrier of the information (the photographic emulsion) is not as important as the image, one approach to conservation is to transfer image content ad infinitum onto new emulsions or displays as the entropy and decay of the original set in. In fact, this approach is already in practice, with the most notable case being Louise Bourgeois's holograms at the Museum of Modern Art in New York, which were remade after the original film had deteriorated.[7] Here, the restoration efforts of the artist's images of architecture, body parts, and other assorted objects went so far as to transfer the images onto a more stable glass substrate! However, this "ship of Theseus" approach to the conservation of active matter, wherein the persistence of the physical materials is not a requirement, diminishes the agency of holographers who have greatly concerned themselves with controlling color, sharpness, and, indeed, the overall visual effects of what they created through the manipulation of the photographic emulsion. Such concerns also extend to the illumination of holograms, in which artists chose a huge variety of approaches that were often optimized for the particular installation of a given artwork. For instance, the first time Simone Forti exhibited her *Angel* in 1976, she used a candle as the source.[8] While beyond the scope of this present essay, these concerns about how holograms were intended to be played back also need better documentation, with an emphasis placed on the nature of the ambient light in the viewing area and the properties of the illuminating element.

In the remainder of this essay we will unpack what constitutes the material structure of a reflection hologram and, through a case study, explore how one holographer, Stephen Benton, crafted a hologram of a toy train that itself paid homage to the history of making holograms.

Holograms As Physical Matter

Invented in 1948 as a way to improve the resolution of electron microscopes, holography wasn't fully realized until the proliferation of the laser in the 1960s and 1970s. Shortly after scientists and engineers developed the first techniques to record analog holograms using lasers in a laboratory setting, the potential of this new medium captured the imagination of artists fascinated with the possibility of working with a new canvas that enabled them to sculpt directly with light. The initial technical draw of the hologram was the futuristic promise of a realistic 3D display. In many respects, such a 3D display (think of the fictitious Princess Leia hologram) couldn't be realized using just analog film—a fact understood by most of the original holographers—but the conceptual and mathematical basis for such a device was in place. Only now, nearly sixty years after the invention of the hologram, is the technology meeting this initial vision with developments like Google's Project Starline.[9] Artists, however, from the start of film-based holography, were fascinated with the new possibilities for abstract expression, 3D composition, color, texture, and motion. The first holographic artists emerged in the late 1960s, shortly after holographic photography was popularized by Emmett Leith and Juris Upatnieks, reaching the greatest numbers in the late 1970s and 1980s, then starting to decline after the popularization of the personal computer and digital graphical media.

The physical material of analog holography is high-resolution photosensitive film specifically designed to record interference fringes with a resolution close to the wavelength of light. When looking at a large-format hologram (e.g., one square meter), the information density is quite obvious: 3D images are displayed with stunning realism and, in some cases, almost indistinguishable from a real object.[10] Even today's massive twenty-to-forty-megapixel digital sensors lack the density to record the information embedded in an analog hologram, thus relegating practical digital holography to the dustbin, at least for now. Given this background, technical study of holograms may include everything from evaluating deterioration of the emulsion to reverse engineering the techniques of making, as well as displaying and documenting, their dynamic visual effects. However, given the large variations in holographic recording setups, processing chemistries, and the sheer density of visual information contained in a hologram (fig. 2), what is the basic information required to document a hologram's state of preservation?

To address this question, we first must consider the key procedures in making and viewing holograms. One simple example of a recording setup is the so-called in-line "Denisyuk" reflection holography as shown in figure 3.[11] The laser beam emitted from the source passes through a beam expander to hit the holographic plate and the object. When this beam (reference beam) and the light scattered from the object (sample beam) interfere, the holographic plate records these interference fringes.

The best display of a hologram, without distortion, requires an external light source to match the reference beam position as it is set in the recording setup. In practice, this beam position is found by trial and error to identify the "sweet spot" when the hologram is at its brightest. As such, reconstruction geometries may vary in each showing of a given hologram and, correspondingly, the reconstruction may change in appearance. Obvious concerns arise from these uncertainties in reconstruction, such as the extent to which the position of the light affects the visual experience of the hologram.

Perhaps the most notable characteristic of holograms is the realistic 3D space they occupy. Because the scene in a hologram moves with the viewing position, the viewer is often found dancing around it in an effort to take in its entirety. Benton once observed that "looking at this magical surface should be like looking through a window: we should see a three-dimensional image of the scene floating behind it with perfect realism, just as it would have looked if we saw the scene itself. We have created a 'window with a memory.'"[12] Obviously, conventional single-shot photodocumentation of a holographic image will fail to capture this "window of memory." Therefore, the documentation of a hologram calls for another approach that is compatible with this dynamic space, as will be discussed in greater detail below.

In addition to the optical appearance of holograms, the material properties relating to development and bleaching are important aspects to consider. Holographic processing chemistries are nearly identical to those used in conventional silver gelatin photography.[13] In fact, early holograms in the United States borrowed from thirty-five-millimeter Kodak films and processing.[14] However, these holograms were visually dissatisfying, possessing low brightness caused by the opaque

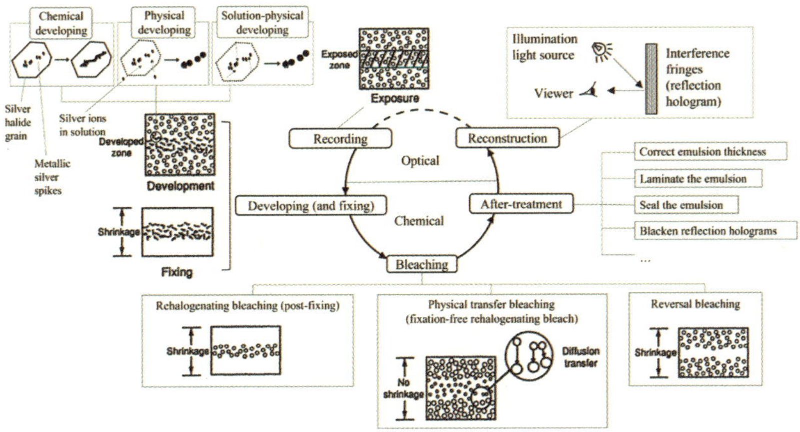

Fig. 2. Making and viewing holograms include optical and chemical steps: recording, developing (and fixing), bleaching, after-treatment, and reconstruction. Holographers may apply different variations of procedures in each step. Image adapted from Bjelkhagen, *Silver-Halide Recording Materials.*

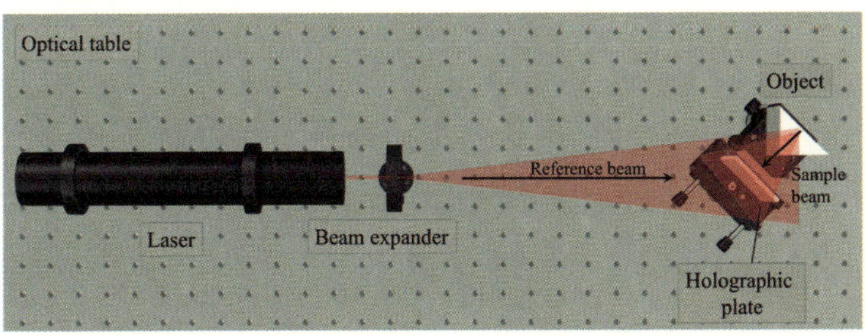

Fig. 3. Diagram of the optical setup of "Denisyuk" reflection holography. The diagram was drawn using the 3DOptix optical simulation and design software.

silver particles and limited angular resolution (parallax) due to the relatively large grain sizes. To overcome these limitations, specialized holographic plates with smaller grains were produced, and developing agents were modified. Likewise, to increase the brightness of reflected or transmitted light, bleaching chemistries were innovated wherein silver halides (AgX, where X = -I, -Br, or -Cl) were used in the finished emulsion rather than silver metal—a so-called phase grating, which produces a wavefront through refraction instead of scattering and absorption inherent to an amplitude grating. Because of these chemistries, conservation concerns for holograms go beyond conventional problems of photographic materials. They include complications associated with AgX particles that remain light-sensitive and will yellow with prolonged exposure to light as well as residual acidity associated with chemical components left over from the bleaching reagents.[15]

As shown schematically in figure 2, the following three types of developers have traditionally been applied to holography:[16]

1. *Chemical (direct) developers* are the most common developing agents that reduce silver halide in contact with exposed silver. The "developed" silver grains have "filament-like" shapes.

2. *Physical developers* function in two steps. Fixation first dissolves and removes all unexposed AgX from the emulsion. Then, exposed particles of silver act as seeds and nucleate fine clusters of metallic silver.

3. *Solution-physical developers* do not contain silver components. Instead, they dissolve the unexposed AgX grains to produce silver ions, which redeposit around the exposed silver to form fine colloidal clusters. More ideal for emulsions with finer grains, these developers were more common in the Soviet Union than in the West.

Bleaches generally fall into these three categories:[17]

1. *Rehalogenating (conventional) bleaches* are used together with a fixing agent. First, unexposed AgX is washed away. Then, developed silver grains are converted into a halide salt. Rehalogenating bleaches cause shrinkage of emulsion since unexposed AgX is lost in fixing.

2. *Physical transfer bleaches* are used without fixation. This type of bleach converts the metallic silver to AgX and migrates the silver ions to the unexposed grains. Physical transfer bleaches conserve the emulsion thickness because the least amount of silver components is lost.

3. *Reversal bleaches* are used without a fixer. With these bleaches, metallic silver is preferentially dissolved, leaving unexposed AgX in the final film. Because metallic silver is lost in bleaching, reversal bleaches also cause shrinkage of emulsion.

Although bleaching improves the brightness of the hologram, the process often causes shrinkage of the emulsion. Accordingly, the fringe spacings also decrease to produce a blue shift in the hologram relative to the wavelength of the laser illuminator. For instance, a reflection hologram recorded using a red He-Ne laser (633 nm) usually retains a green-yellow tint rather than the red of the laser. To compensate for (or indeed, sometimes induce) these color shifts, holographers traditionally tried to optimize such processing chemistries, as will be discussed in more detail in the next section.

Stephen Benton and His *Engine no. 9* Series

Stephen Benton (1941–2003) was a pioneer in holography. Prolific in his invention of novel 3D displays and optics, Benton worked on holography over his entire career at Polaroid and then MIT.[18] Among his fifteen patents, his rainbow (white light transmission) holograms resulted in the ubiquitous embossed security holograms found to this day on credit cards and currency. Likewise, his dynamic holographic video system ushered in the idea of holographic TV. These achievements obscured many other less recognized, yet still key, findings. Also, though he would not have called himself an artist, he was an active promoter of the holographic field and remained a bridge between the engineering and art communities until his death.

In the United States, off-axis transmission holograms had dominated holographic research. Reflection holography, which was invented by Yuri Denisyuk in 1961 in the Soviet Union, was slow to take hold.[19] Yet compared with transmission holograms, reflection holograms held a significant advantage in that they required only a white light source for illumination instead of a laser. These different optical geometries produced shorter fringe spacings resulting in low brightness, inconsistent color shifts in hologram to hologram, and limited viewing angles due to the use of low-resolution emulsion. Such limitations prompted Benton to explore optical setups and processing chemistries to create a new type of reflection hologram with low noise, high brightness, and deep viewing depth, his *Engine no. 9* series.[20] As shown in figure 4, one example from this series depicts a toy train made in homage to the first hologram created by Leith and Upatnieks.

Considering the twelve copies of the hologram studied here, the red color and the appearance of the train (fig. 4A) exhibit considerable variation across the series. Also, evidence of deterioration is found on most of the holograms (fig. 4B), including loss of emulsion on edges and corners, dark patterns in regions of overexposure, and, to be discussed further, whitish "tide lines" along the edges. Although processing chemistries with a similar principle had been developed previously, particularly in the former Soviet Union, Benton pioneered the use of "IEDT (intra-emulsion diffusion-transfer) bleach" for this series.[21] Applying a combination of a chemical developer, a solution-physical developer, and a rehalogenating bleach in sequence resulted in minimal shrinkage to the emulsion during processing.[22] In addition, these steps generated a higher diffraction efficiency and thus brightness "by preserving the polarizable volume of the emulsion."[23] Therefore, by experimenting with IEDT, Benton was able to maintain the color of the reflection holograms and produced one of the brightest reflection holograms ever attained.

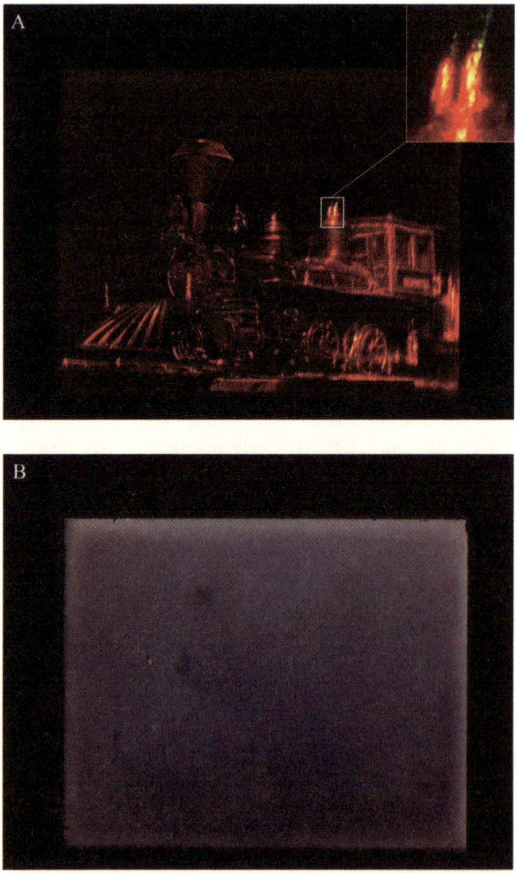

Fig. 4. Stephen Benton and Polaroid staff, *Engine no. 9* (labeled *"Benton 12"*), 1976–77. Hologram, 6½ × 6½ in. (16.51 × 16.51 cm). MIT Museum collection. *A*, with reconstruction. The inset, the whistle of the locomotive in magnification, exhibited colors of the rainbow. *B*, without reconstruction.

To explore the *Engine no. 9* series, we focused on two techniques to characterize the optical properties of the holograms: light-field captures to document the wavefront, or the 3D space occupied by the hologram, and spectral analysis to quantitatively characterize the color of reflection holograms. Additional characterization techniques such as optical coherence tomography (OCT) and X-ray fluorescence (XRF) will be briefly discussed in a later section.

To view the holograms comprised in the *Engine no. 9* series, we used a 3.5-mm-diameter white light LED emitter. Hans Bjelkhagen and David Brotherton-Ratcliffe suggested two general rules for replaying holograms. First, to avoid image blurring, they suggested the use of a small source size and long distance to the hologram to approximate a collimated beam. Second, to avoid aberrations, such as distortion of color and shape, the suggested requirement was to match the illumination geometry to the recording setup.[24] Since the exact recording geometry of the *Engine no. 9* series was unknown (Benton's laboratory records were destroyed), we fixed the position of the light source to allow for its maximum distance to the hologram, the brightest hologram reconstruction, as well as a close-to-forty-five-degree incident angle (fig. 5). This reconstruction geometry was kept unchanged for different holograms and characterization techniques in the following discussions.

With the reconstruction setup described above, the hologram series displayed the expected dynamic 3D view at varying viewing positions. To document the visual appearance, we produced a light field of the holograms using a DSLR (digital single-lens reflex) camera

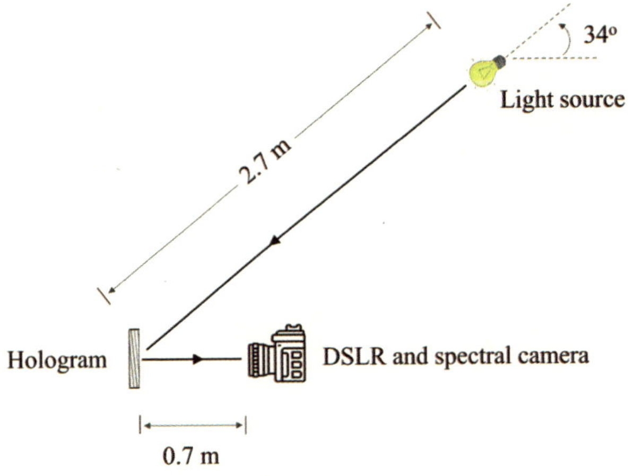

Fig. 5. Diagram of the reconstruction geometry.

attached to an automated *x-y* stage (190 cm by 130 cm). The camera was used to capture a thirteen-by-eight array of viewing angles such that, for each of the 104 images, the camera position was staggered from all surrounding positions by fifteen centimeters in both the vertical and horizontal directions (see figs. 6A and 6B for two of the 104 viewing angles). Once all images were synthesized into a single video or GIF file, we obtained a dynamic rendering of the train. Out of practicality, not all possible angles from the holographic scene were captured using this approach. However, enough of the information is preserved from these light-field arrays to be considered as a promising approach toward conservation documentation.

While the light field records the vertical and horizontal parallax of the holograms, we sought to further analyze the scene to infer Benton's creative processes. In particular, we wanted to estimate the optical arrangement used in recording the holograms and how Benton positioned the object relative to the plane of the hologram. To do this, we constructed a 3D model of the holographic scene using the principles of photogrammetry, which has become more common in the field of cultural heritage, ranging in application from large-scale archaeological sites to small objects.[25] Similarly, here we produced a 3D model from the MIT Museum's *Engine no. 9* copy using commercial software (fig. 6C).[26] Because of the speckle inherent to the hologram combined with the attenuation of light from the small camera aperture, which was required to capture a full depth of field, the 3D model exhibits considerable noise. Despite this limitation, the model records the general holographic space and appears in red with the actual plane of the holographic film demarcated by the black edges of the mount and the fiducial markers. Notably, in figure 6C the reconstructed 3D train straddles this plane, suggesting that Benton utilized a technique known as "image plane holography." Rather than using a real train that would reconstruct behind the hologram in these experiments, Benton projected a master hologram (called H1) onto a holographic film (H2) and controlled the position of the projection, so that the reconstructed space would appear in front of the hologram. The technique augments the visual experience of these holograms by projecting the train out toward the viewer, and it was commonly applied by holographers to more easily control brightness and contrast as well as to create sharper reconstruction in reflection holograms.[27]

When we viewed the *Engine no. 9* series by eye (fig. 7), the varying colors among prints became obvious and prompted us to quantitatively compare the color of each of the twelve copies by spectral analysis. With a spectral camera (Resonon Pika II), we obtained the reflectance spec-

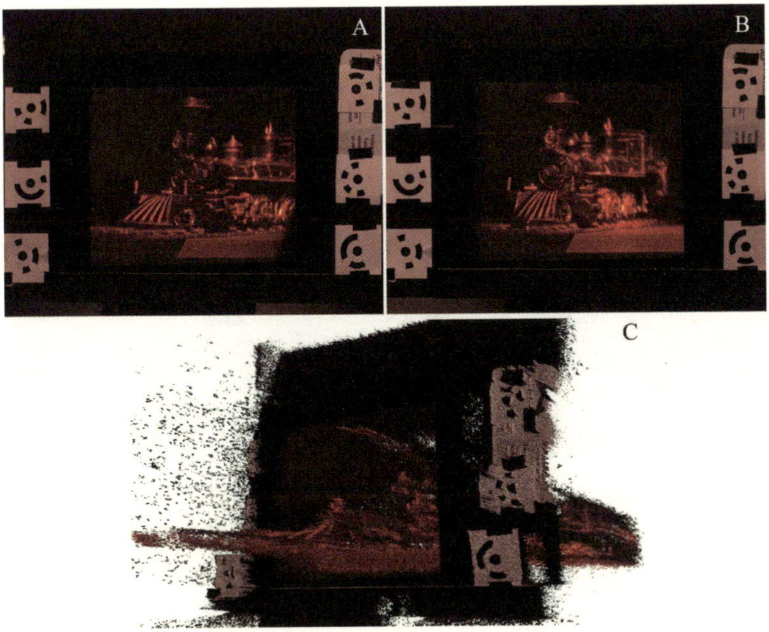

Fig. 6. Registered DSLR images of MIT *Engine no. 9* copy at two camera positions (*A, B*), providing different viewing angles of the train. Six black-and-white fiducial markers along the borders of the hologram assisted with image registration. *C*, the dense cloud 3D model of MIT *Engine no. 9* copy rendered by Agisoft Metashape Professional Edition. The model shows that the plane of the reconstructed wavefront straddles the hologram plane.

tra over the entire hologram, then averaged the spectra at all pixels for each copy (fig. 8). Unlike most reflection holograms, which possessed a greenish tint caused by the shrinkage of the emulsion, nine of the twelve *Engine no. 9* copies exhibited different shades of red (peak centered from 617 nm to 647 nm).[28] This maintenance of the red color close to the wavelength of the He-Ne recording laser at 633 nm is likely the result of Benton's use of the IEDT bleach in this series. It is also interesting to note that in each copy, the noise level (the baseline) and brightness (the peak height) varied as well. Among all copies, the MIT copy possessed the highest brightness with lowest noise—possibly why it was selected for donation to the museum as the best quality from the series.

In addition to the shape and color of the holographic scene, the physical materials of the hologram also matter: the chemical compositions and the microstructures of a hologram that might deteriorate over time. Given these factors, spectral imaging again stood out as a method

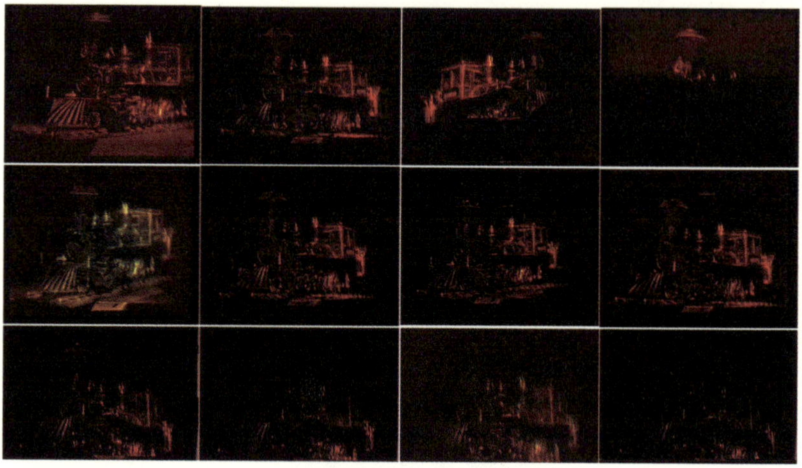

Fig. 7. The *Engine no. 9* series under reconstruction. The twelve copies varied in brightness, noise level, and color. Nine of them displayed different shades of red. While the naked eye could hardly detect the differences, a noninvasive quantitative analysis was possible with spectral imaging.

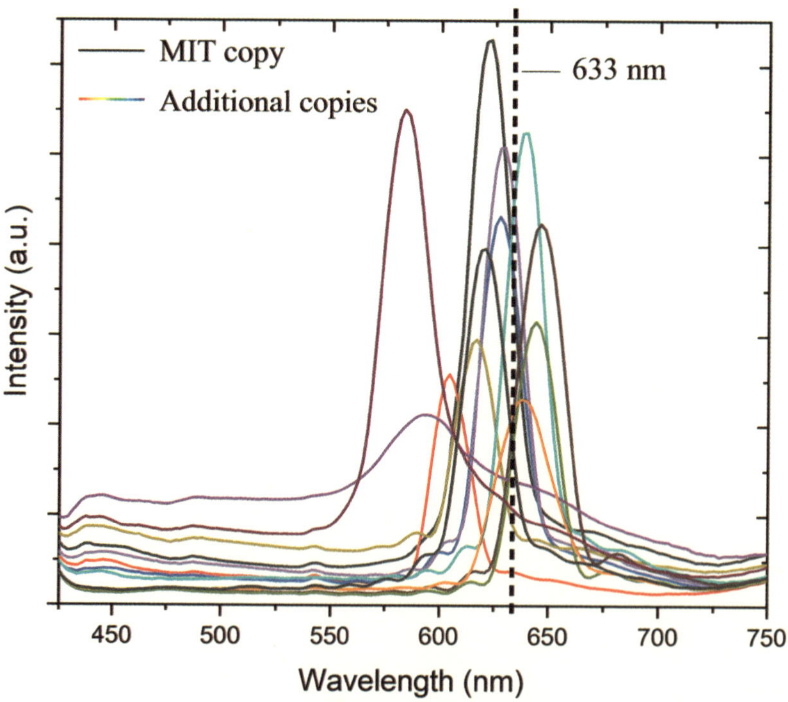

Fig. 8. Average spectra of twelve copies of *Engine no. 9.*

whereby we could noninvasively estimate some of these physical characteristics. In particular, since a hologram's thickness is correlated to color, changes of thickness due to swelling and shrinkage across the image space could be mapped. As observed on one of the copies, the reflectance spectra from distinct regions (fig. 9) indicated inconsistent thickness. Most significantly, the area of tide lines possessed thicker emulsion or larger fringe spacings compared with the rest of the regions, indicated by the seven-nanometer redshift of the peak position.[29] According to the Bragg equation, a spectral shift of seven nanometers corresponds to a change in the emulsion thickness of approximately one percent, or two hundred nanometers, suggesting its ability to detect an approximately seventy-nanometer change of thickness at the resolution limit. In fact, the tide line was hypothetically caused by water vapor diffusion through the edges of the hologram, during which the emulsion swelled. The spectral analysis therefore indicated the possibility of detecting such thickness change with time and across different regions of the hologram; a pixel-wise analysis would help to map this change quantitatively.

Further Thoughts

In this brief case study, we proposed the use of light-field captures and spectral imaging to characterize optical properties of reflection holograms. The light-field captures, a potential documentation approach for holograms, faithfully recorded and displayed the vertical and

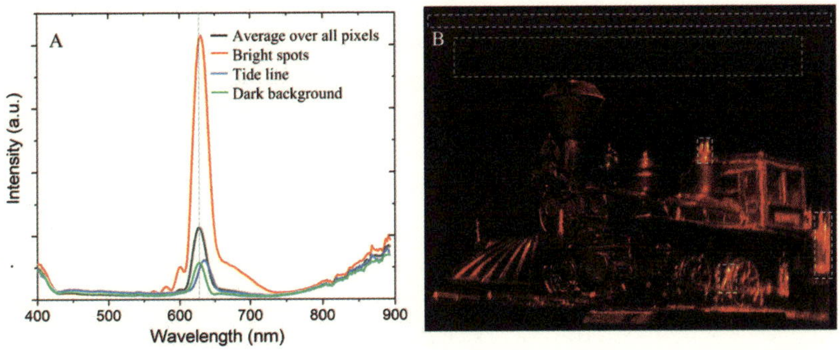

Fig 9. Spectral analysis of *Benton 12*. *A*, the spectra of the hologram averaged in different regions: black—all pixels, red—bright spots, blue—the edges, and green—the dark background area; *B*, a DSLR capture shows the regions corresponding to the averaged spectra.

horizontal parallax of Benton's *Engine no. 9* series as well as preserving their depth of field. Combined with geometric analysis using photogrammetry, the bisecting holographic space with the hologram plane suggested that one of the recording techniques used by Benton was image plane holography. Through spectral imaging, the color variation in different *Engine no. 9* copies suggested that Benton experimented with chemical processing procedures. Particularly, spectral analysis was sensitive to the change of thickness at a submicron scale, thus detecting regional swelling or shrinking of the emulsion due to degradation and/or ingress of water vapor. Thus, the two analytical techniques used here offer a glimpse into Benton's efforts in making reflection holograms—experimenting with chemical and optical techniques and pushing the technical limitations—to produce the MIT copy, which is described as "an exceptionally bright and unusually deep example of its kind."[30]

Nevertheless, the current study requires optimization of experimental procedures to improve the quality of the data. In our current research we are optimizing both illumination and the light-field capture process. For instance, implementing a vibrating light source would theoretically reduce laser speckle. Synthesizing a large depth of field by applying focus stacking, combined with optimizing camera positions, would improve resolution. Also, as this type of photography only captures a small portion of the information in a hologram, we are exploring the use of more data-intensive methods, such as laser ptychography, to capture the full wavefront at a resolution of one to two microns. The resulting data set from these experiments is necessarily large, bordering on hundreds of gigabytes. While such approaches may not be practical to implement or duplicate in many museums' imaging studios, this proof of concept is necessary if full digital holography, with the same resolution as film, is to be realized.

In addition to hyperspectral imaging and light-field captures, analytical techniques such as XRF and OCT could also provide chemical and physical information. For instance, XRF point analysis detected silver in all *Engine no. 9* copies, while bromide and trace elements such as iron were present in some copies. Such elemental information may assist in identifying the type of chemical processing used by the holographer. However, since glass strongly adsorbs X-ray, it blocks the detection of trace elements on the copies layered between glass plates. Therefore, an area without glass substrates or film-based holograms would be more suitable for XRF measurements. OCT, on the other hand, could noninvasively characterize physical layer structures. A preliminary experiment (fig. 10) suggested that a nine-hundred-nanometer OCT system could measure a six-micrometer-thick silver halide emulsion. Since certain brands and types of a holographic film or plate have a defined layered structure, OCT may prove to be valuable for such identification.

Conclusion

Holography was a scientific and cultural symbol from the 1960s to the 1980s. In the case study presented here from Benton's *Engine no. 9*, we characterized the materials and optical properties using hyperspectral imaging and light-field capture and show how these data may be used to recover Benton's experimental methods as he attempted to improve the quality of reflection holograms. While improvements in experimental and analytical procedures are needed, we hope this preliminary work may inspire interest, future studies, and expanded discussions on the conservation of holograms.

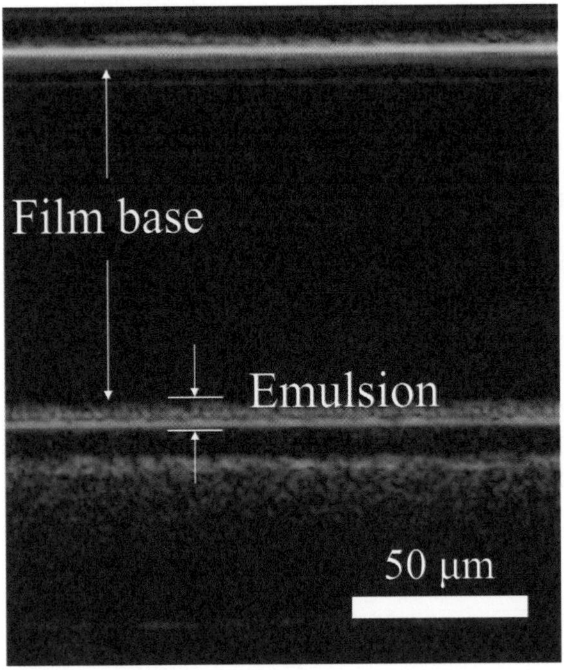

Fig. 10. Traverse scan (B-scan) of a holographic film using OCT generated a cross-sectional image. It noninvasively measures the thickness of the emulsion and the film base. The image was obtained using a Thorlabs Ganymede GAN621 900 nm OCT, with the OCT-LK2-BB lens kit.

NOTES

The authors thank Deborah Douglas, director of collections and curator of science and technology at the MIT Museum; Gloria Martinez, collections manager at the MIT Museum; and Katie Porter, registrar at the MIT Museum, for preparing for lab space and providing access to the hologram collection at the MIT Museum. The authors thank Betsy Connors, holographic artist and previous colleague of Stephen Benton, for consultation on Benton holograms. The research is made possible through the support of the Andrew Mellon Foundation and the National Science Foundation–Partnerships for International Research and Education (NSF–PIRE) grant. This work made use of the MatCI Facility supported by the MRSEC program of the National Science Foundation (DMR-1720139) at the Materials Research Center of Northwestern University.

1. Lucas, *Star Wars*, 25; Johnston, "Introduction," 2.
2. "[Cartier] Hand and Jewels"; Slesin, "Diamond Sleight of Hand."
3. Connors, Benton, and Seamans, "Report from the MIT Museum."
4. "Huddle."
5. "Getty Museum Announces Donation."
6. Wise, "'Transparent Things'" (2013); Wise, "'Transparent Things'" (2016); Serra and Moreno, "Experiencias en la conservación"; Brown and Jacobson, "Archival Permanence"; Jacobson and Brown, "Archival Properties."
7. "Untitled."
8. Nelson, "Phantom Limbs."
9. Bavor, "Project Starline."
10. Slesin, "Diamond Sleight of Hand."
11. Benton and Bove, "In-Line 'Denisyuk' Reflection Holography."
12. Benton and Bove, "Holograms and Perception," 7.
13. Bjelkhagen, *Silver-Halide Recording Materials.*
14. Johnston, "Three-Dimensional Wavefront," 111.
15. Bjelkhagen, "After-Treatment."
16. Bjelkhagen, "Development."
17. Benton and Bove, "In-Line 'Denisyuk' Reflection Holography"; Bjelkhagen, "Bleaching."
18. Spierings, "Chasing the Holoprinter"; "Holography Pioneer."
19. Johnston, "Americanization of Reflection Holography," 200–202.
20. Burns and Jackson, *Holografi*, 61; Benton, "Intra-Emulsion Diffusion-Transfer" (1978).
21. Bjelkhagen, "Development"; Bjelkhagen, "Bleaching."

22. Benton, "Intra-Emulsion Diffusion-Transfer Processing" (1974); Benton, "Intra-Emulsion Diffusion-Transfer" (1978); Benton, Halle, and Bove, "Practical Issues."
23. Bjelkhagen, "Development"; Benton, "Intra-Emulsion Diffusion-Transfer" (1978).
24. Bjelkhagen and Brotherton-Ratcliffe, "Illumination of Colour Holograms."
25. Aicardi et al., "Recent Trends."
26. "Engine #9"; "AgiSoft PhotoScan Professional."
27. Brandt, "Image Plane Holography."
28. Burns and Jackson, *Holografi*, 61.
29. Trochtchanovitch, "Method of Characterization"; Benton and Bove, "In-Line 'Denisyuk' Reflection Holography."
30. Burns and Jackson, *Holografi*, 61.

BIBLIOGRAPHY

AgiSoft PhotoScan Professional. Version 1.4.5. Software. 2016. http://www.agisoft.com/downloads/installer/.

Aicardi, Irene, Filiberto Chiabrando, Andrea Maria Lingua, and Francesca Noardo. "Recent Trends in Cultural Heritage 3D Survey: The Photogrammetric Computer Vision Approach." *Journal of Cultural Heritage* 32 (2018): 257–66.

Bavor, Clay. "Project Starline: Feel Like You're There, Together." *Keyword* (blog). Google. May 18, 2021. https://blog.google/technology/research/project-starline.

Benton, Stephen A. "Intra-Emulsion Diffusion-Transfer." *Holosphere* 7, no. 7 (1978): 4–5.

———. "Intra-Emulsion Diffusion-Transfer Processing of Volume Dielectric Holograms." *Journal of the Optical Society of America* (1974): 1393–94.

Benton, Stephen A., and V. Michael Bove Jr. "Holograms and Perception." In *Holographic Imaging*, 5–13.

———. *Holographic Imaging*. Hoboken, NJ: John Wiley & Sons, 2008.

———. "In-Line 'Denisyuk' Reflection Holography." In *Holographic Imaging*, 173–80.

Benton, Stephen A., Michael Halle, and V. Michael Bove Jr. "Practical Issues in Rainbow Holography." In Benton and Bove, *Holographic Imaging*, 159–72.

Bjelkhagen, Hans I. "After-Treatment." In *Silver-Halide Recording Materials*, 295–320.

———. “Bleaching.” In *Silver-Halide Recording Materials*, 159–213.

———. “Development.” In *Silver-Halide Recording Materials*, 115–58.

———. *Silver-Halide Recording Materials for Holography and Their Processing.* 2nd ed. New York: Springer, 1995.

Bjelkhagen, Hans, and David Brotherton-Ratcliffe. “Illumination of Colour Holograms.” In *Ultra-Realistic Imaging: Advanced Techniques in Analogue and Digital Colour Holography*, 461–92. Boca Raton, FL: CRC Press, 2013.

Brandt, Gerald B. “Image Plane Holography.” *Applied Optics* 8, no. 7 (1969): 1421–29.

Brown, K. C., and R. E. Jacobson. “Archival Permanence of Holograms?” *Journal of Visual Communication in Medicine* 9, no. 2 (1986): 50–55.

Burns, Jody, and Posy Jackson. *Holografi: Det 3-Dimensionella Mediet; 18–28 Mars, 1976, Pa Kulturhuset, Gallereit . . . Stockholm, Sverige.* Stockholm: Stockholm Cultural Administration, 1976.

“[Cartier] Hand and Jewels.” MIT Museum Collections. Accessed December 16, 2020. https://webmuseum.mit.edu/detail.php?term=cartier&module=objects&type=keyword&x=0&y=0&kv=128353&record=0&module=objects.

Connors, Betsy A., Stephen A. Benton, and Warren Seamans. “Report from the MIT Museum.” In *Fifth International Symposium on Display Holography*, edited by Tung H. Jeong, 146–50. Proceedings of SPIE 2333. Bellingham, WA: SPIE, 1995.

“Engine #9.” MIT Museum Collections. Accessed December 16, 2020. https://webmuseum.mit.edu/detail.php?term=engine+%239&module=objects&type=keyword&x=6&y=11&kv=67220&record=0&module=objects.

“Getty Museum Announces Donation of 105 Holograms Created by 20 Noted Artists.” News from the Getty. Press release, March 5, 2019. http://news.getty.edu/getty-museum-announces-donation-105-holograms-created-by-20-noted-artists.htm.

“Holography Pioneer Stephen Benton Dies at 61.” *MIT News*, November 12, 2003. https://www.media.mit.edu/articles/holography-pioneer-stephen-benton-dies-at-62/.

“Huddle.” Art Institute of Chicago Collections. Accessed October 6, 2020. https://www.artic.edu/artworks/244406/huddle.

Jacobson, Ralph E., and Kevin C. Brown. “Archival Properties of Holograms.” *Practical Holography III*, edited by Stephen A. Benton, 60–67. Proceedings of SPIE 1051. Bellingham, WA: SPIE, 1989.

Johnston, Sean F. “The Americanization of Reflection Holography.” In *Holographic Visions*, 200–202.

———. *Holographic Visions: A History of New Science.* New York: Oxford University Press, 2006.

———. "Introduction: Seeking Coherence." In *Holographic Visions,* 1–11.

———. "Three-Dimensional Wavefront Reconstruction." In *Holographic Visions,* 102–19.

Lucas, George. *Star Wars: A New Hope; The Illustrated Screenplay.* New York: Ballantine, 1998.

Nelson, Solveig. "Phantom Limbs: Solveig Nelson on Simon Forti's Holograms." *Art Forum* 57, no. 1 (2018): 260–68.

"Optical Design Simulation: 3DOptix." Accessed December 16, 2020. https://3doptix.com/.

Serra, Rolando, and Alfredo Moreno. "Experiencias en la conservación de hologramas patrimoniales cubanos." *Arquitectura y urbanismo* 35, no. 2 (2014): 93–98.

Slesin, Suzanne. "A Diamond Sleight of Hand." *New York Times,* May 27, 1979. https://www.nytimes.com/1979/05/27/archives/a-diamond-sleight-of-hand.html.

Spierings, Walter. "Chasing the Holoprinter." In "Tribute to Professor Stephen A. Benton," edited by Sunny Bains. Special issue, *Holography* 15, no. 1 (2004): 8.

Trochtchanovitch, Pavel. "Method of Characterization of Effective Shrinkage in Reflection Holograms." *Optical Engineering* 43, no. 5 (2004): 1160–68.

"Untitled." The untitled hologram series, 1998–2014. Louise Bourgeois: The Complete Prints & Books. MoMA (website). https://www.moma.org/s/lb/collection_lb/objbytag/objbytag_tag-vo282442_thumbs.html.

Wise, Andrea. "'Transparent Things, through Which the Past Shines': Investigating Holograms in the Collection of the National Gallery of Australia." *Journal of the American Institute for Conservation* 55, no. 3 (2016): 176–85.

———. "'Transparent Things, through Which the Past Shines': Investigating Holograms in the Collection of the National Gallery of Australia." *Topics in Photographic Preservation* 15 (2013): 142–55.

TWENTY

Making Meiji Red: Semiotic Activity in the Colors of Japanese Woodblock Prints, 1864–1900

Marco Leona and
Henry D. Smith II

Abstract

We propose that the concept of "active matter" be expanded to encompass nonmaterial realms of activity, in particular that of meaning. We propose such "semiotic activity" as a key area of interest, drawing on the work of the French philosopher and art historian Georges Roque, and offering a scientific investigation of the changing colorants of Japanese color woodblock prints in the nineteenth century as a revealing case study of materiality and meaning. In particular, we single out the binaries of expensive/cheap, natural/artificial, native/foreign, and vivid/austere as central to the shifting meanings assigned to newly introduced dyes and pigments.

The Special Nature of the Japanese Color Woodblock Print

The Japanese color woodblock print emerged in its fully developed form in 1765 with the invention of the "brocade print" (*nishiki-e*), which built on earlier experiments with printing in two or three colors in selective areas. The new method, led by the artist Suzuki Harunobu (1725–70), added more blocks and covered a far larger surface area with color, in particular the background. Several generations of increasingly skilled carvers and printers had emerged in the century and a half since the woodblock printing of books had suddenly exploded in the early seventeenth century.

The nishiki-e technique was distinguished first by the use of a sophisticated rubbing tool known as a *baren*, brought from China but improved by Japanese printers, which enabled the application of great pressure at multiple points, firmly pressing the colorants into the wetted paper. The other device, invented in Japan, was a simple but remarkably precise system of registration, using two notches in the keyblock known as *kentō*, one in the lower right corner and the other along the lower edge to the left. Both *baren* and *kentō* remain in use today in the large community of printers who follow the traditional technique of Japanese color woodblocks, now known simply as *mokuhanga*, "woodblock prints."

The crucial material variable was color, the use and variety of which set Japanese color printing apart in a class unmatched anywhere in the early modern world of printing on paper, for its precision, vibrant colors, and a remarkably low cost per sheet. The natural sources of nishiki-e color, both organic and mineral, acting as both dyes and pigments in ways that are not yet fully understood, are now well known for the first century of the medium, until the introduction of the wholly new synthetic organic dyes that began with the accidental discovery by William Henry Perkins (1838–1907) of mauve in 1864.[1]

The one truly important change in this interval was the introduction of Prussian blue as the dominant source of that hue, previously provided by a duller indigo and a fugitive dayflower (*tsuyukusa*). This crucial inorganic dye, accidentally discovered by Johann Jacob Diesbach (dates unknown) in 1704, made its way to Japan in limited quantities in the late eighteenth and early nineteenth centuries but suddenly became dominant after a Prussian blue factory was built in Canton in 1822, providing the pigment at a far lower cost than that supplied by the Dutch trade from Europe. This became the striking blue of the landscapes of Katsushika Hokusai (1760–1849) and Utagawa Hiroshige (1797–1858) and has remained the dominant blue colorant of Japanese-style woodblocks until the present.[2]

The great remaining mystery has been the identity of the new synthetic dyes that appeared in Japanese nishiki-e from the 1860s and exactly how they were used. In the absence of any scientific or historical research, a variety of narratives emerged over time, the most extreme alleging that *all* the existing vegetable and mineral colorants were displaced by imported synthetics, typically described as "aniline," and often indicating Germany as the primary source.

The conspicuous absence of serious research on the actual colors of Meiji prints enabled a proliferation of opinions that reflected not objective knowledge about matter, but rather an assortment of subjective biases about what were thought to be cheap, imported, and unnatural colorants. These biases were from the start aligned with an array of aesthetic judgments that dismissed the visual qualities of the perceived colors as "vulgar," "garish," "harsh," and "strident."

The apotheosis of this kind of meaning-full activity for material colors came with James Michener's influential *The Floating World: The Story of Japanese Prints* (1954), by which time the synthetic colors had become not just aesthetically but even morally repugnant: "ugly," "grotesque," "violent," and "cruel." He opined that "[w]hen vile German aniline dyes reached Japan, all restraint disappeared, and some of the most grotesque prints ever issued flooded the shops. . . . No pre-aniline print is quite the complete horror that most of the post-aniline prints are."[3] Over six decades later, Michener's vituperative tone has given way to more lenient and even positive assessments of the colors of Meiji woodblock prints among collectors and scholars, but the notion of a wholesale replacement of "soft" older dyes by stronger "aniline" colors remains entrenched in the literature.[4]

Conservation professionals today would be unlikely to sympathize with Michener's attitude, but they might well voice a general concern about the stability of aniline dyes, many of which were known to be light fugitive and prone to bleeding when exposed to moisture.[5] We are not concerned, however, with the chemical or physical activity of these dyes. Meiji printers managed to limit bleeding through skillful techniques of mixing with existing dyes and pigments, both vegetable and mineral. What rather concerns us is the identity of the new dyes that were introduced to Japan after the opening of free trade at Yokohama in 1858. Where did they come from, how much did they cost, and were they indeed all "aniline" dyes? Beyond the precise identification of the historical introduction of new dyestuffs, we are concerned with evolving subjective judgments about the nature of such dyes as cheap, foreign, artificial, and gaudy. How did these opinions, most of them lacking in objective scientific evidence, culminate over decades in Michener's fulminations not only over the new synthetic dyes, but also the use of strong, bright colors in the final decades of the Edo period? To carry out such an investigation, we must combine scientific and art historical findings with a broader investigation into culture, aesthetics, and language, engaging in a truly interdisciplinary collaboration of the sciences and the humanities.

The New Story That Science Tells

Our task here is first to introduce wholly new research, both scientific and historical, about the material activity of the changing colorants actually used in nishiki-e from the 1860s. With this new evidence in hand, we can then look back to reinterpret their semiotic activity.

The pervasive notion that aniline dyes are typical of Meiji prints was curiously met by the conspicuous absence at the time we began our own investigation of studies in the scientific literature describing the identification of any such colorants. This is not at all surprising, given that organic dyes are extremely difficult to identify in works of art. To start with, these substances have extremely high tinting strength; that is, the actual amount of dye used is extremely small. Analysis of works of art should preferably be carried out by noninvasive methods, and samples, if one is allowed to remove any at all, should be microscopic. In the case of organic dyes, all the noninvasive methods available to cultural heritage scientists are basically useless.[6] Elemental analysis by X-ray fluorescence spectrometry is of no value, since dyes all uniformly contain the same carbon, hydrogen, oxygen, and nitrogen atoms (in addition to sometimes chlorine, bromine, and sulfur). Fiber optics reflectance spectroscopy (commonly known as FORS) is extremely useful to discriminate between indigo and Prussian blue, but the spectra obtained from red and violet dyes are not as characteristic.[7]

Our work on surface-enhanced Raman spectroscopy (or SERS), a technique that uses silver nanoparticles to amplify the vibrational response of molecules exposed to intense laser radiation, provided an ideal tool for the study of synthetic dyes.[8] This technique is both highly specific and extremely sensitive. While it requires a sample, accurate molecular signatures can be detected from fragments as small as a few tens of micrometers, such as a section of a single paper fiber from a print (for reference, human hair is between fifty and one hundred micrometers in diameter).

Armed with the right tool, we proceeded to attempt a chronological survey of woodblock prints, spanning the time period from the opening of the port of Yokohama to international trade in 1859 until the end of the century. We analyzed over sixty prints, datable with precision to the year and often to the month of issue. Our survey shows that a relatively small number of new dyes were introduced at key moments and that most of the traditional colorants of the late Edo period continued in use throughout the Meiji period.[9] The full timeline for the introduction of foreign dyes in Meiji woodblock prints is given in figure 1.

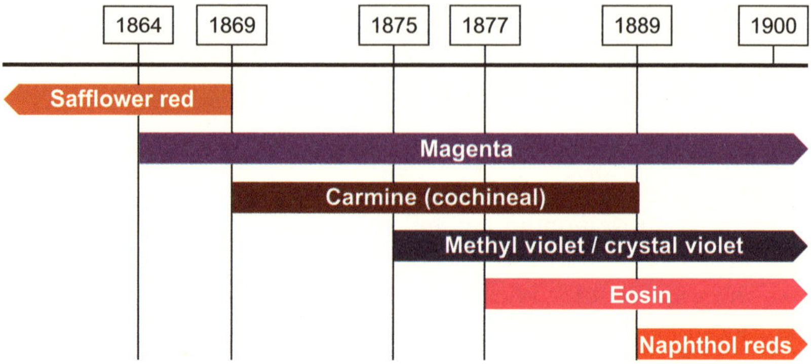

Fig. 1. Timeline for the use of major purple and red natural and synthetic dyes in Japan from the late Edo to the late Meiji period.

The use of aniline dyes in Meiji prints began in early 1864, when the purple dye rosaniline began to appear, at first in combination with Prussian blue, to give a deep violet blue (fig. 2). From 1875, Prussian blue was usually mixed with methyl violet for a stronger purple. Rosaniline (also known as fuchsine or magenta) and methyl violet are members of the triarylmethane dye family, arguably the first synthetic dyes to enjoy widespread commercial success and continued use (crystal violet, a dye closely related to methyl violet, is still in use to this day as an ingredient in blue rollerball inks).

The most significant discovery we made, however, was the introduction in early 1869 of a new red dye with a long history of use in other parts of the world: cochineal, which rapidly replaced safflower as the dominant red in woodblock prints. Although the new cochineal "carmine" (the term for the lake pigment produced from the dried insects) enabled a more generous use of solid red through its lower cost and great ease of application, it was not until the next year that Toyohara Kunichika (1835–1900), the leading designer of actor prints at the time, began regularly to use solid red backgrounds (fig. 3). Carmine remained the primary red of nishiki-e prints for the next two decades, often combined with vermilion, much as had frequently been the case with safflower.

By 1875 a change in the purple areas becomes evident. Two purple aniline dyes belonging to the triarylmethane family were introduced into the market in the later nineteenth century: methyl violet (basic violet 1) in 1866 (invented in 1861) and crystal violet (basic violet 3) in 1883. These bright hues were put to excellent use in "Newspaper

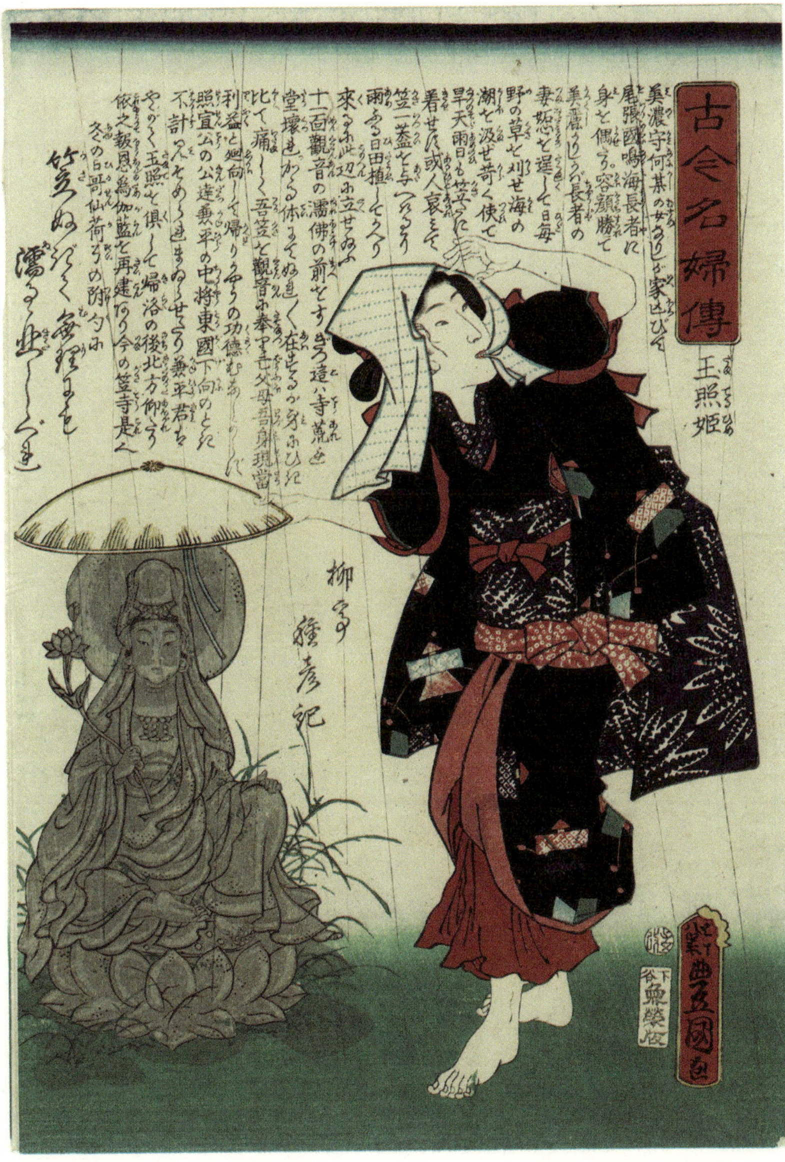

Fig. 2. Utagawa Kunisada (1786–1865), *Princess Tamateru* (*Tamateru-hime*), from the series *Biographies of Famous Women Ancient and Modern* (*Kokin meifu den*), Fifth Month, 1864. Color woodblock print, 10 × 14⅛ in. (25.2 × 35.8 cm). Collection of the authors. This is one of the earliest appearances in nishiki-e of the new synthetic dyes from Europe in early 1864. This is magenta (known by chemists as rosaniline), the purple seen on the *obi* sash of Princess Tamateru. It was muted by mixture with Prussian blue but marks a new stage in the history of the colorants used in Japanese woodblock prints.

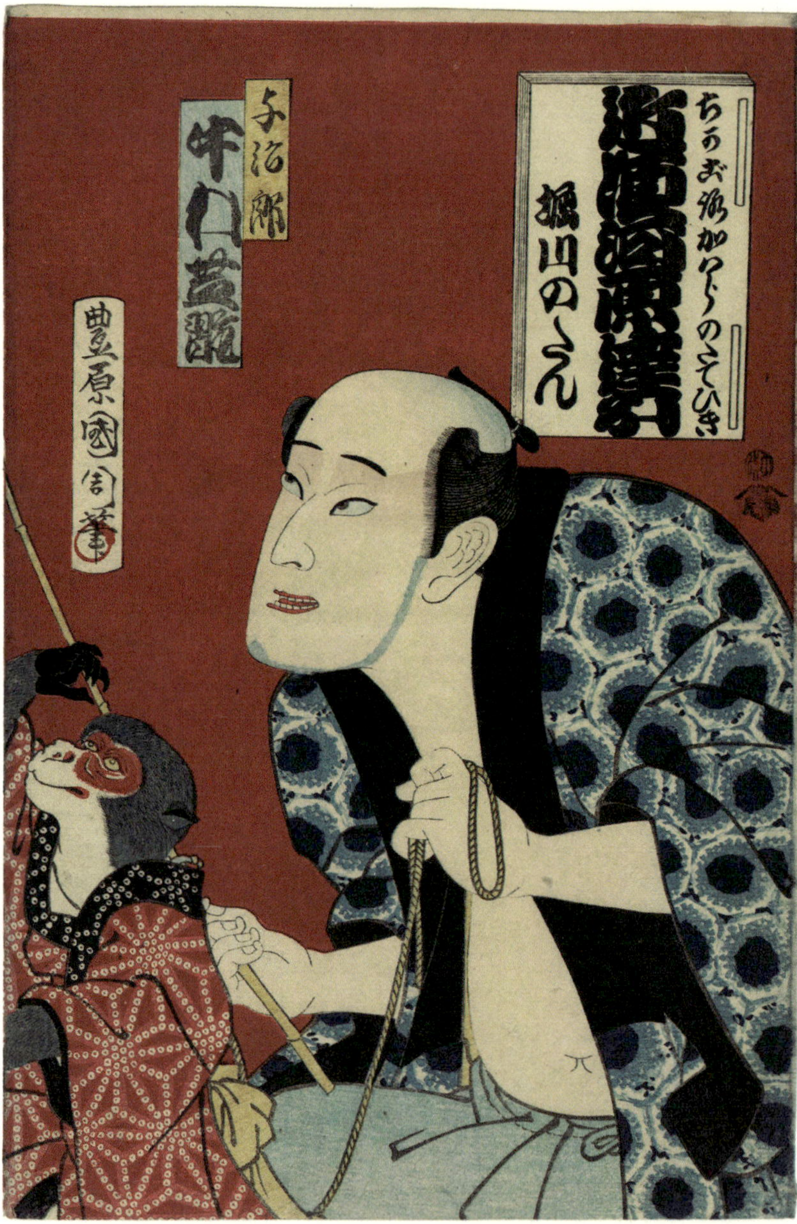

Fig. 3. Toyohara Kunichika (1835–1900), Nakamura Shikan IV as Yojirō the monkey handler in Horikawa scene from kabuki play *Chikagoro kawara no tatehiki*, right panel of triptych, Tenth Month, 1870. Color woodblock print, 9⅝ × 14¼ in. (24.3 × 36.0 cm). Collection of the authors.

nishiki-e" (fig. 4), which plucked sensational crime stories from the new daily press (in this case from an issue of the *Yūbin Hōchi Shinbun* newspaper published two months earlier on February 15) and illustrated them in full-size prints. The brighter purple in the border frame is the new synthetic methyl violet, probably a very early example. The red is pure cochineal carmine, with no mixing. The blue is Prussian blue (mixed with shell white), the yellow is orpiment, and the green is a mixture of the two.

From 1877, we see the introduction of eosin (acid red 87/solvent red 43), characterized by a brilliant pink color, which was first synthetized by Heinrich Caro (1834–1910) in 1871 by brominating fluorescein. It was widely used in the West as an artists' material, often in the form of the lake pigment known as geranium lake, and it was a common lithographic ink. In the very print that began our research in May 2014 (fig. 5), we detected eosin pink on the right (the dress of one lady, trim of the other) and the two-tone mix of carmine and eosin in the dramatic bolt of red fabric to the left. It is an exquisite example of the use of new reds and pinks to evoke a more feminine ethos.

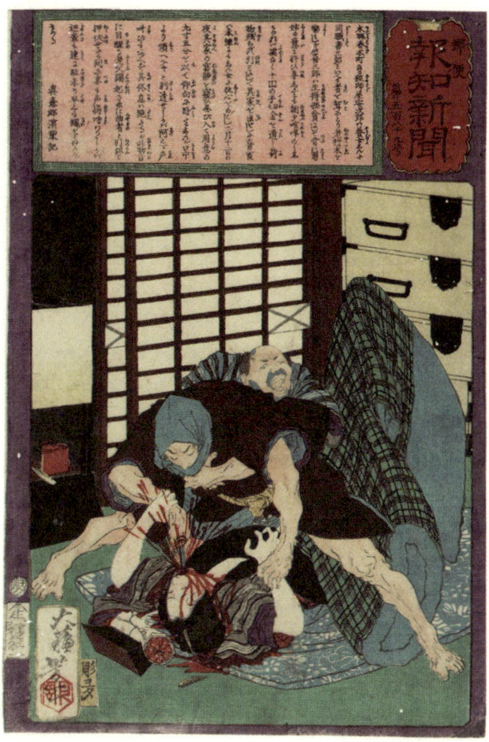

Fig. 4. Tsukioka Yoshitoshi (1839–92), *Yūbin Hōchi Shinbun, dai 589-gō* (*Yūbin Hōchi News, no. 589*), April 1875. Color woodblock print, 13¾ × 9⅜ in. (35 × 23.7 cm). Collection of the authors.

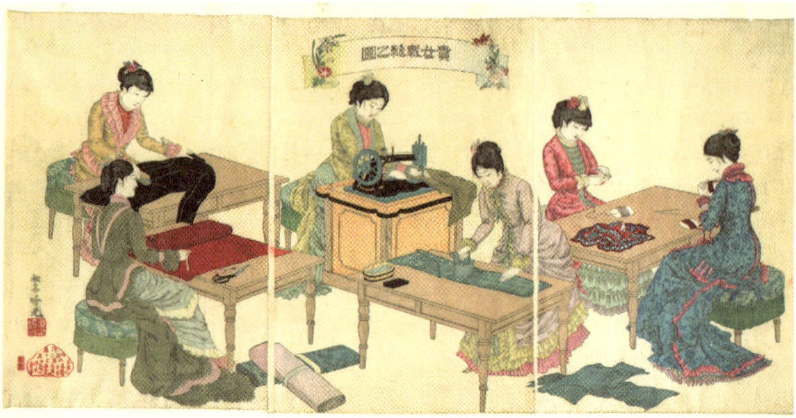

Fig. 5. Adachi Ginkō (active 1874–97), *Ladies Sewing* (*Kijo saihō no zu*), September 1887. Color woodblock print, 14 × 28½ in. (35.6 × 72.4 cm). Metropolitan Museum of Art (JP3272). Gift of Lincoln Kirstein, 1959.

From 1889, a succession of red naphthol dyes of more striking color suddenly appeared, quickly replacing cochineal carmine as the dominant red in nishiki-e prints. Just about this time, however, a tendency to more restrained use of color and more painterly effects began to emerge in color woodblocks, and with the exception of a burst of dynamic color in prints depicting the Sino-Japanese War (1894–95), the use of strong colors in Meiji prints receded. For a dramatic example of such war prints, see figure 6, in which the bright fireball seems to have an organic life of its own, with deep naphthol scarlet along the bottom and rings of the familiar orpiment yellow above. The eerie orange reflection off the tree is simply a mix of the same two colorants. The experience of Kobayashi Kiyochika (1847–1915) in 1876–81, depicting night landscapes of Tokyo, is reprised here in the hazy moon and the distant silhouettes of trees and mounted cavalry.

Historical Change and Semiotic Activity in Nishiki-e Colors

How much of the new story told by scientific analysis, shown in the timeline in figure 1, can be corroborated in Japanese documents of the Meiji period? Why were these changes not widely recognized at the time? Looking back at the contemporary record, we find much concrete information, but no coherent narrative about their identity and what they meant. Those involved in the production of Meiji nishiki-e were of course well aware of the nature of the changing dyes and pigments they were using. The publishers purchased the colorants themselves, through retail dealers in Tokyo, and the printers they employed would have been intimately familiar with the material nature and qualities of the new dyes. References to these dyes may be found in newspaper articles and in the growing technical literature on the dyes used for textiles: magenta was "*majenta*," and cochineal carmine was known as *yōkō*, "western safflower," which Japanese-English dictionaries from the 1870s confirmed to be "carmine" or "cochineal." Eosine (known as *yōshin*) and naphthol "scarlet" (*sukaretto*) also appeared in both the daily press and in numerous technical manuals.

Much more difficult to detect in contemporary records, however, is any recognition that these imported dyes signaled a new era of color in popular woodblock prints, for better or for worse: no overall story emerged to describe or evaluate the changes. The only exception unearthed so far is an account in the daily newspaper *Yomiuri shinbun* in August 1890 by an unidentified author, "Trying to Reverse the Decline of Azuma Nishiki-e." But it is a confused and misleading tirade by a collector of contemporary prints who laments the use of *yoko* (still thought to be synthetic), *yōshin*, and *sukaretto*, claiming that the resulting colors "would not last more than a few months."[10] Note that the objection is not to colors too strong or bright, but rather those especially prone to fade (which was not necessarily the case).

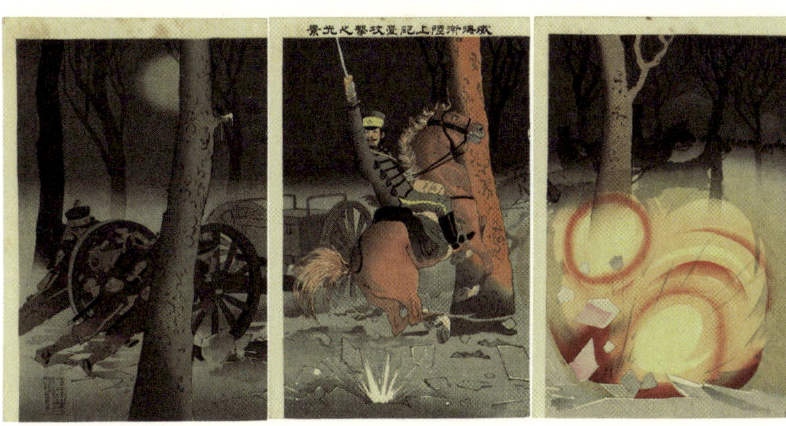

Fig. 6. Kobayashi Kiyochika (1847–1915), *Japanese Attacking the Cannon Base at Weihaiwei* (*Ikaiei rikujō hōdai kōgeki no kōkei*), February 1895. Color woodblock print, 14¾ × 30 in. (37.5 × 76.0 cm). Collection of the authors.

Meanwhile, wholly apart from the specific identification of the new dyes, there emerged from the mid-1880s among Western writers on Japanese art a coherent narrative about the changes of the colors of woodblock prints, one based not on scientific knowledge but on purely subjective aesthetic judgements. The key figure was William Anderson (1842–1900), an English physician who went to Japan in 1873 as a foreign expert to serve in the Imperial Naval College in Tokyo. During his stay of over six years, Anderson developed a keen interest in Japanese paintings and amassed a collection that he sold to the British Museum in 1881, a total of some three thousand works.[11] Back in London in the early 1880s, Anderson devoted himself to the production of a monumental work titled *The Pictorial Arts of Japan,* mostly on the sort of paintings he had collected, but which now included a section on "chromoxylography" (the standard Victorian term for color woodblock printing), only six pages long but offering a clear scheme for the history of nishiki-e.[12]

The reigning conception of the historical arc of Japanese painting at the time had been put forth by the French art historian Louis Gonse in his *L'art japonais* of 1883, in which he concluded his history of Japanese painting from its origins by asserting that

> Hokusai . . . is the last and most brilliant figure of a progress of more than ten centuries. . . . At his death [in 1849] began an irreparable decadence . . . [when] Japanese imitated, they no longer created. The revolution of 1868 is the gulf which separates the purely Japanese art from this hybrid art, occupied only with the demands of exportation, and which sacrifices itself to European tastes as to the golden calf. Such art has nothing which appeals to our eyes or our imagination.[13]

The new narrative that Anderson provided was a radical rejection of the unilinear scheme of pure native development over centuries culminating in Hokusai. He proposed rather a biological model of the sort dominant in Europe from Giorgio Vasari's *Lives* of 1550, by which art evolved through continuing life cycles, each with an organic procession from birth to florescence and on to decay and death. Anderson applied this paradigm specifically to the history of the Japanese color woodblock print (whereas Gonse had been concerned only with painting), which he tracked from its late seventeenth-century origins (the period that would come to be known as "Primitive" in the stock narrative), on to a sudden florescence when the art was brought to perfection between 1765 and 1785, from Harunobu's creation of nishiki-e on to Utagawa Toyokuni (1769–1825). But then, he wrote,

> from 1830, a decadence sets in, the colours become crude, and were no longer balanced with the happy daring of Toyokuni. It was the adoption of cheap European pigments that gave the coup-de-grace, and the colour sense of the people . . . was first to be shocked, then perverted by the chromatic discords that in the present day have transformed the shops of the single-sheet vendors into places to be passed with averted head.[14]

So now the quality and materiality of color itself—an issue that Gonse never considered—became the prime indicator of the progression into the final stages of decadence and decline. In time, as the paradigm was further developed by successive Western writers, the era of the 1890s in which this discourse flourished would come to be seen as the "death of ukiyo-e." This narrative established by Anderson of a passage into decadence and decline in color woodblock printing from 1830 would be sustained and elaborated by subsequent Western writers for a century and more.

To grasp the structure of Anderson's assertion that from 1830, "colours become crude," and that "[i]t was the adoption of cheap European pigments that gave the coup-de-grace," we turn to a provocative essay by the French philosopher and art historian Georges Roque, who has proposed that the basic meanings of color in art are determined by a series of binary oppositions independent of hue.[15] Some of these are peculiar to the history of Western art, but two in particular are central to the key distinctions that emerged in the discourse about color in nishiki-e, namely, "expensive vs. cheap" and "natural vs. artificial." For the case of Japanese pigments, a third crucial opposition must be taken into account, that of "native vs. foreign" (or "indigenous vs. imported"). In this scheme, the new synthetic dyes from the 1860s were always imagined as "synthetic, cheap, and foreign." The "foreign" was always seen in this context as "Western," overlooking the centuries of active international trade in pigments throughout Asia.

It is especially revealing that these conceptions of Meiji colorants began not in Japan but in the often uninformed opinions of Western observers, which were then taken up by the Japanese themselves. Anderson's *Pictorial Arts of Japan* of 1886 was translated into Japanese in 1891 and attracted considerable publicity in the daily press.[16] It now seems clear that the emerging community of Japanese connoisseurs and scholars of the woodblock prints that were now called "ukiyo-e" in deference to Western usage rather than the customary Japanese term "nishiki-e" tended to follow Western opinion, rather than engaging in any research of their own about the actual colorants used in woodblock prints.

Our precise identification of the changing colorants of nishiki-e calls into question all these negative qualities, particularly in view of the revelation that the dominant red for the first two decades of the Meiji period was cochineal carmine, a lake pigment derived from the wholly natural product of an insect from the New World and that was actually higher in price than the traditional safflower but that because of its greater tinting strength and ease of printing was in the end more economical. The allegation of a "foreign" origin overlooks the centuries over which Japanese had imported colorants from China and Southeast Asia, including the closely related insect dye of lac, widely used as a painting pigment in late Tokugawa Japan.[17]

Beyond these objective distinctions of cost, level of human intervention ("synthesis"), and place of origin must be added a cluster of murkier but no less important subjective oppositions, both aesthetic and ideological, which Roque specified as "vivid" versus "austere," but which encompass a richly varied vocabulary that is heavily dependent on language and culture. For the dominant Western view of the color red in Meiji nishiki-e in particular, this subjective language tended to "gaudy" on the "vivid" pole rather than the equally plausible native attributions of "festive," "auspicious," or "healing."

All these distinctions reveal a lively pattern of semiotic activity in the modern understanding of the colors of Japanese woodblock prints. In short, the "activity" of materials must always go beyond materiality itself and take into account the changing meanings assigned to those materials and the ways in which they are chronicled historically. It is essential that conservation experts be aware of this realm of the malleable meanings of color, shifting over time in accord with new findings about the materiality of color, but also constantly sensitive to new meanings.

NOTES

1. The pioneering scientific study of the traditional plant-derived colorants of ukiyo-e, focused on their fading rates, is Feller, "Identification of Traditional Organic Colorants." See also Sasaki, "Materials and Techniques," in which the section "Colourants," 333–37, includes a table

listing all known ukiyo-e colorants at that time, but noting that nishiki-e of Meiji "have yet to undergo in-depth study."

2. For the introduction of Prussian blue into Japanese prints, see Smith, "Hokusai and the Blue Revolution." For the Canton factory, see Bailey, "Note on Prussian Blue," and for the exact date of 1822, see "Manufactory of Prussian Blue."
3. Michener, *Floating World*, 228.
4. See the recent assertion in an authoritative reference work that "[b]y the 1870s these imported colours were popular, inexpensive, and . . . with the 1880s aniline dyes had practically replaced the older vegetable and mineral colourants." Merritt, "Woodblock Prints," 243.
5. Tristan, "Analysis and Conservation."
6. The one crucial exception is the use of the noninvasive technique of excitation–emission matrix fluorescence spectroscopy, pioneered in Japan by Shimoyama Susumu and used most creatively and provocatively by Yamato Asuka in her "Kōki ukiyo-e hanga ni shiyō sareta shikizai no hensen ni kansuru kenkyū," in which she was the first to hypothesize from scientific evidence the sudden appearance of cochineal carmine in Japanese color prints in early 1869. Her results were made public in Yamato, Tsuchiya, and Yonemura, "Investigation of Colorants" (2013), and then in Yamato, Yonemura, and Tsuchiya, "Investigation of Colorants" (2014).
7. Leona and Winter, "Fiber Optics Reflectance Spectroscopy."
8. Leona, "Microanalysis of Organic Pigments."
9. Cesaratto, Luo, Smith, and Leona, "Timeline."
10. "Azuma nishiki-e no suisei o bankai sen to su." "*Azuma nishiki-e*" or "Nishiki-e of the East" is a stock phrase for Edo color woodblock prints, said to have been coined by the author and inventor Hiraga Gennai.
11. For Anderson's style of collection, see Akiko of Mikasa, "Specimen or Art?"
12. Anderson, *Pictorial Arts*, 153–58.
13. Gonse, *Art japonais*, 307–8. The quotation here is from the English translation, *Japanese Art*, 94–95.
14. Anderson, *Pictorial Arts*, 154.
15. Roque, "Meaning of Pigments." The "table of oppositions" appears on p. 275. Roque's table also included the two dimensions of color other than hue, bright/dark and saturated/unsaturated. An updated version of Roque's article may be found in "Esquisse d'une sémiotique des pigments," chapter 4 of his *La cochenille*, 83–98.
16. The Japanese translation of Anderson's *Pictorial Arts* was, like the original, a lavish and costly publication with many high-quality reproductions

in color lithography, so it is unclear how widely it circulated. But the opinions about woodblock colors that he espoused certainly became known to Japanese specialists in the history of ukiyo-e.

17. For the use of lac (known in Japan as *shōenji*) in late Edo painting, see Hokusai's painting manual *Ehon saishiki tsū* (A treatise on colors) of 1848, in which it is recommended frequently as a glaze over vermilion, precisely the way in which cochineal carmine lake was widely used in early modern European oil paintings.

BIBLIOGRAPHY

Akiko of Mikasa, Princess. "Specimen or Art? Collecting Japan in the Victorian Age: Siebold, Anderson and Fenollosa." *Japan Society Proceedings* 46 (2009): 89–108.

Anderson, William. *The Pictorial Arts of Japan.* London: Sampson Low, Marston, Searle, & Rivington, 1886.

"Azuma nishiki-e no suisei o bankai sen to su" [Trying to reverse the decline of Azuma nishiki-e]. *Yomiuri shinbun,* August 18, 1890, 2.

Bailey, Kate. "A Note on Prussian Blue in Nineteenth-Century Canton." *Studies in Conservation* 51, no. 2 (2012): 116–21.

Cesaratto, Anna, Yan-Bing Luo, Henry D. Smith II, and Marco Leona. "A Timeline for the Introduction of Synthetic Dyestuffs in Japan during the Late Edo and Meiji Periods." *Heritage Science* 6, no. 22 (2018): https://doi.org/10.1186/s40494-018-0187-0.

Feller, Robert. "Identification of Traditional Organic Colorants Employed in Japanese Prints and Determination of Their Rates of Fading." In *Japanese Woodblock Prints: A Catalogue of the Mary A. Ainsworth Collection,* edited by Roger Keyes, 253–66. Oberlin, OH: Allen Memorial Art Museum, Oberlin College, 1984.

Ficke, Arthur Davison. *Chats on Japanese Prints.* New York: Frederick A. Stokes, 1917.

Gonse, Louis. *L'art japonais.* Paris: A. Quantin, 1883.

———. *Japanese Art.* Translated by M. P. Nickerson. Chicago: Belford-Clarke, 1891.

Leona, Marco. "Microanalysis of Organic Pigments and Glazes in Polychrome Works of Art by Surface-Enhanced Resonance Raman Scattering." *Proceedings of the National Academy of Sciences of the United States of America* 106, no. 35 (September 2009): 14757–62.

Leona, Marco, and John Winter. "Fiber Optics Reflectance Spectroscopy: A Unique Tool for the Investigation of Japanese Paintings." *Studies in*

Conservation 46 (2001): 153–62.

"Manufactory of Prussian Blue Established in the Vicinity of Canton." *Asiatic Intelligence and Monthly Register*, March 1822, 298.

Merritt, Helen. "Woodblock Prints of the Meiji Era." In *The Hotei Encyclopedia of Japanese Woodblock Prints*, edited by Amy Reigle Newland, 241–60. Amsterdam: Hotei, 2005.

Michener, James. *The Floating World: The Story of Japanese Prints*. New York: Random House, 1954.

Roque, Georges. *La cochenille, de la teinture à la peinture: Une histoire matérielle de la couleur*. Paris: Gallimard, 2021.

———. "The Meaning of Pigments." In *Los itinerarios de la imagen: Prácticas, usos y funciones*, edited by Linda Báez Rubí, Eilie Carreon Blaine, and Deborah Dorotinsky Alperstein, 265–86. Mexico City: Universidad Nacional Autónoma de México, 2010.

Sasaki, Shiho. "Materials and Techniques." In *The Hotei Encyclopedia of Japanese Woodblock Prints*, edited by Amy Reigle Newland, 324–50. Amsterdam: Hotei, 2005.

Smith, Henry D., II. "Hokusai and the Blue Revolution in Edo Prints." In *Hokusai and His Age: Ukiyo-e Painting, Printmaking, and Book Illustration in Late Edo Japan*, edited by John T. Carpenter, 234–69. Amsterdam: Hotei, 2005.

Tristan, Pamela De. "The Analysis and Conservation of Aniline-Dyed, Nineteenth-Century Japanese Prints." In *Proceedings of the Fourth International Conference of the Institute of Paper Conservation*, 107–17. London, April 1997.

Yamato, Asuka. "Kōki ukiyo-e hanga ni shiyō sareta shikizai no hensen ni kansuru kenkyū" [A study of changes in the colorants used in late ukiyo-e prints]. MA thesis, Tohoku University of Art and Design, 2014.

Yamato, A., A. Tsuchiya, and S. Yonemura. "Investigation of Colorants Used in Ukiyo-e Woodblock Printings during the End of Edo Period and Meiji Era Using Nondestructive Analysis Method—Red and Purple Colorants." [In Japanese.] *Proceedings of the 35th Congress of the Japan Society for the Conservation of Cultural Property*, 216–17. Tokyo, 2013.

Yamato, A., S. Yonemura, and A. Tsuchiya. "Investigation of Colorants Used in Ukiyo-e Woodblock Printings during the End of Edo Period and Meiji Era Using Nondestructive Analysis Method—the Second Report: Transition of Red Colorants, and Yellow Colorants." [In Japanese.] *Proceedings of the 36th Congress of the Japan Society for the Conservation of Cultural Property*, 226–27. Tokyo, 2014.

Appendix: Events of the Research Project Conserving Active Matter

October 31, 2017
Trinidad Rico (Rutgers University), "Heritage, Secrecy, and Failure: The Atomic Project Huemul"

November 2, 2017
Samantha Alderson (American Museum of Natural History) and Peter M. Whiteley (American Museum of Natural History), "Over Land and Sea: The Journey of a Chief's Chest Back to Haida Gwaii from the American Museum of Natural History"

November 27, 2017
Conserving Active Matter: A Cultures of Conservation Research Project
What Is Active Matter?
Welcome
Peter N. Miller (Bard Graduate Center)
Introduction
Jessica Walthew (Cooper Hewitt, Smithsonian Design Museum; Bard Graduate Center)
Presentations
Admir Masic (Massachusetts Institute of Technology), "Natural Fibers: Preservation of the Old and Inspiration for the New"
Glenn Wharton (New York University), "Active Matter in the Conservation of Contemporary Art"
Justin Broackes (Brown University), "Philosophical Matter in the New Modern World: Atomists, Identity, and the 'Plastic Life of Nature'"

Moderator
Alicia Boswell (Bard Graduate Center)

November 28, 2017
What Is Conserving Active Matter?
Introduction
Peter N. Miller (Bard Graduate Center); Peter Fratzl (Humboldt University); Wolfgang Schäffner (Humboldt University); Robert van Langh (Rijksmuseum)
History
Ittai Weinryb (Bard Graduate Center); Ann-Sophie Lehmann (University of Groningen); Spike Bucklow (University of Cambridge)
Philosophy
Ivan Gaskell (Bard Graduate Center); A. W. Eaton (University of Illinois at Chicago); Carolyn Korsmeyer (University at Buffalo); Sherri Irvin (University of Oklahoma)
Indigenous Ontologies
Aaron Glass (Bard Graduate Center); Jolene Rickard (Cornell University); Kelly McHugh (National Museum of the American Indian, Smithsonian Institution)
Materials Science
Jennifer L. Mass (Bard Graduate Center); Marc Walton (Northwestern University); Paul Messier (Yale University)
Response
Admir Masic (Massachusetts Institute of Technology); Glenn Wharton (New York University); Justin Broackes (Brown University)

January 29, 2018
Jennifer L. Mass (Bard Graduate Center), "Reverse Engineering the Past to Secure a Future for the 1876 Dry Tortugas National Park Harbor Light"

February 1, 2018
Carolyn Riccardelli (The Metropolitan Museum of Art), "After the Fall: The Treatment of Tullio Lombardo's *Adam*"

February 19, 2018
Noémie Étienne (University of Bern) "Exoticizing in the Enlightenment: A Global History of Switzerland in Objects"

April 18, 2018

Conserving Active Matter: Indigenous Ontologies

Presentations

Jamie Jacobs (Rochester Museum and Science Center), "The Orator's Dilemma: Wampum as Material, Media, and Memory"

Rose Evans (The ObjectLab, Auckland, New Zealand), "Making Our Past Visible: Toi Moko and Our Connection with Whakapapa (Genealogy)"

Sanchita Balachandran (The Johns Hopkins Archaeological Museum; Johns Hopkins University), "The Museum Was Like Heaven, All the Gods Being There: Approaches to the Preservation of Archaeological-Devotional Objects at the Madras Government Museum, India, in the Late 19th and Early 20th Centuries"

Moderator

Aaron Glass (Bard Graduate Center)

November 1, 2018

Conserving Active Matter: History

Welcome and Introduction

Peter N. Miller (Bard Graduate Center); Ittai Weinryb (Bard Graduate Center)

Magic

Frank Klaassen (University of Saskatchewan); Nicolas Weill-Parot (École Pratique des Hautes Études)

Buddhism

Fabio Rambelli (University of California, Santa Barbara); Wen-shing Chou (Hunter College, City University of New York)

Subjectivities

Surekha Davies (The John Carter Brown Library); Daniel Garber (Princeton University)

Traditions

Moshe Idel (The Hebrew University of Jerusalem); André Laks (Panamerican University)

Modernism

Joyce Tsai (University of Iowa); Claudia Wedepohl (The Warburg Institute)

March 28–29, 2019

Conserving Active Matter: Materials Science

Welcome and Introduction

Peter N. Miller (Bard Graduate Center); Jennifer L. Mass (Bard Graduate Center)

Metamorphosis and the Modern
James Coddington (Institute of Fine Arts, New York University); Jessica Walthew (Cooper Hewitt, Smithsonian Design Museum); Margo Delidow (Whryta Contemporary Art Conservation; Whitney Museum of American Art); Clara Rojas-Sebesta (Whitney Museum of American Art); Chris McGlinchey (The Museum of Modern Art)
Undead Objects: Preservation of the Source Code
Marc Walton (Northwestern University); Deena Engel (New York University); Catherine S. Mallinckrodt (Virginia Museum of Fine Arts); Ashley Duhrkoop (Virginia Museum of Fine Arts); Pepe Karmel (New York University); James Coddington (Institute of Fine Arts, New York University)
The Death of a Painting: Metal Soaps and Molecular Self-Assembly
Francesca Casadio (The Art Institute of Chicago); Barbara Berrie (National Gallery of Art); Silvia Centeno (The Metropolitan Museum of Art); Petria Noble (Rijksmuseum)
The Afterlife: Materials Research and Reanimating Objects
Matthew Collins (Natural History Museum of Denmark); Glenn Gates (The Walters Art Museum); Tim Wess (University of the Sunshine Coast); Marco Leona (The Metropolitan Museum of Art)

November 8, 2019
Conserving Active Matter: Philosophy—Degradation as an Aesthetic Value
Welcome and Introduction
Peter N. Miller (Bard Graduate Center); A. W. Eaton (University of Illinois at Chicago); Ivan Gaskell (Bard Graduate Center)
Growth in Decay
Erich Hatala Matthes (Wellesley College), "What Can Conservation Conserve?"
Alva Noë (University of California, Berkeley), "Entanglement and Death"
Wabi-Sabi, Kintsugi, and the Joys of Deformation and Rupture
Yuriko Saito (Rhode Island School of Design), "Aesthetics of Repair"
Rumiko Handa (University of Nebraska) "Time and Imperfection"
Puzzles in Modern and Contemporary Art and Material Culture
Francesca Esmay (Solomon R. Guggenheim Museum) and Jeffrey Weiss (independent curator), "Iteration, Example, Version: The Variable Realization of Minimal Art"
Sherri Irvin (University of Oklahoma) "The Expressive Import of Degradation and Decay in Contemporary Art"

Worlds in Ruins
Carolyn Korsmeyer (University at Buffalo), "Degradation: Appearance and/or/vs. Reality"
Elisabeth Schellekens (University of Uppsala), "Beauty, Destruction, and the Temple of Bel"

September 22, 2020
Jennifer L. Mass (Bard Graduate Center), "Nineteenth-Century Black Patinas on Ancient Bronzes: Chemical Innovations and Asian Influences"

February 2, 2021
Soon Kai Poh (Bard Graduate Center), "Introducing *Conserving Active Matter*"

April 13, 2021
Screening and Panel Discussion—*Those That, at a Distance, Resemble Another*
Jessica Sarah Rinland (director); Grace Kim-Butler (Vanderbilt University); Soon Kai Poh (Bard Graduate Center)

May 3, 2021
Julia M. Brennan (Caring for Textiles), "From Trauma to Tenderness: Following a 'Middle Path' for Conservation Work at Memorial Sites"

Index

Contributors

Spike Bucklow is professor of material culture and teaches conservation at the Hamilton Kerr Institute, University of Cambridge.

Francesca Casadio is associate vice president and Grainger Executive Director of Conservation and Science, the Art Institute of Chicago, IL.

Oliver Cossairt is associate professor in the Computer Science and Electrical and Computer Engineering departments at Northwestern University, Evanston, IL.

A. W. Eaton is professor in the Philosophy Department and associate dean, Liberal Arts and Sciences, University of Illinois–Chicago.

Rose Evans (Te Āti Awa and European descent) is an objects conservator and director of Objectlab Ltd, a heritage consultancy based in Auckland, Aotearoa (New Zealand), having specialized in the conservation of Maori and Pacific taonga and contemporary sculpture across a range of institutions from Te Papa Museum of New Zealand (1990–2005) to the British Museum (2002).

Ivan Gaskell is professor of cultural history and museum studies at Bard Graduate Center, New York City.

Guido Giglioni is associate professor of history of philosophy at the University of Macerata, Italy.

Aaron Glass is associate professor of anthropology, museum studies, and Indigenous arts at Bard Graduate Center, New York City.

Sven Haakanson Jr., Sugpiaq from the community of Old Harbor, Alaska, is curator of North American anthropology and associate professor in anthropology at the University of Washington, Seattle.

Pengxiao Hao is postdoctoral scholar in the Institute for the Conservation of Cultural Heritage of Shanghai University, Shanghai, China.

Sherri Irvin is presidential research professor of philosophy and senior associate dean of the Graduate College at the University of Oklahoma.

Jamie Jacobs, an enrolled member of the Tonawanda Band of Senecas, is of the Turtle Clan and acts as cultural and ceremonial custodian for the Tonawanda Ceremonial Longhouse. He is also curator for all Iroquoian archeological and anthropological material owned by the Rock Foundation, whose Haudenosaunee Collections are housed at the Rochester Museum of Science in Rochester, NY.

Carolyn Korsmeyer is research professor of philosophy at the University at Buffalo, NY.

André Laks was professor of Greek and of ancient philosophy at the University of Lille, at Princeton University, and at the University of Paris-Sorbonne. Since his retirement in 2011, he has been teaching at the Universidad Panamericana, Mexico City.

Marco Leona is David H. Koch Scientist in Charge, Department of Scientific Research, Metropolitan Museum of Art, New York City.

Jennifer L. Mass is the Andrew W. Mellon Professor of Cultural Heritage Science at Bard Graduate Center, New York City.

Chris McGlinchey is research scholar and adjunct professor of conservation science, Conservation Center of the Institute of Fine Arts, New York University.

Kelly McHugh is head of collections care and stewardship, Collections Management Department, National Museum of the American Indian, Smithsonian Institution, Washington, DC.

Peter N. Miller is dean and professor at Bard Graduate Center, New York City.

Alva Noë is professor of philosophy at the University of California, Berkeley, where he is also a member of the Institute for Cognitive and Brain Sciences, the Center for New Media, and the Program in Critical Theory.

Soon Kai Poh is a conservator and Andrew W. Mellon Foundation Postgraduate Fellow at Bard Graduate Center, New York City.

Yuriko Saito is professor emerita of philosophy at the Rhode Island School of Design, Providence.

Henry D. Smith II is professor emeritus of Japanese history, Department of East Asian Languages and Cultures, Columbia University, New York City.

Marc Vermeulen is research associate at the Center for Scientific Studies in the Arts at Northwestern University, Evanston, IL.

Marc Walton is codirector of the Center for Scientific Studies in the Arts and research professor of materials science and engineering at Northwestern University, Evanston, IL.

Lee Palmer Wandel is WARF Michael Baxandall and Linda and Stanley Sher Professor of History, University of Wisconsin–Madison.

Ittai Weinryb is associate professor at Bard Graduate Center, New York City.

Florian Willomitzer is research assistant professor in the department of Electrical and Computer Engineering at Northwestern University, Evanston, IL.